Subsurface Transport
and
Fate Processes

Robert C. Knox
David A. Sabatini
Larry W. Canter

LEWIS PUBLISHERS
Boca Raton   Ann Arbor   London   Tokyo

Library of Congress Cataloging-in-Publication Data

Knox, Robert C.
   Subsurface transport and fate processes / Robert C. Knox, David A.
Sabatini, Larry W. Canter.
     p. cm.
   Includes bibliographical references and index.
   ISBN 0-87371-193-9
   1. Soil pollution. 2. Soil physics. 3. Water, Underground —
Pollution. I. Sabatini, David A., 1957– . II. Canter, Larry W.
III. Title.
TD878.K59 1993
628.5'5—dc20   92-24931
                                                       CIP

Direct all inquiries to CRC Press, Inc., 2000 Corporate Blvd., N. W., Boca
Raton, Florida, 33431.

PRINTED IN THE UNITED STATES OF AMERICA
 2 3 4 5 6 7 8 9 0
Printed on acid-free paper

To my mother and my friends, especially Betty and Janice and their families, for their loyalty and positive influence.

To Frances, Caleb and Peggy,
and to Donna, Bill, Mike and Lily
for their support, encouragement, sacrifices, and inspiration

To Donna, Doug and Carrie, Steve, and Greg

Robert C. Knox, P.E., is an Associate Professor of Civil Engineering and Environmental Science at the University of Oklahoma in Norman, Oklahoma. Dr. Knox received B.S. (1978), M.S. (1979), and Ph.D. (1983) degrees in civil engineering from the University of Oklahoma. After completing his doctoral studies, he spent one year as a research engineer for the Environmental and Ground Water Institute at the University of Oklahoma, then took a faculty position at McNeese State University in Lake Charles, Louisiana. He returned to the University of Oklahoma as an Assistant Professor in 1986.

Dr. Knox's research interests include subsurface transport and fate processes, innovative remediation technologies for soils and ground water, solute transport and multiphase flow modeling, and solid waste disposal systems. He has worked on research for the U.S. Environmental Protection Agency, the Department of Defense, the U.S. Geological Survey and several state agencies in Oklahoma. Dr. Knox has focused research and consulting activities on oilfield-related contaminants including brine movement through the subsurface, disposal of drilling muds, and remediation of hydrocarbon contamination. Currently, he is conducting research on methods for improving the hydraulic efficiency of ground water extraction operations for pump and treatment remediation schemes using chemical agents.

Dr. Knox previously co-authored three Lewis Publishers books entitled *Septic Tank Systems Effects on Ground Water Quality*, *Ground Water Quality Protection*, and *Ground Water Pollution Control*. He recently co-edited and contributed chapters to *Transport and Remediation of Subsurface Contaminants: Colloidal, Interfacial, and Surfactant Phenomena*, published by the American Chemical Society. Dr. Knox has also published numerous technical reports and papers dealing with ground water quality management, ground water pollution control, and environmental impact assessment. He has received several awards including the DOW Outstanding Young Faculty Award for the Gulf-Southwest Section of the American Society for Engineering Education in 1986; an Oklahoma University Distinguished Lectureship in 1989; and the Outstanding Young Engineer of the Year from the Oklahoma Society of Professional

Engineers in 1991. His research on energy-related environmental impacts resulted in being named a University of Oklahoma Energy Center Fellow in 1991.

Dr. Knox has taught short courses to industry and government personnel on ground water quality management, ground water quality protection, site assessment and closure of underground storage tanks, ground water containment and removal systems, and environmental impact assessment. He has also given invited lectures and presentations in the U.S. and abroad. He has extensive experience as an expert witness on subsurface contamination cases and has testified before a U.S. Senate Committee and the U.S. Environmental Protection Agency at public hearings. Dr. Knox is president of Knox Engineers Inc. which focuses on siting and design of solid waste disposal facilities and removal of underground storage tanks.

David A. Sabatini is an Assistant Professor of Civil Engineering at the University of Oklahoma, Norman, Oklahoma. Dr. Sabatini received his B.S. in civil engineering from the University of Illinois, Urbana, IL in 1981, his M.S. in civil engineering from the Memphis State University, Memphis, TN, in 1985 and his Ph.D. in civil engineering from Iowa State University, Ames, IA, in 1989. He is also a licensed professional engineer. From 1981 to 1983, Dr. Sabatini worked as a Civil Engineer for the Illinois Central Gulf Railroad (Chicago, IL and Memphis, TN). He joined the faculty at the University of Oklahoma in 1989.

Dr. Sabatini's research interests include development of an improved understanding of subsurface contaminant transport, development of advanced technologies for remediation of subsurface contamination, development of effective ground water protection strategies, and development of innovative processes for water and wastewater treatment (with an emphasis on resource recovery). Past and present sponsors of his research include the National Science Foundation, Environmental Protection Agency, Department of Defense, and the Department of Agriculture. Dr. Sabatini recently co-edited and contributed chapters to the monograph "Transport and Remediation of Subsurface Contaminants: Colloidal, Interfacial and Surfactant Phenomena" published by the American Chemical Society. He is author or co-author of numerous journal articles, papers, research reports, etc. on research topics as outlined above. Examples of journals in which he has published and/or served as referee include: *Environmental Science and Technology, Ground Water, Journal of Contaminant Hydrology, Journal of Environmental Quality, Journal of the Water Pollution Control Federation, Environmental Pollution, Journal of Environmental Engineering Division — ASCE, Journal of Irrigation and Drainage Division — ASCE.* He has participated in teaching numerous short courses on ground water-related topics in the United States, Scotland, and Thailand, including a Subsurface Transport and Fate short course for the U.S. Environmental Protection Agency.

Dr. Sabatini has received numerous honors and awards. In particular, his doctoral program was sponsored by a USDA National Needs Fellowship, he received a Premium for Academic Excellence Award from Iowa State University, he received a Research Excellence Award for his dissertation from Iowa State University, and he received a Junior Faculty Summer Research Fellowship from the University of Oklahoma. Dr. Sabatini is also a member of the following honorary societies: Chi Epsilon, Tau Beta Pi, Phi Kappa Phi, and Sigma Xi.

Larry W. Canter, P.E., is the Sun Company Professor of Ground Water Hydrology and Director, Environmental and Ground Water Institute, University of Oklahoma, Norman, Oklahoma. Dr. Canter received his Ph.D. in environmental health engineering from the University of Texas in 1967, M.S. in sanitary engineering from the University of Illinois in 1962, and B.E. in civil engineering from Vanderbilt University in 1961. Before joining the faculty of the University of Oklahoma in 1969, he was on the faculty at Tulane University and was a sanitary engineer in the U.S. Public Health Service. He served as Director of the School of Civil Engineering and Environmental Science at the University of Oklahoma from 1971 to 1979.

Dr. Canter's research interests include ground water pollution source evaluation, aquifer vulnerability mapping, ground water protection strategies, and soil and ground water remediation technologies. Currently, he is conducting research on methods for prioritizing ground water contamination sources and evaluating ground water protection programs. Dr. Canter previously co-authored three Lewis Publishers ground water-related books entitled Septic Tank Systems Effects on Ground Water Quality, Ground Water Pollution Control, and Ground Water Quality Protection. He has also written several books on environmental impact assessment. He is also the author or co-author of numerous refereed papers and research reports related to environmental impact studies or ground water pollution evaluation. In 1982 he received the Outstanding Faculty Achievement in Research Award from the College of Engineering, and in 1983 the Regent's Award for Superior Accomplishment in Research.

Dr. Canter served on the U.S. Army Corps of Engineers Environmental Advisory Board from 1983 to 1989. He has conducted research, presented short courses, or served as advisor on environmental impact assessment and/ or ground water pollution to institutions in Mexico, Panama, Colombia, Venezuela, Peru, Scotland, The Netherlands, France, Germany, Italy, Greece, Turkey, Kuwait, Thailand, Saudi Arabia, and the People's Republic of China.

PREFACE

Subsurface contamination is arguably man's most complex environmental problem. The movement of chemical and biological constituents in the subterranean environment can involve multiple phases with a myriad of potential reactions in an inherently nonhomogeneous, anisotropic porous media. The ultimate fate of constituents introduced to the subsurface is difficult to predict; altering the subsurface to control the fate of these constituents in order to effect remediation is a formidable challenge. In order to predict or control the fate of any substance in the subsurface environment, one must possess a thorough understanding of the basic processes that can influence the transport and fate of the substance. This will require knowledge and expertise from a variety of professions. No other problem truly requires a multi-disciplinary approach more than remediation of the subsurface.

This book is intended as a basic introductory text for the very broad field of subsurface transport and fate processes. The individual chapters are comprehensive reviews of the state of the knowledge on specific topics. Detailed descriptions of the individual chapter topics would require more than a single text; each chapter could easily be expanded into an entire text. No single manuscript could provide detailed coverage of these ever expanding topics. To that end, we have attempted to present a level of information that provides a working knowledge of the basic principles in each topic and allows perusal of the refereed literature for more detailed information on a subject of particular interest.

The manuscript is well suited for use as a text for upper level undergraduate or introductory graduate level instruction; however, a certain level of basic training is assumed. The text is written to an audience that has a working knowledge of the fields of subsurface hydrology, a firm grasp of basic chemistry, and some training in the area of environmental impact assessment. After the introductory chapter, the next three chapters cover the subsurface processes that can influence constituents. The processes are grouped as hydrodynamic processes, abiotic processes, and biotic processes to facilitate discussion; however, it must be emphasized that all of the subsurface processes are interrelated. Chapters 5 and 6 deal with methods for measuring and assessing the impact of contaminants moving through the subsurface, while Chapter 7 discusses the deterministic computer methods used to mathematically model movement of contaminants. The final chapter is a capstone chapter that attempts to integrate information from the previous chapters through two case studies.

Even as this manuscript goes to press, the state of knowledge in subsurface contamination is growing dramatically. In particular, information related to remediation technologies for subsurface contamination and understanding of complex transport processes is growing quite rapidly. The reader should consider this text as only the first rung on the ladder. The references cited at the end of each chapter can give insight as to who is working in specific areas and where certain information is being published; for the most recent data the reader is directed to these sources.

The authors wish to express their appreciation to several individuals and groups who indirectly contributed to the development of this book. First, we would like to express our appreciation to the external reviewers including Dr. Candida West, Dr. James Weaver, Mr. Randall Ross, and Mr. Jerry Thornhill, all from the U. S. Environmental Protection Agency's Robert S. Kerr Environmental Research Laboratory (USEPA RSKERL) in Ada, Oklahoma; and Dr. Joe Suflita, Department of Botany and Microbiology, University of Oklahoma. Several other personnel from USEPA RSKERL have contributed to the formulation of this manuscript through years of cooperative research including Marion Scalf, James McNabb, Dr. Carl Enfield, and Jack Keeley. Our thanks also go out to several faculty colleagues who have participated on research projects related to the subject matter including Dr. Leale Streebin, Dr. Joakim Laguros, and Dr. James Robertson from Civil Engineering and Environmental Science; and Dr. Jeffrey Harwell, Chairman, Chemical Engineering and Materials Science. In addition, we would like to acknowledge the investments of our mentors, including Dr. John W. Smith (Memphis State University) and Dr. T. Al Austin (Iowa State University). We also wish to acknowledge the many graduate students who undertook research endeavors that resulted in stimulating discussion and useful insight (especially Thomas Soerens and Joseph Rouse). The authors are extremely grateful to the technical typists who have worked diligently in preparing this manuscript; specifically, Mrs. Ginger Geis and Mrs. Mittie Durham of the Environmental and Ground Water Institute, University of Oklahoma, and Mrs. Betty Craig. The authors also want to express their appreciation to Lewis Publishers, especially to Ms. Kathy Walters, Associate Editor, for their efforts in conjunction with this book.

The authors gratefully acknowledge the support and encouragement of Dr. Ronald Sack, Chairman of Civil Engineering and Environmental Science, and the College of Engineering at the University of Oklahoma relative to faculty writing endeavors. Most important, the authors wish to thank their families for their patience and understanding.

Robert C. Knox

David A. Sabatini

Larry W. Canter

CONTENTS

1

<u>BACKGROUND</u>

1.1 INTRODUCTION

With the twentieth anniversary of "Earth Day" receiving much notoriety, many in the media took the opportunity to look back at the emphases placed on different aspects of "our environment" over the past two decades. Following passage of the National Environmental Policy Act, early emphasis was placed on cleaning up the nation's air and surface water resources. Subsequently, land-based waste disposal and uncontrolled hazardous waste sites received much attention. With the beginning of a new decade, tremendous energy is now being exerted to deal with the millions of underground storage tanks nationwide.

The interesting aspect of all of these environmental issues is that they have all had a direct effect on our access to, and awareness of, high quality subsurface resources. Environmental professionals are now, more than ever before, addressing the impacts of man's activities on the subsurface soils, gases, microflora, and most importantly, water. In order to assess the impacts of our actions, we must first understand how the subsurface environment functions.

The evolution of our understanding of subsurface transport and fate processes is an interesting study in itself. For many years the soils beneath us were thought to be the "perfect filter", i.e., movement of contaminants through the subsurface would be confined to the uppermost reaches of the soil horizon. Then we slowly began to accept the fact that aqueous and nonaqueous fluids could readily migrate through the subsurface. This realization is best exemplified by the outlawing of saltwater (brine)

1

"evaporation" pits in the oil and gas industry. The term evaporation was a misnomer as it became obvious that most of the saltwater infiltrated rather than evaporated.

As it became more obvious that we had heavily impacted the subsurface and contaminated some of our ground water resources, we began the knee-jerk response of pump-and-treat remediation. The only solution was to pump out our mess and deal with it at the surface remediation. If it got down there, we could get it out if we pumped long and hard enough. That was our only solution. As we started extracting large quantities of water from our aquifers, we noticed strange behavioral patterns in the contaminants. It seemed as though each chemical behaved differently in the subsurface. Very few of the chemicals moved as fast as we liked, and some of them were somehow transformed from their original composition. We even started finding gases associated with some of the contamination episodes. Hence, three realities came to light: (1) the porous media could physically alter migration of contaminants; (2) the subsurface was decidedly not abiotic; and (3) multiple phases had to be considered.

Given the above realities, it can be proposed that the subsurface is the most complex of all the physical environments and that subsurface contamination represents our most difficult environmental problem. In order to address this problem, large numbers of trained professionals with specialized expertise in subsurface transport and fate processes will be required. Remediation of subsurface contamination requires multiple technical disciplines; however, the key component will always be expertise in contaminant hydrogeology.

1.2 OBJECTIVES

The overall objective of this book is to give the reader a basic understanding of the predominant processes affecting the subsurface transport and fate of chemical constituents in aqueous and nonaqueous fluids. With this understanding, a series of goals or applications of the acquired information can be identified. First, the material presented herein will allow professionals to develop more definitive predictions for contaminant migration in the subsurface. Second, the information provided will also allow for development of more accurate and more comprehensive transport models. Third, an improved understanding of subsurface processes should result in improved pollution prevention, monitoring, and remediation programs.

It is important to qualify the stated objectives. To achieve the goals listed above, one must incorporate skills and knowledge from areas outside this

Table 1.1 Predominant Ionic Species of Natural Waters

Major Cations	Major Anions
Calcium – Ca^{+2}	Bicarbonate – HCO_3^-
Magnesium – Mg^{+2}	Carbonate – CO_3^{-2}
Sodium – Na^+	Sulfate – SO_4^{-2}
Potassium – K^+	Chloride – Cl^-
Minor Metals	**Minor Anions**
Silicon – Si	Nitrate – NO_3^-
Iron – Fe	Phosphate – PO_4^-

text. Most importantly, the knowledge gained from field experience must be incorporated. There is no such thing as an office hydrogeologist. A truly complete understanding of the subsurface involves grasping the theoretical considerations presented in this text and observing and analyzing the real world results of these phenomena in the field.

1.3 GROUND WATER QUALITY

A primary concern when addressing subsurface transport and fate processes is the potential for ground water contamination. When considering ground water contamination, it is necessary to first know what constitutes uncontaminated ground water. As one might guess, the natural or background quality of ground water does not include any synthetic organics or high levels of heavy metals. In fact, natural waters, including ground water, have their chemical composition dominated by selected cations and anions as shown in Table 1.1. It is important to emphasize that the constituents listed in Table 1.1 are only the predominant ionic species, not the entire list of naturally occurring ions. Many other ions, metals, and radioisotopes can be found in ground water. However, these minor constituents are usually found only at very low concentrations.

Because ionic species involve electrical charges, concentration data in the traditional units of mass per unit volume is not always useful. In some instances, the amount of "electrical charge" a given ion represents is a needed measure. This is especially true when doing comparative analysis of different ions or ion ratios. The amount of electrical charge a given ion contributes to

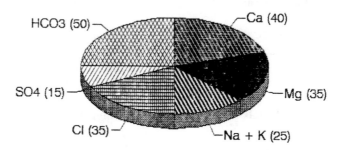

Figure 1.1.   Pie chart of water quality.

a solution is a function of both its mass concentration and valence. The units used for comparing relative concentrations of ionic species are equivalents per unit volume, usually milliequivalents per liter.

1.3.1 Qualitative Analyses

Owing to the fact that the quality of natural waters is dominated by a few select ions, much effort has been put forth for qualitatively and quantitatively addressing and analyzing concentration data for these (and other) constituents. Tabular comparisons of data for multiple components is too cumbersome. Several methods have been developed for graphically depicting water quality information. These methods are discussed individually below.

Figure 1.1 is a pie chart depicting the ionic composition of a natural water. The size of the "slice" associated with each ion is proportional to the percentage of the total charge that the ion contributes to its ionic group, i.e., positive or negative. The total charge for each group is depicted on the "x-axis." (Note, the total meq/ℓ of all cations must equal the total meq/ℓ for all anions. For the predominant ions, a charge difference of 15% or more would not be acceptable.) The pattern produced by the pie chart is indicative of the composition of the water. It is the *patterns* of the pie charts that are used in comparing different waters.

Figure 1.2 is another graphical means of presenting water quality information and is called a Stiff diagram (Stiff, 1951). The Stiff diagram plots ionic concentrations, with cations on the left and anions on the right. The plotted values are connected in a dot-to-dot fashion resulting in a distinctive figure. The size of the diagram is proportional to the ionic strength of the

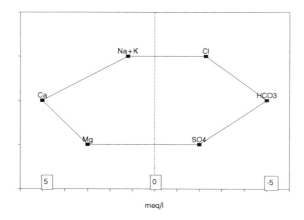

Figure 1.2. Stiff diagram of water quality.

natural water; the shape of the diagram is indicative of it's composition. Stiff diagrams are commonly used on water atlases to depict the water quality from different formations within a given area. Similar diagrams are usually color coded.

Figure 1.3 is an ionic ratio, sometimes called a Schoeller diagram (Schoeller, 1962). The Schoeller diagram is a simple plot of the ratios of different ion pairs. The ratio values are once again connected in a dot-to-dot fashion and result in a distinctive pattern. Waters of similar quality will have similar Schoeller diagrams. The Schoeller diagram has an advantage over the previous two graphs in that multiple waters can be plotted on a single graph. This allows for direct comparison of different waters and can show the effects of mixing of two different waters.

Figure 1.4 is a trilinear diagram also referred to as a Piper plot (Piper, 1944). The Piper plot consists of two triangles and an upper diamond. The left triangle plots the relative contribution of the major cationic species; the right triangle plots the relative contribution of the major anionic species. The upper diamond plots the relative contribution of the major ion pairs. Hence, a complete analysis of the major ions in a given water will plot as three distinct points. Water of similar quality will plot in the same general area on all three figures. The Piper plot also allows for multiple waters to be plotted for comparative analysis.

The major advantage of the Piper plot, however, is the ability to show the effects of mixing waters of different composition and the effects of retardation on plume migration. Morris et al. (1983) outlines the procedure by which the results of binary (two waters) or tertiary (three waters) mixing are depicted on a Piper plot. In simple terms, the effects of mixing of two waters can be represented by a mixing line; the mixing of three waters is a mixing triangle. Any mixture of two waters will fall on the mixing line; any mixture of three

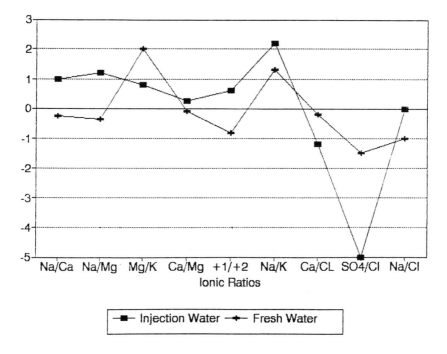

Figure 1.3.   Schoeller diagram of water quality.

waters will fall within the mixing triangle. It is important to note that true mixing of different composition waters involves no other reactions. The effects of reactions on plume migration using Piper plots is discussed in detail in Chapter 8.

1.3.2 Quantitative Analyses

The graphical procedures discussed above are not meant for quantitative analyses. In order to interpret what the actual concentrations of the various species found in a ground water mean in relation to different samples or some numerical standard, quantitative techniques need to be utilized. Most of the quantitative techniques now available are based on statistical analyses.

The development of statistical techniques for analyzing ground water quality data can be traced back to the Resource Conservation and Recovery Act (RCRA), which mandated ground water monitoring at waste disposal facilities. Originally, EPA promulgated standards that required the use of t-tests for analyzing data from different wells. Concerns were voiced that the proposed

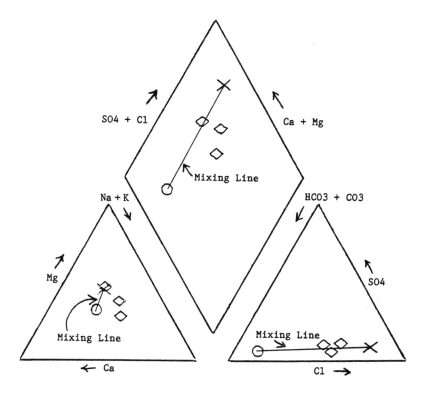

KEY

$\bigcirc$ – Background (uncontaminated) Well

$\times$ – Brine Leachate

$\diamondsuit$ – Brine Impacted Monitoring Well

Figure 1.4. Piper plot of water quality.

methods could result in high rates of both "false positives" and "false negatives". In 1988, EPA amended the proposed methods with five different statistical methods that are more appropriate for ground water applications (U.S. EPA, 1989).

The statistical methods now available for RCRA facilities are divided into categories based on the compound being tested and the type of comparison to be completed. Different methods are now accepted for comparing data from background and detections wells, comparing data from detection wells with fixed standards, intra-well comparisons, and analysis of data sets with many entries below the detectable limits. These methods are summarized in Table 1.2 and discussed briefly below.

Table 1.2 Summary of Statistical Methods

Compound	Type of Comparison	Recommended Method	Section of Guidance Document
Any compound in background	Background vs compliance well	ANOVA	5.2
		Tolerance limits	5.3
		Prediction intervals	5.4
	Intra-well	Control charts	7
ACL/MCL specific	Fixed standard	Confidence intervals	6.2.1
		Tolerance limits	6.2.2
Synthetics	Many nondetects in dataset	See "below detection limit" Table 8.1	8.1

Source: U.S. Environmental Protection Agency, 1989.

1. Analysis of Variance — ANOVA techniques are used to compare the mean values of different groups of observations. The techniques are used to separate differences due to random errors from statistical differences in the observations. ANOVA methods can include parametric (use of statistical parameters) and nonparametric (rank ordering of observations) techniques.
2. Tolerance Intervals — The tolerance interval approach establishes a concentration range that is constructed to contain a specified proportion of observations with a specified confidence coefficient. The tolerance interval approach, as with many statistical approaches, assumes the population from which observations are being made is normally distributed. Methods are available for checking this assumption.
3. Control Charts — Control charts are graphical interpretation methods in which the ground water quality data is plotted over time. This allows for detection of long term trends and/or seasonality associated with the data.

1.4 SUBSURFACE AND CONTAMINANT PROPERTIES AFFECTING FATE AND TRANSPORT

Two basic elements affecting the transport and fate of contaminants in the subsurface are properties of the subsurface materials or the subsurface environment and physicochemical and biological properties of the contaminant. Nonreactive (conservative) chemicals will move through the subsurface environment with the ground water (hydrodynamic processes) and will not be affected by abiotic or biotic processes that may be active in the subsurface. Conversely, contaminants that have the potential to be reactive (nonconservative) will not be affected during ground water transport if the subsurface environment is not conducive to the reactions that affect the contaminant (e.g., a contaminant that is susceptible to aerobic degradation but is in an anaerobic subsurface environment). Thus, for the fate and transport of the contaminant to be altered from flow with the ground water (for interactions between the subsurface environment and the contaminant to occur), it is necessary that both the contaminant property and the subsurface environment be conducive to these interactions. The goal of this section is to provide an overview of several subsurface and contaminant properties which may affect the transport and fate of the contaminants in the subsurface and indicate the nature of the impacts caused by these properties. The approach to be utilized will involve an overview of processes that affect the fate and transport of contaminants in the subsurface and subsequently a discussion of subsurface and contaminant properties affecting these processes.

1.4.1 Overview of Interactions

For purposes of this discussion, the general categories of processes affecting subsurface fate and transport are hydrodynamic processes, abiotic (nonbiological) processes, and biotic processes. Hydrodynamic processes affect contaminant transport by impacting the flow of ground water (in terms of quantity of flow and flow paths followed) in the subsurface. Examples of hydrodynamic processes are advection, dispersion, preferential flow, etc. Abiotic processes affect contaminant transport by causing interactions between the contaminant and the stationary subsurface material (e.g., sorption, ion exchange) or by affecting the form of the contaminant (e.g., hydrolysis, redox reactions). Biotic processes can affect contaminant transport by metabolizing or mineralizing the contaminant (e.g., organic contaminants) or possibly by utilizing the contaminant in the metabolic process (e.g., nutrients, nitrate under denitrifying conditions). Examples of biotic processes are aerobic, anoxic, and anaerobic biodegradation. Table 1.3 gives an expanded list of subsurface processes and corresponding subsurface and contaminant properties influencing these processes.

Subsequent chapters will be devoted to discussing the three categories of processes (hydrodynamic, abiotic, and biotic) in detail, and at such time, the interrelation of subsurface and contaminant properties will be discussed for all the processes listed in Table 1.3. For introductory purposes, an example from each process category will be discussed here.

Table 1.3 Subsurface Processes and Corresponding Subsurface and Contaminant Properties and Interactions Affecting the Fate and Transport of Contaminants

Process	Subsurface Property
Hydrodynamic Solute Transport	
Advection	Ground water gradient, hydraulic conductivity, porosity
Dispersion	Dispersivity, pore water velocity
Preferential flow	Pore size distribution, fractures, macropores
Abiotic Solute Transport	
Adsorption	Organic content, clay content, specific surface area
Volatilization	Degree of saturation
Ion Exchange	Cation exchange capacity, ionic strength, background ions
Hydrolysis	pH, competing reactions
Precipitation/Dissolution	pH, other metals
Cosolvation	Types and fraction of cosolvents present
Redox	pE, pH
Colloid Transport	pH, ionic strength, flow rate, mobile particle size, aquifer and particle surface chemistry
Biotic	
Metabolisim/Cometabolism	Microorganisms, nutrients, pH, pE (electron acceptors), trace elements
Multiphase Flow	
	Intrinsic permeability, saturation, porosity

The hydrodynamic process of dispersion (spreading of the contaminant about the mean ground water velocity) is attributed to the distribution of flow paths for the subsurface system and the diffusion of the contaminant. Dispersivity is defined as a soil parameter that describes the spreading due to the distribution of flow paths (when Fickian dispersion is assumed). However, for low ground water flow velocities, molecular diffusion may dominate the contaminant transport. Thus, the dispersivity and ground water velocity of the subsurface system, and the molecular diffusion coefficient of the contaminant, are properties that control the dispersion process.

The abiotic process of sorption (accumulation of the contaminant at the surface of a solid interface — typically the stationary subsurface material) will slow down the movement of the contaminant as it accumulates on the subsurface medium. For neutral organic contaminants and subsurface materials with organic carbon content present, sorption is commonly of hydrophobic (water disliking) chemicals into the soil organic carbon content. The more hydrophobic a chemical, the greater the water disliking characteristic. Thus, as the solubility of a chemical decreases, the potential for the chemical to sorb at an interface is expected to increase. Also, as the organic carbon content of the subsurface material increases, the total capacity of the soil to sorb the contaminant increases. This will result in additional ground water passing through the subsurface material before the sorptive capacity is exceeded and the contaminant appears downgradient. Thus, the organic

Table 1.3 (Continued)

Process	Contaminant Property	Interactions
Hydrodynamic Solute Transport		
Advection	Independent of contaminant	
Dispersion	Diffusion coefficient	Dispersion coefficient
Preferential flow		
Abiotic Solute Transport		
Adsorption	Solubility, octanol-water partition coefficient	
Volatilization	Vapor pressure, Henry's constant	
Ion Exchange	Valency, dipole moment	
Hydrolysis	Hydrolysis half life	
Precipitation/Dissolution	Solubility versus pH, speciation reactions	
Cosolvation	Solubility, octanol-water partition coefficient	
Redox	pK_a	
Colloid Transport	Sorption, reactivity, speciation, solubility	Colloid stability
Biotic		
Metabolisim/Cometabolism	BOD, COD, degree of halogenation, etc.	
Multiphase Flow		
	Solubility, volatility, density, viscosity	Relative permeability, residual saturation, wettability, interfacial tension (surface tension), capillary pressure

Source: Sabatini and Knox, 1992. With permission.

carbon content of the subsurface system and the solubility (hydrophobicity) of the contaminant are two properties that affect the sorption process.

The biotic process of aerobic biodegradation may convert an organic contaminant to another form (metabolites) or to harmless end products (e.g., CO_2 and H_2O). For aerobic biodegradation (metabolism) to function, aerobic microorganisms must be present in the subsurface system. The microorganisms require free oxygen as an electron acceptor (free oxygen is present only at high values of pE), nutrients such as nitrogen and phosphorous, certain trace elements and an acceptable environment (pH, temperature, etc.). These conditions must be satisfied in the subsurface system. For aerobic microorganisms to biodegrade organic contaminants, the process must be energetically favorable for the microorganisms. The contaminant must be readily metabolized (e.g., highly halogenated organic contaminants are typically refractory under aerobic conditions — difficult for microorganisms to metabolize) and present at sufficient concentrations to make the metabolism energetically favorable. The biochemical oxygen demand (BOD) is a test that quantifies the biochemical oxygen equivalent of the organics present in a contaminant and thus indicates if the contaminant is susceptible to aerobic biodegradation. Thus, the aerobic biodegradation process is affected by properties of the subsurface environment (presence of microorganisms, nutrients, free oxygen, etc.) and properties of the contaminant (concentration, BOD, etc.).

REFERENCES

Morris, M. D., J. A. Berk, J. W. Krulik, and Y. Eckstein. "A Computer Program for a Trilinear Diagram Plot and Analysis of Water Mixing Systems," *Ground Water* 21(1, January-February):67–78 (1983).

Piper, A. M. "A Graphic Procedure in the Geochemical Interpretation of Water Analyses," *Trans. Am. Geophys. Union* 25:914–928 (1944).

Sabatini, D. A. and R. C. Knox, *Transport and Remediation of Subsurface Contaminants: Colloidal, Interfacial, and Surfactant Phenomena*, ACS Symposium Series 491, American Chemical Society, Washington, D.C. (1992).

Schoeller, H. *Les Eaux Souterraines* (Paris: Masson and Cie, 1962).

"Statistical Analysis of Ground-Water Monitoring Data at RCRA Facilities: Interim Final Guidance," U.S. EPA Report-530-SW-89-016 (February, 1989).

Stiff, H. A. "The Interpretation of Chemical Water Analysis by Means of Patterns," *J. Pet. Technol.* 3(10):15–17 (1951).

2

HYDRODYNAMIC PROCESSES

2.1 INTRODUCTION

Grouping the various processes affecting the transport and fate of fluid substances in the subsurface is somewhat arbitrary in that most all of the processes are interrelated. However, in order to develop an understanding of the basic mechanics of each of the various processes, some sort of categorized approach is necessary. In this chapter we will address the hydrodynamic processes that affect the subsurface transport and fate of fluids and dissolved constituents. Hydrodynamic processes include those phenomena that result from the physical movement of fluids in the subsurface.

2.1.1 Darcy's Law

The basic equation governing fluid flow through a saturated porous medium is known as Darcy's Law and can be written as shown in Equation 2.1.

$$V = -KS*$$
(2.1)

where V =   specific discharge (Darcy velocity) of fluid (L/t)
         K =   hydraulic conductivity tensor (L/t)
         S^* =   hydraulic gradient (dimensionless)

Darcy's Law is covered in detail in most texts dealing with ground water flow (e.g., Bouwer, 1978; Driscoll, 1986; Davis and DeWiest, 1966; Fetter, 1980; Freeze and Cherry, 1979; Bear, 1979). However, in order to understand the basic hydrodynamic processes affecting subsurface transport and fate, several subtleties related to ground water flow and Darcy's Law need to be highlighted. First and foremost the Darcy velocity, V, (Equation 2.1) is not the actual velocity of the fluid in the porous medium. The Darcy velocity is an average or discharge velocity based on volumetric flow. To calculate the actual (seepage, pore water) velocity of the fluid one can use Equation 2.2.

$$V_s = \frac{V}{\eta} = \frac{-KS*}{\eta} \qquad (2.2)$$

where V_s = seepage (pore water) velocity (L/t)
       η = porosity of the formation (dimensionless)

In most transport and fate analyses, the seepage velocity is the required velocity term.

The second point to note about Equation 2.1 is that the hydraulic conductivity (K) is a function of both the saturated formation and the flowing fluid (water). Hydraulic conductivity is a tensor quantity, meaning it possesses vector qualities (magnitude and direction) for each of the orthogonal (x,y,z) directions. A property specific to the porous medium is commonly referred to as the intrinsic permeability and is calculated as shown in Equation 2.3.

$$k = K\frac{\mu}{\gamma} = \frac{K\mu}{\rho g} \qquad (2.3)$$

where k = intrinsic or physical permeability of the porous medium (L^2)
       μ = dynamic viscosity of fluid (M/Lt)
       γ = unit weight of fluid ($M/L^2/t^2$)
       ρ = mass density of fluid (M/L^3)
       g = gravitational acceleration (L/t^2)

In subsurface transport and fate studies involving immiscible phases or highly contaminated water, the density and viscosity of the fluids will be different than that of water and can have significant effects on transport of the fluids. For these situations, the coefficient of permeability will be needed to assess the fluid conductive properties of the formation.

The third point to note about Equation 2.1 is that, in the strictest sense, the hydraulic gradient (S^*) is applicable only along streamlines (Figure 2.1). The hydraulic gradient represents the rate of change in total (elevation plus pressure)

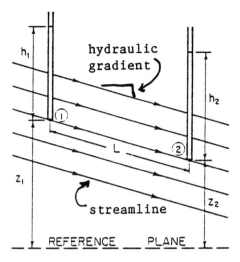

Figure 2.1. Hydraulic gradient and streamlines.

head with respect to position. Often times, Equation 2.1 is written and applied in its finite difference form as shown in Equation 2.4.

$$V = -K \frac{\left(P_2 / \gamma_2 + Z_2\right) - \left(P_1 / \gamma_1 + Z_1\right)}{L} \qquad (2.4)$$

where P/γ = pressure head (L)
       Z = elevation head (L)
       L = distance from point 1 to point 2 (L)

Strictly speaking, L must be taken along a streamline.

The more general form of Darcy's Law is written with differentials as

$$V_\ell = -K_\ell \frac{\partial h}{\partial \ell} \qquad (2.5)$$

where V_ℓ = Darcy velocity in the "ℓ" direction
         K = hydraulic conductivity in the "ℓ" direction
    $\partial h/\partial \ell$ = hydraulic gradient in the "ℓ" direction

Equation 2.5 helps explain the often misunderstood convention of using a minus sign (–) in Darcy's Law. The minus sign is needed because derivatives are taken in the direction of flow, i.e., lower total head minus higher total head. The minus sign assures a positive velocity from high to low total head.

Although Darcy's Law governs flow of a fluid at a point in a saturated porous medium, other more complex equations govern the behavior of the aquifer system as a whole unit. Chapter 7 discusses some of the most common equations governing ground water behavior. In all cases, one of the equations will have to be solved in order to develop the velocity distributions within the porous media. Ground water velocity tends to dominate the hydrodynamic subsurface contaminant transport processes and can influence the degree to which the other processes and/or reactions can occur.

2.1.2 Variations in Hydraulic Conductivity

The influence of hydraulic conductivity on contaminant transport is evident in Darcy's Law. The structure of subsurface formations with respect to hydraulic conductivity is most often anisotropic and heterogeneous. In order to explain anisotropy and heterogeneity, a new function is introduced as

$$\Phi_\ell = \left(-K_\ell h\right) \tag{2.6}$$

where Φ_ℓ = velocity potential in the "ℓ" direction (L^2/t). Applying Darcy's Law to the velocity potential results in

$$V_\ell = \frac{\partial \Phi_\ell}{\partial \ell} \tag{2.7}$$

Anisotropy occurs when a formation has a preferential direction of flow, i.e., when "K_x" does not equal "K_y". Consider the case of an alluvial aquifer formed by the deposition of alternating sequences of coarse materials (sand) and fine materials (silt and clay). It is easy to picture the layering of the formation as depicted schematically in Figure 2.2. Flow of water in the "sand layer" sandwiched between two "clay layers" will be less restricted in the lateral (K_x) directions than in the vertical (K_y) directions.

A formation is referred to as isotropic if the hydraulic conductivity does not vary with orientation within the formation ($K_x = K_y = K_z$). In certain cases, the assumption of isotropic conditions is considered to be sufficient. In those instances Darcy's Law becomes

$$V_\ell = \frac{\partial \Phi_\ell}{\partial \ell} = -\frac{\partial(Kh)}{\partial \ell} \tag{2.8}$$

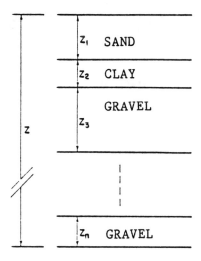

Figure 2.2.   Anisotropic formation due to alternating sequences of deposition.

In essence, the assumption of isotropic conditions allows us to drop the subscript on the hydraulic conductivity tensor.

Homogeneous conditions exist when the hydraulic conductivity is independent of the position of the reference point within the formation. Under homogeneous conditions Darcy's Law can be written as

$$V_\ell = -K_\ell \frac{\partial h}{\partial \ell}$$ (2.9)

The assumption of homogeneity allows us to move the hydraulic conductivity tensor outside the differential because it is now constant in space. Heterogeneity would keep the hydraulic conductivity tensor inside the derivative, i.e., the conductivity changes with position. (Note that Equation 2.9 makes no assumption of isotropy.)

The various combinations of anisotropy and heterogeneity are best explained schematically as in Figure 2.3. In Figure 2.3, the hydraulic conductivity of the formation in two directions at two different locations within the aquifer is depicted as a vector (arrows). The length of each arrow is proportional to the magnitude of the hydraulic conductivity. Also, the appropriate form of Darcy's Law is shown in each case.

Because hydraulic conductivity has a direct influence on the seepage velocity in a saturated formation, it will also have a direct impact on the transport of contaminants in the subsurface. Spatial variability in hydraulic

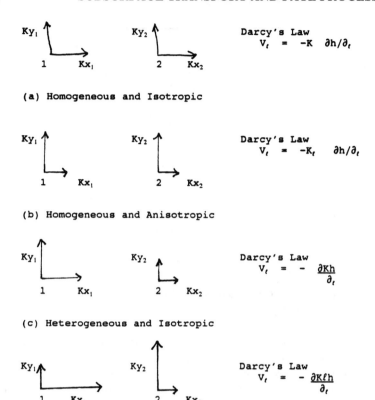

(a) Homogeneous and Isotropic

Darcy's Law
$$V_\ell = -K \ \partial h / \partial_\ell$$

(b) Homogeneous and Anisotropic

Darcy's Law
$$V_\ell = -K_\ell \ \partial h / \partial_\ell$$

(c) Heterogeneous and Isotropic

Darcy's Law
$$V_\ell = - \frac{\partial Kh}{\partial_\ell}$$

(d) Heterogeneous and Anisotropic

Darcy's Law
$$V_\ell = - \frac{\partial K\ell h}{\partial_\ell}$$

Figure 2.3. Variations in hydraulic conductivity.

conductivity is a critical factor controlling mass transport (Smith and Schwartz, 1981). Sposito et al. (1986), in their review of certain mass transport equations, point out that the spatial variability of subsurface solute movements is derived mainly from the variability of the hydraulic conductivity. Smith and Schwartz (1981) point out that uncertainties in the velocity field are due to variations in the hydraulic conductivity and to the random nature of the hydraulic gradient. Additionally, it was noted that hydraulic conductivity determinations do not lead to large reductions in the uncertainty in the velocity field and measurements in homogeneous media are significantly more useful than measurements in heterogeneous media. They noted that hydraulic conductivity values are most effective when determined for areas with large hydraulic gradients, such as those near the water table. Finally, the study noted the structure of the hydraulic conductivity field created the potential for significant errors in transport predictions based on limited data sets. Molz et al. (1983) summarize the importance of hydraulic conductivity on contaminant transport by noting that the inability to specify and understand the hydraulic conductivity distribution in natural aquifers is the most important problem in contaminant transport studies. They also point out that the vertical hydraulic conductivity profile is very important in determining large scale contaminant dispersion properties and the dispersion process in general. This topic will be discussed later in detail.

Heterogeneity in an aquifer system has a pronounced effect on contaminant migration. The difficulty in characterizing subsurface heterogeneities severely limits our abilities to model the transport process. Philip (1980) proposes two types of heterogeneity. The first, deterministic heterogeneity, is the condition where various soil properties (including hydraulic conductivity) vary spatially, and possibly temporally, in a known way. The second type, stochastic heterogeneity, is the condition where the spatial variations of soil properties is irregular, may involve many scales, and is imperfectly known, i.e., it is essentially random. Figure 2.3 attempts to depict deterministic heterogeneity in an aquifer system. Stochastic heterogeneities are the subject of much recent research as will be discussed later.

2.2 SOLUTE TRANSPORT EQUATION

Similar to many of the ground water flow equations, several forms of the solute transport equation can be found in the literature. The basic equation in each case is usually developed by utilizing a conservation of mass approach for a control volume and employing Fick's Law of diffusion. The equation in statement form is

net rate of	flux of	flux of	loss or gain	
change of	= solute out	– solute into	+ of solute mass	
solute mass within	of the	the	due to	
the volume	volume	volume	reactions	(2.10)

Given that this chapter is dealing only with the hydrodynamic processes affecting transport and fate, the third term on the right hand side of Equation 2.10 will not be addressed. The changes in solute mass due to reactions is covered in Chapters 3 and 4.

By postponing consideration of solute reactions, Equation 2.10 becomes simply an accounting of solute fluxes. To date, the literature has focused on two solute fluxes in porous media flow: advection (convection) and hydrodynamic dispersion. Advection is the bulk movement of solute at a velocity equal to the mean velocity of flow in the aquifer system. As noted by Davis et al. (1985) convection may be caused by differences in the water density (natural convection), regional movement of the water (advection), and pumping (forced convection). The terms advection and convection are sometimes used interchangeably; however, the differences between the two should be noted.

Dispersion is the spreading of a solute as it moves through a porous media. Dispersion can be characterized as macroscopic or microscopic, as will be discussed later. The net effect of the dispersive phenomena is to decrease the concentration gradient between two different solutions of the same solute. In essence, dispersion reduces the sharp interface between the two solutions.

In order to grasp the concepts of advective flow and microscopic dispersion, it is necessary to consider an idealized condition (Figure 2.4). Figure 2.4.a shows a soil column initially ($t \leq t_0$) saturated with solute free water (concentration of solute = 0). At some time "t_0" a tracer solution containing a solute (concentration of solute = C_0) is introduced to the top of the soil column. As the solution moves through the column it displaces (flushes out) the solute free water. The effluent concentration of the solute (C) is then monitored over time.

If flow of the tracer through the column were solely advective, a sharp interface between the two fluids would exist continuously as the tracer flushes out the solute free water. In essence, the tracer would travel through the column as a plug. A breakthrough curve, illustrating the relative effluent concentration (C/C_0) vs time, would take the form of the dashed line in Figure 2.4.c. The sharp jump in the breakthrough curve indicates that the solute in the tracer fluid travels along with the bulk motion of the fluid, i.e., the solute moves because the solution is moving. This is known as advection.

In reality, the breakthrough curve for the column effluent would look more like the solid curve in Figure 2.4.c. This curve shows an earlier first arrival of solute and a more gradual increase up to the input concentration, C_0. The difference between the dashed line and solid curve in Figure 2.4.c shows the effects of the phenomenon termed hydrodynamic dispersion.

The microscopic dispersive phenomenon depicted in Figure 2.4.c is the result of two processes: mechanical dispersion and molecular diffusion. Mechanical dispersion is the result of velocity variations in the pore channels and the tortuous nature of flow in the porous medium. The two mechanical dispersion processes are depicted graphically in Figures 2.5.a and 2.5.b.

Molecular diffusion (Figure 2.5.c) is the random movement of molecules in a fluid due to concentration gradients. Molecular diffusion is governed by Fick's Law which states that the diffusion rate is proportional to the concentration gradient or

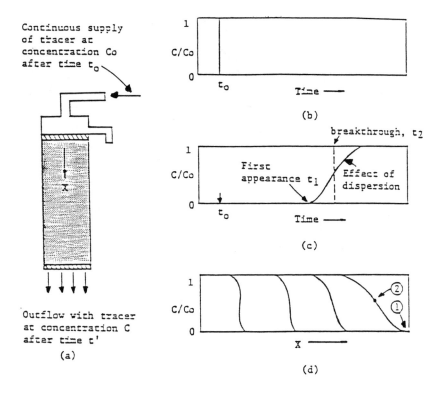

Figure 2.4. Longitudinal dispersion of a tracer passing through a column of porous medium: (a) column with steady flow and continuous supply of tracer after time t_0; (b) step-function-type tracer input relation: (c) relative tracer concentration in outflow from column (dashed line indicates plug flow condition and solid line illustrates effect of mechanical dispersion and molecular diffusion); and (d) concentration profile in the column at various times.

$$\frac{\partial C}{\partial t} = \nabla\left(D_o \nabla C\right) \tag{2.11}$$

where D_o = diffusion coefficient for a given solute species whose concentration is C.

In developing the solute transport equations for porous media flow, Fick's Law is assumed to govern the hydrodynamic dispersion phenomenon or

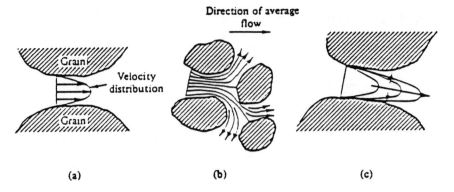

Figure 2.5. Spreading due to mechanical dispersion (a, b) and molecular diffusion (c). (Bear and Verruijt, 1987. Reprinted by permission of Kluwer Academic Publishers.)

$$\frac{\partial C}{\partial t} = \nabla(D\nabla C) \qquad (2.12)$$

where D = the coefficient of hydrodynamic dispersivity or the dispersion coefficient (L^2/t).

The dispersion coefficient has been related to the diffusion coefficient and the mean flow velocity (V_s) by the expression

$$D = D* + \alpha(V_s)^n \qquad (2.13)$$

and

$$D* = \tau D_o \qquad (2.14)$$

where α = medium property called dispersivity (L)
         τ = tortuosity of the medium
         n = constant

Most often, the diffusion coefficient (D^*) in Equation 2.13 is assumed to be negligible, relative to the mechanical dispersion. For systems with low pore water velocities (low hydraulic conductivities or low hydraulic gradients) or low dispersivities, the molecular diffusion may become relatively significant.

Although the above development is based on the one-dimensional soil column shown in Figure 2.4, the dispersive phenomenon is decidedly multidimensional. In simple terms, solutes in flowing ground water will be

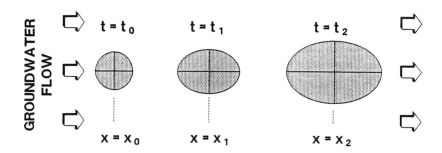

Figure 2.6. Transport of a contaminant slug through a porous aquifer (Palmer and Johnson, 1989).

dispersed both in the direction of flow (longitudinally) and normal (transverse) to the direction of flow. The two-dimensional dispersion of a contaminant slug is depicted schematically in Figure 2.6. For most flow systems, the longitudinal dispersivity tends to be significantly larger (10 to 30 times) than the transverse dispersivity in either the lateral or vertical direction.

By utilizing the mass balance approach for an elemental volume (assuming Fick's Law to govern dispersion and ignoring solute reactions), the two-dimensional equation for a nonreactive, dissolved chemical species in flowing ground water can be written as

$$\frac{\partial C}{\partial t} = \frac{\partial}{\partial x_i}\left(D_{ij}\frac{\partial C}{\partial x_j}\right) - \frac{\partial C V_i}{\partial x_i}$$

$$(2.15)$$

$$i,j = 1,2$$

where i,j correspond to the coordinate directions.

Equation 2.15 is one form of the (Fickian) solute transport equation. The first term on the right side of Equation 2.15 represents the changes due to (microscopic) hydrodynamic dispersion. The second term on the right side represents the changes due to convection. Hence, Equation 2.15 is also referred to as the convection-dispersion equation (CDE).

It is extremely important to be cognizant of the assumptions used in developing the solute transport equation. The solute transport equation is applicable to those situations involving dissolved constituents. The behavior of two miscible fluids of differing solute concentrations is often referred to as miscible displacement. Miscible displacement refers to the contact of two fluids of differing solute concentrations with no interfacial tension effects, i.e., the two fluids are miscible. The basic assumption used in miscible displacement studies is that the solute concentration does not appreciably affect the density or viscosity of the fluid. Hence, the two fluids, although of different solute concentrations, behave as a single phase. When density and viscosity differences are significant, multiphase flow theory is used. Multiphase flow, as discussed in the latter part of this chapter, involves immiscible fluids.

Equation 2.15 can be expanded to include solute reactions, artificial sources and/or sinks, and three-dimensional flow systems. However, for the remainder of this chapter, Equation 2.15 will serve as the basis for our discussion of hydrodynamic processes.

2.2.1 Problem Scale

Subsurface contaminant transport processes are interesting in that they can be approached and analyzed, and they manifest themselves, at various scales. These scales range from molecular (e.g., surface adsorption sites) to regional (e.g., the long term accumulation of nitrates in a regional aquifer). In analyzing the hydrodynamic subsurface transport and fate processes, it is important to be aware of the scale at which the various processes are occurring.

Bhattacharya and Gupta (1983) give a good comparative description of three different space-time scales regarding dispersion during solute transport. The first scale is the kinetic scale, where molecular displacements are governed by interactions of the solute molecules with the liquid and solid phases. The second scale, the microscopic scale, is where the liquid and solid phases are considered to form a heterogeneous continuum, and solute displacements are represented by considering the stochastic dynamics of the (small) kinetic scale. The Darcy scale is the largest of the three scales, and solute movements at this scale are represented as averages of the microscopic scale.

At this point it is necessary to discuss the averaging of values at various scales. Bear and Verruijt (1987) describe the concept of a representative elementary volume (REV) which can be used in a universally applicable averaging procedure. The basic concept of the REV is to select a volume of the subsurface such that the averaged values of all geometrical characteristics at any point in the medium will be a single valued function of that point, i.e., the medium beomes a continuum. The importance of the REV concept is revealed in most theoretical works involving stochastic analyses (averaging) of solute transport phenomena.

2.2.2 Microscopic Dispersion

As noted above, the dispersive phenomenon can be analyzed at various scales. The three mechanisms contributing to dispersion depicted in Figure 2.5 occur at the "pore channel" or "microscopic" level. However, these mechanisms are discussed only in terms of the freely flowing or bulk water in the pores. Taking the scale of the problems down even further, to the "soil particle" level, allows us to analyze the solute transport phenomena at a smaller scale. As discussed by Crittenden et al. (1986), the pore channel is occupied by both the freely flowing or bulk solution, the fixed film or stationary water surrounding the soil particles, and aggregates. Analyzing the system at this scale, as shown in Figure 2.7, requires that we include additional processes. In addition to the advective and dispersive phenomena in the bulk solution, there is also mass (solute) transport from the bulk solution to the stationary phase. Once in the stationary phase, the solute is then subjected to adsorption, surface and pore diffusion, and/or transfer through the film back to the bulk solution. If the stationary phase processes are ignored, the dispersive phenomenon reverts back to the microscopic scale depicted in Figure 2.5. The net effect of these microscopic mechanisms is to cause additional spreading of the solute.

If one considers the problem at the "laboratory scale", the effects of dispersion can be analyzed with breakthrough curves, as shown in Figure 2.4. Soil columns are traditionally considered to be one-dimensional; however, the dispersive phenomenon is always multi-dimensional. Figure 2.8 is a schematic depiction of the multi-dimensional dispersion of a solute in a one-dimensional flow field. The tracer slug is being advected to the right. Dispersion occurs both in the direction of flow (longitudinally) and transverse to the direction of flow. Wang and Anderson (1982) note that there is greater dispersion in the direction of flow, but transverse dispersion does occur. Figure 2.8 really depicts the macroscopic outcome of the microscopic mechanisms.

A final point relative to the problem of scale involves comparing the Darcy and seepage velocities as outlined in Equation 2.2. The Darcy velocity itself represents a volumetric average velocity; hence, the seepage velocity is itself an averaged value at the Darcy scale. Referring to Figure 2.8, one can see the effects of the averaging process. Although the average seepage veloc-

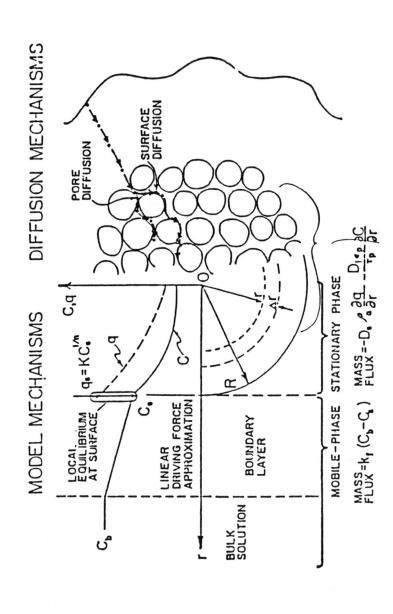

Figure 2.7. Microscopic Intra-aggregate Mass Transport Mechanisms. (Crittenden, J. C., N. J. Hutzler, D. G. Geyer, J. L. Oravitz, and G. Friedman, *Water Resour. Res.*, 22:271–284, 1986. Copyright by the American Geophysical Union.)

Figure 2.8. Statistical distribution of flow paths around local heterogeneities leads to dispersion. The process is shown here at
a microscopic scale where pore space surrounds gravel-
sized grains. (R. A. Freeze and J. A. Cherry, GROUND-
WATER©, 1979, p. 384. Reprinted by permission of
Prentice Hall, Englewood Cliffs, New Jersey.)

ity is some value representing one-dimensional (x-direction) solute transport,
it is apparent from the figure that the seepage velocity is an average of the
two-dimensional (x and y) pore channel velocities.

If one were to examine the individual pores (an unrealistic option) it
would be noted that each pore channel has a unique velocity distribution.
Heller (1963) notes that Darcy's Law is simply not valid at this scale. The
velocity distribution within the pores would have to be developed by applying
the Navier-Stokes equation to the fluid within the actual microscopic pore
boundaries.

2.2.3 Macroscopic Dispersion (Macrodispersion)

Earlier discussions of the dispersive phenomenon focused on the velocity
variations at the microscopic or pore channel level. As outlined in the
discussion of hydraulic conductivity variations, aquifers tend to be
heterogeneous at the macroscopic or field level (i.e., there exists significant
variations in hydraulic conductivity with respect to position within an aquifer).
Smith and Schwartz (1981) point out that these heterogeneities in hydraulic
conductivity create the spatial variation in the velocity field that actually
causes macroscopic dispersion or macrodispersion, Figure 2.9.

Understanding macrodispersion does not necessarily make its quantification

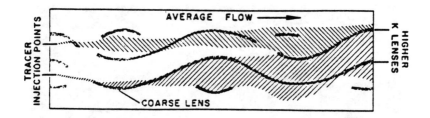

Figure 2.9.   Macroscopic Dispersion (Reprinted with permission from *Groundwater Contamination,* **1984. Published by National Academy Press, Washington, D.C.).**

an easy task. Variations in hydraulic conductivity (heterogeneities) are not easy to assess. Davis et al. (1985) note that local differences in hydraulic conductivity can cause local flow directions to be distorted and the actual directions of flow will diverge from the directions predicted on the basis of spaced wells used in aquifer analyses, such as aquifer pumping tests.

2.2.4 Dispersion Coefficients

Of all the parameters discussed in this chapter that enter into contaminant transport calculations, the dispersion coefficient tends to be the most difficult to determine. The dispersion coefficient has been promoted as being a function of the aquifer dispersivity and the mean flow velocity (Equation 2.13) raised to some power "n". However, the value of the exponent "n" is the subject of much research and disagreement.

Taylor (1953) used one-dimensional capillary tube analysis to show the dispersion coefficient to be proportional to the second power of the velocity. Aris (1956) extended Taylor's analysis to straight tubes of any cross-section. He showed the dispersion coefficient to be linearly proportional to the mean velocity of flow.

Scheidegger (1957) used a statistical approach to analyze dispersion in porous media. He summarized his analysis by promoting two possible relationships between the dispersion coefficient and the velocity, i.e., "n" to be either 1 or 2. Saffman (1959), in his statistical analysis of dispersion, showed that the longitudinal dispersion coefficient (D_L) and the transverse dispersion coefficient (D_T) are related to the mean flow velocity by:

$$D_L = V_s \ln(V_s A) \qquad (2.16a)$$

$$D_T = V_s \qquad (2.16b)$$

where A is a constant which depends on the molecular diffusion coefficient.

De Jong (1958) used the Markov process, which is equivalent to having a completely mixed cell. He also showed that the dispersion coefficient is linearly related to the mean flow velocity. Bear (1961), in his one-dimensional interconnecting array of small cells, found the dispersion coefficient to be linearly proportional to the average flow velocity.

Brigham et al. (1961) used the model of a bundle of capillary tubes and experimentally determined the dispersion coefficient in packs of granular material (beads and sandstones) and consolidated Berea sandstone. They found the dispersion coefficient to be proportional to a power of the velocity. The exponent "n" in these studies ranged from 1.24 to 1.19. They also concluded that the exponent is dependent on the porous media and lies between 1.00 and 2.00.

Rumer (1962) used quartz gravel and glass beads as a porous media for measuring the dispersion coefficient in steady and unsteady state flow. He reported the dispersion coefficient to be proportional to the flow velocity. For unsteady flow the proportionality was linear; for steady flow it was found to be nonlinear.

Perkins and Johnston (1963) conducted dispersion tests on unconsolidated sands and glass bead packs and found the longitudinal dispersion coefficient to vary with the ground water velocity. Figure 2.10 illustrates the ratio of dispersion coefficient to diffusion coefficient vs the dimensionless Peclet number. Perkins and Johnston (1963) concluded that at sufficiently low interstitial velocities (low Peclet numbers), molecular diffusion will equalize the concentration of solute within each pore space and have a significant effect on the dispersion coefficient. However, at high flow rates (high Peclet numbers), time for diffusion to equalize solute concentrations is insufficient and the dispersion coefficient is dominated by advection. In regions where both diffusion and dispersion are present, the total dispersion coefficient is represented by the sum of the two individual coefficients, as shown in Equation 2.13.

Klotz and Moser (1974) conducted about 2500 tracer injection tests on columns packed with different sands. They used the statistical model of Scheidegger (1957) to calculate the dispersion coefficient. The results of their experiments showed that the dispersion coefficient depends linearly on the velocity as long as the mixing effect is primarily due to dispersion.

Klotz et al. (1980) extended their laboratory work to a field study, where they used the results of 4,000 tests. They reported that the linear correlation of the dispersion coefficient with the flow velocity is not valid below a certain discharge velocity.

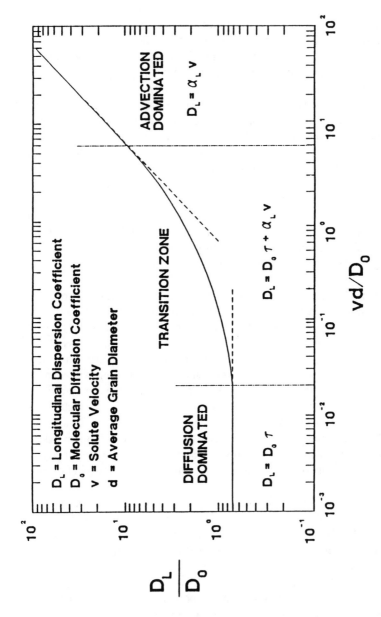

Figure 2.10.   Dispersion coefficient as a function of Peclet number (Palmer and Johnson, 1989).

Peters et al. (1984) used the continuum approach to calculate the dispersion coefficient by unequal viscosity miscible displacements. They found that the dispersion coefficient is proportional to the "nth" power of the flow velocity. In their experiments with consolidated rocks, they reported the exponent n of velocity to be 1.05.

Lee and Okyiga (1986), in their statistical model of thin permeable porous media, showed the dispersion coefficient to be proportional to the second power of the flow velocity.

Sahimi et al. (1986) used a Monte Carlo simulation of a continuous time, random walk model of the dispersion process in chaotic porous media. They showed the dispersion coefficient to be proportional to the velocity with the exponent being 1.27.

The vast majority of dispersion coefficients are determined for longitudinal dispersion using laboratory column experiments exemplified by Blackwell (1962), Ebach and White (1958), Klotz and Moser (1974), Brigham et al. (1961), Harleman and Rumer (1963), Legatski and Katz (1967), and Salter and Mohantz (1982). These experiments have traditionally been conducted on either glass beads or unconsolidated sand-packed columns.

Recently, Menzie et al. (1988) have conducted longitudinal dispersivity tests on consolidated Berea sandstone cores. These experiments were conducted to analyze the dispersive phenomenon and to obtain any correlations between the dispersion coefficient and the rock properties (e.g., porosity, grain size, permeability, etc.). The dispersivity reported based on the laboratory experiments falls in the range of 10^{-4} to 10^{-2} m.

Dispersion coefficients have also been calculated by application of a numerical method. In this case, the contaminant distribution is known and the numerical model is used to identify the dispersion coefficient (the inverse problem). It has been found that the dispersion coefficient is a function of the modeling procedure (Robson, 1974). The values for dispersivity based on modeling studies at the field scale fall into the range of 1 to 100 m (Anderson, 1979).

Dispersion coefficients have also been obtained from various field tests. These tests entail the injection of a tracer at one well and the observation of the tracer at several observation wells. Detailed methods of analysis for these types of tests are contained in Smith (1978), Sudicky, et al. (1983), Santz (1980), and Dieulin (1980). The values of dispersivity from these field tests have been reported in the range of centimeters to meters.

Pickens and Grisak (1981) provide an overview of dispersivity values obtained by different methods. In addition to tabulating numerous values from a variety of sources, they also identify the range of values reported. They report laboratory dispersivity values ranging from 0.01 to 1 cm, while values obtained from computer modeling studies at the field scale range from 12 to 61 m. Intermediate values obtained from various types of field tracer tests ranged from 0.012 to 15.2 m.

Knopman and Voss (1987) studied the sensitivity of the dispersion coefficient and found it to be at least an order of magnitude less sensitive than the sensitivity of the average linear velocity. This has significant implications for monitoring requirements, in that it implies that more effort is required to estimate the dispersion coefficient. The study also noted that designs aimed at optimizing determination of a parameter by reducing a certain variance may do little to reduce other variances; hence, sampling design requires a multiobjective approach.

2.2.5 Scale Effects

As noted in the previous discussion, the values for dispersion coefficient or dispersivity tend to be a function of the type of test used. As a general rule, dispersivities determined in the laboratory tend to be much less than those obtained from field studies. The apparent increase of either the dispersion coefficient or the dispersivity with the overall dimension of the region through which solute transport occurs is called the scale effect (Sposito et al., 1986).

The microscopic pore scale velocity differences depicted in Figure 2.5 are not the dominant dispersive processes in stratified, heterogeneous (real world) aquifers. It appears that most field scale dispersion is caused by larger scale velocity variations between the numerous sublayers and other heterogenieties that occur in an aquifer, a process called differential advection (Wheatcraft and Tyler, 1988). Schwartz (1977) noted that macroscopic dispersion in a porous medium results from large-scale contrasts in hydraulic conductivity. This study was also early to promote the idea that dispersivities for some media may not be unique. The study emphasized the need for identifying the true nature of the hydraulic conductivity distribution. Without a clear understanding of this distribution, dispersivity loses its identity as either a spatially dependent coefficient or a unique constant reflecting features of the structure of the porous medium.

Guven et al. (1984) found that nonuniformities in the hydraulic conductivity profile may produce large values of longitudinal macrodispersivity in a stratified aquifer. They also note that the longitudinal macrodispersivity, unlike its microscopic counterpart, will generally not be a unique physical property of an aquifer, but will depend on the travel distance and flow conditions.

Gelhar et al. (1979) note that scale problems are attributable to both the variations in hydraulic conductivity and the fact that large distances are required for the solute transport phenomena to become Fickian in their behavior. In a subsequent work, Gelhar and Axness (1983) note that the three-dimensional statistical anisotropy of aquifer heterogeneity, and the asymmetry associated with its orientation, are essential elements of the natural dispersion process.

2.2.6 Stochastic Convection Dispersion Models

Realizing that both deterministic and stochastic heterogeneities in the hydraulic conductivity distribution exist, several researchers have developed stochastic convection-dispersion models for solute transport. Detailed descriptions of these models, along with applications, can be found in Gelhar et al. (1979), Matheron and de Marsily (1980), Gelhar and Axness (1983), and Dagan (1984). Basically, the stochastic models involve coupling the concepts from the theory of random processes with the physical laws governing the flow of water and solutes in the subsurface. The variability of solute movement is assumed to be governed by variations in hydraulic conductivity. These variations are assumed to be random processes that are amenable to stochastic analyses.

Sposito et al. (1986) review the fundamental problems with stochastic models. They note that most of the stochastic models involve averages for a collection of aquifers rather than for a specific real aquifer. Molz et al. (1986) also point out that the specific real aquifer must be statistically homogeneous and ergodic, a condition that does not allow for the real world variability patterns of hydraulic conductivity. Sposito et al. (1986) considered the above limitations and concluded that "much more theoretical research is required and the stochastic convection-dispersion model does not yet warrant unqualified use as a tool for physically based, quantitative applications of solute transport theory to the management of solute movement at field scales."

2.2.7 Non-Fickian Models

An outgrowth of the realization of the scale dependence of dispersion and the use of stochastic analyses is the promotion of non-Fickian behavior. Simply stated, many researchers have found that contaminant transport does not adhere to the model developed by assuming validity of Fick's Law for describing hydrodynamic dispersion.

Examination of Equations 2.11, 2.12, and 2.13 can give possible insight as to the problems underlying contaminant transport models developed under Fickian assumptions. The basic assumption of Fick's Law is that the driving force for diffusion is the concentration gradient, i.e., Equation 2.11 applies. Fick's Law is then assumed to be applicable to hydrodynamic dispersion as shown in Equation 2.12. The dispersion coefficient D is assumed to be a function of both the chemical species D^* and the porous medium αV_s^n, as shown in Equation 2.12. However, the term D^*, which is the diffusion coefficient extracted from Fick's Law, is then assumed to be negligible. This, in essence, is assuming the pore channel velocities and their variations to be driven by concentration gradients. This questionable assumption suggests that the development of the basic mathematical expression for contaminant transport through porous media discussed above may be invalid.

Several studies have shown non-Fickian behavior of dispersion in porous media flow. Gelhar et al. (1979) used stochastic analyses to identify several facts related to non-Fickian behavior. Their results include

1. The mean transport process in a stratified aquifer becomes Fickian after long travel times.
2. The approach to Fickian flow is asymptotic, resulting in significant non-Fickian transport early in the process.
3. During development of the dispersion process there are significant departures from the classical normal concentration distribution associated with Fickian processes.

The study proposed a modification of the traditional convection-dispersion equation as

$$\frac{\partial C}{\partial t} + U \frac{\partial C}{\partial x} = (A + \alpha_L)U \frac{\partial^2 C}{\partial x^2} - B \frac{\partial^3 C}{\partial t \partial x^2} - BU \frac{\partial^3 C}{\partial x^3} \qquad (2.17)$$

where U = mean flow velocity (L/t)
           α_L = longitudinal dispersivity (L)
           A, B = coefficients

Although this study promotes the use of a non-Fickian model, it was focused on the relation of dispersion coefficients to heterogeneities in the porous medium.

In a related study, Gelhar and Axness (1983) examined three-dimensional macrodispersion using stochastic analyses. A significant conclusion of this study was that the classical gradient transport (Fickian) relationship was valid only for large-scale displacements.

Matheron and de Marsily (1980) examined the flow of solutes parallel to planes of stratification. This study identified a non-Fickian type of transport that never conforms to the usual convection-dispersion equation. Their modification of the equation allowed the dispersion coefficient D to be a function of time; however, they noted that this solution was an artifact. The study concluded "a better mathematical formulation for the transport process in porous and fractured media, for all times, seems necessary."

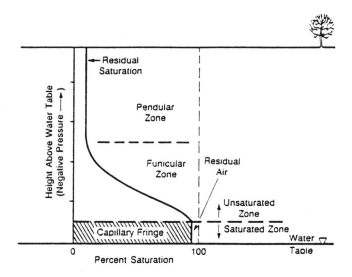

Figure 2.11. Schematic of the unsaturated and saturated subsurface zones (Abdul, 1988).

2.3 PERMEABILITY, SATURATION, AND HYSTERESIS

To grasp the concepts of transport in the vadose zone and multiphase flow, it is first necessary to understand the physical nature of the subsurface environment in terms of its ability to hold and transport fluids. Figure 2.11 is a depiction of moisture distribution in the vadose and saturated zones of the subsurface; it is called a water characteristic curve. The water characteristic curves shows the volumetric water content (abscissa) vs the pressure head (ordinate) for the subsurface. The pressure heads are negative, and they are equal in magnitude to the height of the water above the water table. The negative value of the pressure heads represents the fact that the water in the vadose zone is held in the pores by surface tension effects and could only be removed by applying work to the fluid.

The pressure head value at which the water content first begins to decrease is called the air-entry value of the soil. The air entry value is approximately equal to the equilibrium height of the capillary fringe. This is the point at which air becomes continuous in the soil pores.

The volumetric water content of soils above the water table approaches a constant value, below which the water content will not fall. This value is

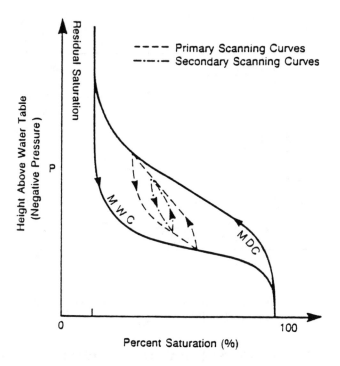

Figure 2.12. Hypothetical curves for a soil showing the main drainage curve (MDC), the main wetting curve (MWC), and primary and secondary curves (Abdul, 1988).

referred to as the irreducible saturation. The fact that the soils contain water at or above the finite irreducible saturation value (including some saturated pores) shows the inappropriateness of the term "unsaturated zone" to describe soils above the water table.

A few subtle points relative to Figure 2.11 need to be highlighted. First, the capillary fringe above the water table contains some (but not all) saturated pore spaces. The water in the capillary fringe is held in place by capillary forces; hence, the absolute pressure of the fluid is less than atmospheric. Second, the water table, located below the capillary fringe, is actually defined as the level at which the fluid is at atmospheric pressure. Third, the unsaturated pore spaces are occupied by both water and air. The air in the pore spaces will affect the movement of infiltrating fluids.

Another concept that needs to be explained is the phenomenon of hysteresis. Figure 2.12 is a hypothetical water characteristic curve for a soil under two different conditions. The upper curve depicts the water content distribution above the water table after the water table has been lowered. This is referred to as the main drainage curve (MDC). The lower curve is the main wetting curve (MWC) and it depicts the water content distribution above a rising

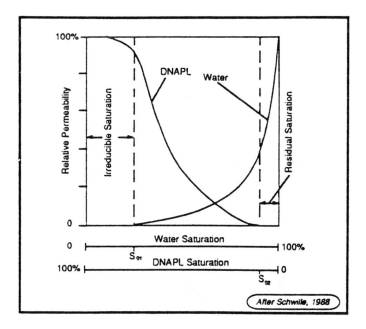

Figure 2.13. Two-phase flow relative permeability (Huling and Weaver, 1991).

water table. The term "hysteresis", as applied to water characteristic curves, simply means that a soil will "wet" and "drain" differently. Hysteresis is due in part to entrapped air that remains for some time in the soil pores after the soil has been wetted (Bouwer, 1978).

As described in the beginning of this chapter, the intrinsic permeability is the characteristic of a porous medium that is indicative of its ability to transmit fluids. In van Dam (1967), two important points are made relative to permeability in multiphase flow systems.

1. The permeability of a given porous medium for one fluid in the presence of another fluid is reduced with respect to the single-phase permeability.

2. The reduction in permeability of the porous medium is dependent on the wetting of the medium by one of two fluids.

In other words, because the porous medium is now transmitting two immiscible fluids, it must now share its capacity (flow area) among different phases. Also, the capacity given to each fluid is less than if a single fluid were present. This reduced value is called the relative permeability and is defined as the ratio of the permeability at a given saturation to the permeabil-

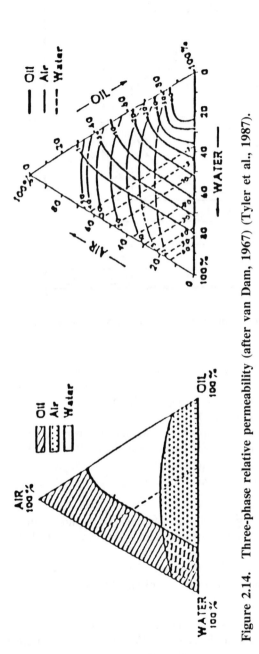

Figure 2.14. Three-phase relative permeability (after van Dam, 1967) (Tyler et al., 1987).

ity at 100% saturation. Figure 2.13 shows the relative permeability of a two-phase (oil-water) system. Figure 2.14 shows the relative permeability of a three-phase (oil-water-air) system.

The saturation percentage of the pore spaces by the various fluids represents one of the most important parameters in developing multiphase flow theory. Simply stated, the saturation is a volumetric percentage of the pore spaces occupied by a given fluid. Saturation affects the relative permeabilities (Figures 2.13 and 2.14) and the capillary pressures of the various fluids in the multiphase flow system.

Figures 2.15 and 2.16 depict hydrocarbons migrating freely through both the saturated and vadose zones; however, certain conditions must be met before this flow can occur. van Dam (1967) noted that there must be a minimum saturation for a fluid in a porous medium. Some of the fluid will be retained by the medium as "residual saturation". If enough fluid is introduced to the formation to exceed the needs of residual saturation, the excess fluid will continue to flow through the medium.

The effects of residual saturation can be seen in Figure 2.14. The graph shows that the relative permeability of both fluids is zero until enough fluid is available to overcome the residual saturation needs. At this point, the relative permeability becomes a positive value and flow can then occur.

2.4 EFFECTS OF DENSITY

The solute transport equation assumes that there are no significant density differences between the fresh water and the solute moving through the formation. In some cases, the concentration of dissolved constituents are large enough to create densities higher than that of fresh water. The effect of the increased density is to impart an increased vertical (downward) component to the ground water velocity. Frind (1982) used the concept of equivalent fresh water head to calculate vertical velocities induced by density. Density differences and vertical velocities are important when considering highly concentrated solutes such as seawater or oilfield brines, and when dealing with heavy immiscible organic fluids.

2.5 MULTIPHASE FLOW

The convective-dispersive solute transport equation discussed previously is only applicable to single-phase flow. The solute transport equation applies

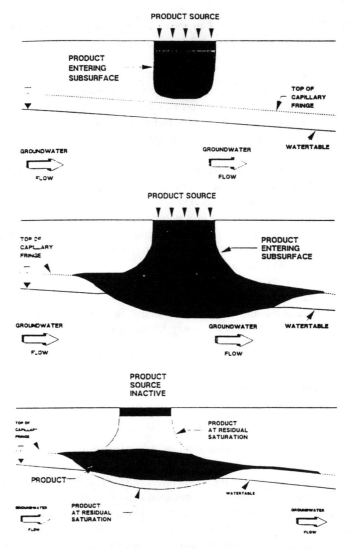

Figure 2.15.   Movement of LNAPLs into the subsurface: (a) distribution
                of LNAPL after small volume has been spilled; (b)
                depression of the capillary fringe and water table; (c)
                rebounding of the water table as LNAPL drains from
                overlying pore spaces (Palmer and Johnson, 1989).

to two fluids which differ only in the concentration of a solute. Basic to the
development of the solute transport equation is the assumption that the two
fluids behave as a single (usually aqueous) phase. The aquifer system therefore consists of only soil particles and a single liquid phase in the pore
channels.

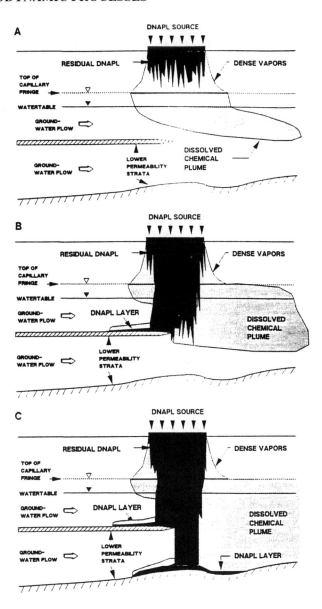

Figure 2.16.   Movement of DNAPLs into the subsurface: (a) distribution
                of DNAPL after small volume has been spilled; (b) distri-
                bution of DNAPL after moderate volume has been spilled;
                (c) distribution of DNAPL after large volume has been
                spilled; (Palmer and Johnson, 1989).

Increasingly, organic chemicals with limited aqueous solubilities are being released to subsurface environments and threatening ground water resources. Chemicals with limited aqueous solubilities are often referred to as immiscible with the aqueous phase. The terms miscible and immiscible actually refer to the absence or presence of capillary pressures between two fluids, respectively. Immiscible fluids would be those that have significant capillary pressures resulting from interfacial tension effects.

An example of two immiscible fluids is oil and water. In most cases, oil can be assumed to have a negligible solubility in water and vice versa. In general, hydrocarbons as a whole tend to be immiscible with water. Most of the current work dealing with multiphase flow in ground water systems is an outgrowth of petroleum industry interest in enhanced hydrocarbon recovery through chemical flooding. The reader is referred to Buckley and Leverett (1942), Fatt and Dykstra (1951), Leverett and Lewis (1941), Muskat et al. (1937), and van Dam (1967) for more information on hydrocarbon recovery.

Because of the widespread occurrence of fluids immiscible with water, general terms and acronyms have recently been promulgated. The general term for the immiscible fluids is non-aqueous phase liquids (NAPLs). With the increased interest in remediation of subsurface hydrocarbon contamination, extensive information regarding the transport and fate of solutes and NAPLs in the subsurface has been developed. Several excellent summary articles on subsurface transport and fate can be found in the recent literature. The interested reader is referred to Mackay, et al. (1985), McCarthy and Zachara (1989), Mackay (1989), Johnson, et al. (1989), Mercer and Cohen (1990), and Huling and Weaver (1991). Generally, NAPLs are divided into two classes based on density: those lighter than water (LNAPLs), and those denser than water (DNAPLs). Most natural and refined hydrocarbons are LNAPLs, while most chlorinated organics and solvents are DNAPLs.

In considering the transport of NAPLs in the subsurface, density and viscosity are of prime interest (Corapcioglu and Hossain, 1986). Density of an immiscible organic fluid is the parameter which delineates LNAPLs from DNAPLS, i.e., the "floaters" from the "sinkers". Viscosity is a measure of fluid resistance to flow. A less viscous fluid will more readily penetrate a porous media. From Equation 2.3 we see that fluids with higher densities and lower viscosities than water will be more mobile in the subsurface than water (Huling and Weaver, 1991)

NAPL migration in the subsurface is affected by the characteristics of the release scenario (volume of release, area of infiltration, duration), properties of both the LNAPL and subsurface media, and subsurface flow conditions (Feenstra and Cherry, 1987). The general migration pattern of LNAPLs released to the subsurface is depicted in Figure 2.15. The product will migrate downward only if a sufficient amount is available to meet residual saturation needs of the soil material. The product continues downward migration until the capillary

Table 2.1 Common Hydrocarbon Contaminants in Ground Water

Hydrocarbon	Solubility (g/l of Water)	Density at 4°C (g/cm)	Viscosity at 4°C (cP)
Tetrachloroethylene	0.15	1.623	0.89
1,1,2,2-Tetrachloroethane	2.90	1.595	1.84
Carbon tetrachloride	0.80	1.594	0.97
Choloroform	8.22	1.483	0.58
Trichloroethylene (TCE)	1.10	1.464	0.58
1,1,1-Tricholoroethane (TCA)	4.40	1.339	1.18
Methylene chloride	20.00	1.327	0.513
1,2-Dichloroethane	8.69	1.2351	0.836
Hexacholoroethane	0.05	2.2091	2.26
Benzene	1.78	0.88	0.653
Toluene	0.52	0.86	0.588
Water	—	1.00	1.0019

Source: Corapcioglu and Hussain, 1986.

fringe is reached, at which point the product will tend to mound. The increased head will depress the water table. The product can move laterally due to capillary forces in the vadose zone (Tyler et al., 1987).

The general migration pattern of DNAPLs released to the subsurface is depicted in Figure 2.16. The product will move downward through the subsurface in a non-uniform pattern. The "viscous fingering" depicted in Figure 2.16 occurs due to the unstable flow of DNAPLs through water. At the capillary fringe, the DNAPL will mound slightly to build enough head to displace the water held by capillary forces to the soil medium. The DNAPL will continue downward through the saturated zone until residual saturation is reached or an impermeable soil zone is encountered. After ponding on the impermeable layer, the DNAPL will spread out laterally due to gravitational forces (head differences) and viscous drag exerted by the flowing ground water (Parker et al. 1986).

It should be noted that in Figures 2.15 and 2.16, both the LNAPLs and DNAPLs are depicted as not totally immiscible. Some of the product goes into solution in the saturated zone. Although their solubilities are limited (Table 2.1), hydrocarbon constituents do have a small propensity to go into aqueous solution.

In subsurface flow of immiscible fluids, a gaseous phase could occupy part of the available pore spaces. Obviously, in the vadose zone overlying a fresh water aquifer, the pore spaces are occupied by both air and tightly held water in the form of residual saturation. However, the gaseous phase does not have to be air. For the case of gasoline in the subsurface, volatilization of the free product leads to gasoline vapors in the pore spaces. Therefore, in the unsaturated zone, the transport of a contaminant could include three phases: solutes in the aqueous phase, vapors in the air phase, and unaltered constituents in the immiscible phase (Corapcioglu and Hossain, 1986).

In the saturated zone below the water table, immiscible contaminant transport is usually assumed to involve two phases: the aqueous phase and the immiscible phase. In most work, these two phases are assumed to be distinct and each behaves as a separate continuum (Reible et al., 1986). However, as discussed in Chapter 7, recently developed models now allow for interaction between phases. The complexity of the mathematical statement is increased because the number of phases increases.

2.6 FLOW AND TRANSPORT IN THE UNSATURATED ZONE

When considering the structure of the porous medium above the water table, it is important to identify the fluids existing in the pore spaces of the solid matrix. Theoretically, unsaturated flow should be treated as a two phase system of water and air. The usual approach is to analyze only the flow of water and consider the air as part of the solid phase (Bouwer, 1978).

The movement of fluids through the unsaturated zone is analyzed using Darcy's Law. However, the analysis is complicated by the fact that the hydraulic conductivity is dependent on the water content, which is related to the pressure head as shown in the water characteristic curves discussed earlier. The volumetric water content also directly affects the transport of solutes in the unsaturated zone. The solute velocity and the coefficient of dispersivity are both functions of the volumetric water content.

The equations governing flow of water and solute transport through the vadose zone are similar to those discussed previously for saturated flow. The equations are complicated by the dependence of hydraulic conductivity, velocity, and the dispersion coefficient on the volumetric water content. The one-dimensional form of the flow equation is

$$\frac{\partial}{\partial z}\left(K(\psi)\frac{\partial \psi}{\partial z}\right) + \frac{\partial}{\partial z}\left[K(\psi)\right] = \Gamma(\psi)\frac{\partial \psi}{\partial t} \qquad (2.18)$$

where $K(\psi)$ = the hydraulic conductivity
         $\Gamma(\psi)$ = the specific water capacity
         ψ = the pressure head

Equation 2.18 is a nonlinear partial differential equation (Palmer and Johnson, 1989), which increases the complexity of its solution.

The one-dimensional form of the advection-dispersion equation for solute transport in the vadose zone is (Palmer and Johnson, 1989)

$$\frac{\partial}{\partial z}\left(\Theta_w D_v \frac{\partial C}{\partial z}\right) - \frac{\partial(qC)}{\partial z} = \frac{\partial(\Theta_w C)}{\partial t} \qquad (2.19)$$

where C = solute concentration
 D_v = vadose zone dispersion coefficient
 Θ_w = volumetric water content
 q = volumetric water flux

The dispersion coefficient is assumed to be analogous to the coefficient in the saturated solute transport equation

$$D_v = D_o \tau + \alpha V(\Theta_w) \qquad (2.20)$$

where $V(\Theta_w)$ = solute velocity.

2.7 FRACTURED MEDIA FLOW

The previous discussions on solute transport and multiphase flow focused on continuous porous media. Another type of subsurface formation that transmits fluids is a fractured formation. Fractured formations can be viewed as a solid rock matrix, typically with a low permeability, that is cracked or fractured. Owing to the low permeability of the rock matrix, most fluid flow through the formation will be through the interconnected fracture channels.

The study of contaminant transport in fractured rocks is in its infancy, yet it is of considerable importance (Anderson, 1984). Certain types of fractured rocks are being considered as repositories for high-level radioactive waste. Clays, which are also susceptible to fracturing, are currently used to contain a variety of municipal and industrial wastes.

The velocity of fluids moving through fractures has traditionally been approximated using the so-called cubic law. Grisak and Pickens (1980) note that the cubic law is developed from solution of the Navier-Stokes equation for nonturbulent flow of a viscous, incompressible fluid between two parallel plates. The equation is developed by considering the conceptual model shown in Figure 2.17. The average velocity parallel to the planar surface is devel-

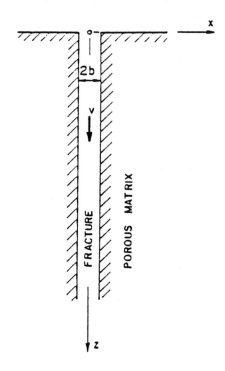

**Figure 2.17.   Fracture-matrix system (Tang., D. H., Frind, E. O., and
Sudicky, E. A., *Water Resour. Res.*, 17:555–564, 1981.
Copyright by the American Geophysical Union).**

oped by using Darcy's Law and the equivalent hydraulic conductivity in the
form of

$$V = -K_f \frac{dh}{dz} \tag{2.21a}$$

$$K_f = \left(\frac{\rho g}{12\mu} \right) (2b)^2 \tag{2.21b}$$

where V = average velocity in the fracture (L/t)
    K_f = equivalent hydraulic conductivity in the fracture (L/t)
    dh/dz = hydraulic gradient
    ρ = fluid density (M/L^3)
    g = acceleration of gravity (L/t^2)
    μ = dynamic viscosity of fluid (M/Lt)
    $(2b)$ = fracture aperture (L)

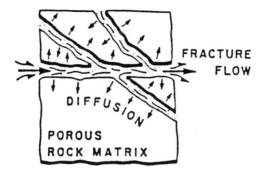

Figure 2.18. Schematic diagram representing flow through fractures and diffusion of contaminants from fractures into the rock matrix of a dual porosity medium (Reprinted with permission from *Groundwater Contamination*, 1984. Published by National Academy Press, Washington, D.C.).

The cubic law equation is developed by multiplying Equation 2.21 by the cross-sectional area of the fracture normal to the flow or

$$A = W(2b) \tag{2.22}$$

and

$$Q = AV = -W\left(\frac{\rho g}{12\mu}\right)\frac{dh}{dz}(2b)^3 \tag{2.23}$$

where W = characteristic width of the fracture (L).

Witherspoon et al. (1980) used a laboratory investigation on artificially induced tension fractures in granite, basalt, and marble to study the validity of Equation 2.23. Their results showed the cubic law to be valid for fractured media flow if a correction factor "f" is used to account for roughness.

Given that determining the velocity of fluids within a fracture can be approximated using Equation 2.23 or modifications thereof, the mechanics of solute transport within the fractured media can then be analyzed. Referring to Figure 2.17, it is important to note that, although the permeability of the fractured formation is dominated by the fracture channels, the impermeable rock matrix still possesses a definite porosity. Figure 2.18 shows that fluid transport occurs in the fracture channel, but significant solute transport can occur within the rock matrix, due mostly to molecular diffusion through the pore channels. The level of diffusion into the rock matrix will be a function of solute concentration, pore water velocity, and duration of the solute source.

A variety of equations can be developed to describe the solute transport phenomena in a fractured formation. Sudicky and Frind (1982) list these phenomena as (1) advective transport in the fractures, (2) molecular diffusion and mechanical dispersion along the fracture axes, (3) molecular diffusion from the fracture to the porous matrix, (4) adsorption onto the face of the matrix, (5) adsorption within the porous matrix, and (6) radioactive decay. The majority of previous work in developing the governing equations simply apply solute transport theory to account for some or all of the above processes (Sudicky and Frind, 1982; Grisak and Pickens, 1980; Tang et al., 1981; and Neretnieks et al., 1982). Applying the principles developed previously for solute transport to flow in the fracture yields

$$\frac{\partial C}{\partial t} = \frac{1}{R}\frac{\partial}{\partial z}\left(\eta D_{zz}\frac{\partial C}{\partial z}\right) + \frac{\partial}{\partial z}\left(\eta D_{xx}\frac{\partial C}{\partial x}\right) - q_z\frac{\partial C}{\partial z} - \lambda R C \qquad (2.24)$$

where
η = porosity
D_{zz}, D_{xx} = hydrodynamic dispersion coefficients (L^2/t)
λ = first-order reaction constant (t^{-1})
R = retardation factor due to adsorption

Some of the subtleties regarding Equation 2.24 should be noted. First, the retardation factor R is developed by assuming a given type of equilibrium adsorption process, usually linear (see Chapter 3). Second, the transverse dispersive phenomena in the "x" direction is assumed to be dominated by molecular diffusion. Hence, D_{xx} in Equation 2.24 is just equal to D^* as described by Equation 2.14. The porosity of the fracture is usually assumed to be one; therefore the Darcy velocity (Equation 2.5) is exactly equal to the velocity of the fluid.

Based on similar versions of Equation 2.24 several researchers have examined solute transport in fractured media. Grisak and Pickens (1980) conducted a numerical and analytical modeling study and noted that solute diffusion into the rock matrix is important in controlling solute transport. These results were subsequently verified by Grisak et al. (1980), who performed laboratory tracer experiments using calcium and chloride as tracers. Tang et al. (1981) developed an analytical model that showed longitudinal dispersion in the fracture to be significant for low velocities. Sudicky and Frind (1982) also used analytical solutions to show that fracture spacing can have a significant effect on the advance rate and ultimate penetration of a contaminant.

In summary, although still in its infancy, solute transport in fractured media has been successfully modelled using the concept of a "dual porosity" medium. The equations governing fractured media transport have been solved analytically and numerically and verified in the laboratory. However, as noted by Anderson (1984), difficulties in characterizing the geometry of a fractured rock system in the field complicates testing of the theory.

REFERENCES

Abdul, A. S. "Migration of Petroleum Products Through Sandy Hydrogeologic Systems," *Ground Water Monitoring Rev.* VIII(4 Fall):73–81 (1988).

Anderson, M. P. "Using Models to Simulate the Movement of Plumes Through Groundwater Flow Systems," *Crit. Rev. Environ. Control* 9:97–156 (1979)..

Anderson, M. P. "Movement of Contaminants in Groundwater: Groundwater Transport — Advection and Dispersion," in *Groundwater Contamination* (Washington, D.C.: National Academy Press, 1984) pp. 37–45.

Aris, R. "On the Dispersion of a Solute in a Fluid Flowing Through a Tube," *Proc. R. Sci. Soc.* 235:67–77 (1956).

Bear, J. "On the Tensor Form of Dispersion in Porous Media," *J. Geophys. Res.* 66(4 April):1185–1197 (1961).

Bear, J. *Hydraulics of Groundwater* (New York: McGraw-Hill, 1979).

Bear, J. and A. Verruijt "Modeling Groundwater Pollution," in *Modeling Groundwater Flow and Pollution,* (Dordrecht, Holland: D. Reidel, 1987.)

Bhattacharya, R. N. and V. J. K. Gupta "A Theoretical Explanation of Solute Dispersion in Saturated Porous Media at the Darcy Scale," *Water Resour. Res.* 19(4 August): 938–944 (1983).

Blackwell, R. J. "Laboratory Studies of Microscopic Dispersion Phenomena," *Soc. Pet. Eng. J.* 2(1): 1–8 (1962).

Bouwer, H., *Groundwater Hydrology,* (New York: McGraw-Hill, 1978).

Brigham, W. E., P. W. Reed, and J. N. Dew "Experiments on Mixing During Miscible Displacement in Porous Media," Soc. Pet. Eng. J. 1:1-8 (1961).

Buckely, S. E. and M. S. Leverett "Mechanics of Fluid Displacement in Sands," *Trans. Am. Inst. Min. Metall. Pet. Eng.* 146:107–116 (1942).

Corapcioglu, M. Y. and M. A. Hossain "Migration of Chlorinated Hydrocarbons in Groundwater," in *Hydrocarbons in Ground Water: Prevention, Detection and Restoration,* (Worthington, OH: National Water Well Association, 1986), pp. 33–52.

Crittenden, J. C., N. J. Hutzler, D. G. Geyer, J. L. Oravitz, and G. Friedman "Transport of Organic Compunds with Saturated Groundwater Flow: Model Development and Parameter Sensitivity," *Water Resour. Res.* 22(3 March):271–284 (1986).

Dagan, G. "Solute Transport in Heterogeneous Porous Formations," *J. Fluid Mech.* 145:151–177 (1984).

Davis, S. N. and R. J. M. De Wiest *Hydrogeology,* (New York: John Wiley & Sons, 1966).

Davis, S. N., D. J. Campbell, H. W. Bentley, and T. J. Flynn "Ground Water Tracers," (Worthington, OH: National Water Well Association, 1985).

De Jong, G. J. "Longitudinal and Transverse Diffusion in Granular Deposits", *Trans. Am. Geophys. Union* 39(1):67–74 (1958).

Dieulin, A. "Propogation de pollutin dans un aquifere alluvial, L'effect de parcours," Doctoral Dissertation, Université Pierre et Marie Curie, Paris (1980).

Driscoll, F. G. *Groundwater and Wells,* 2nd ed. (St. Paul, Minnesota: Johnson Division, 1986).

Ebach, E. A. and R. R. White "Mixing of Fluids Flowing Through Beds of Packed Solids," *Assoc. Ind. Chem. Eng. J.* 4(2):161–169 (1958).

Fatt, I. and H. Dykstra "Relative Permeability Studies," *Trans. Am. Inst. Min. Metall. Pet. Eng.* 192:249–256 (1951).

Feenstra, S. and J. A. Cherry "Dense Organic Solvents in Ground Water: An Introduction," Progress Report 0863985, Institute for Ground Water Research, University of Waterloo, Canada (1987).

Fetter, C. W. *Applied Hydrogeology,* Columbus, OH: Charles E. Merrill Publishing Company, 1980).

Freeze, R. A. and J. A. Cherry *Groundwater,* (Englewood Cliffs, NJ: Prentice-Hall, 1979).

Frind, E. O. "Simulation of Long-Term Transient Density-Dependent Transport in Ground Water," *Adv. Water Resour.,* 5:73–88 (1982).

Gelhar, L. W., A. L. Gutjahr, and R. L. Naff "Stochastic Analysis of Macrodispersion in a Stratified Aquifer," *Water Resour. Res.* 15:1387–1897 (1979).

Gelhar. L. W. and C. L. Axness "Three-Dimensional Stochastic Analysis of Macrodispersion in Auifers," *Water Resour. Res.* 19(1 February):161–180 (1983).

Grisak, G. E. and J. F. Pickens "Solute Transport Through Fractured Media 1. The Effect of Matrix Diffusion," *Water Resour. Res.* 16(4August):719–730 (1980).

Grisak, G. E., J. F. Pickens, and J. A. Cherry "Solute Transport Through Fractured Media 2. Column Study of Fractured Till," *Water Resour. Res.* 16(4 August):731–739 (1980).

Guven, O., F. J. Molz, and J. G. Melville "An Analysis of Dispersion in a Stratified Aquifer," *Water Resour. Res.* 20(10 October):1337–1354 (1984).

Harleman, D. R. F. and R. R. Rumer "Longitudinal and Lateral Dispersion in an Isotropic Porous Medium," *J. Fluid Mech.* 16(3):385–394 (1963).

Heller, J. P. "The Interpretation of Model Experiments for the Displacement of Fluids Through Porous Media," *J. Am. Inst. Chem. Eng.* 9(4 July):452–459 (1963).

Huling, S. G. and J. W. Weaver "Dense Nonaqueous Phase Liquids," U.S. Environmental Protection Agency, Ada, OK,Ground Water Issue, EPA/540/4-91-002 (March, 1991).

Johnson, R. L., C. D. Palmer, and W. Fish *Transport and Fate of Contaminants in the Subsurface,* U.S. Environmental Protection Agency, Ada, OK, EPA/625/4-89/019 (1989).

Klotz, D. and N. Moser "Hydrodynamic Dispersion of Aquifer Characteristic, Model Experiment with Radioactive Tracers," *Isot. Techniques Groundwater Hydrol.* II(March):341–355 (1974).

Klotz, D., K. P. Seiler, H. Moser, and F. Neumaier "Dispersivity and Velocity Relationship from Laboratory and Field Experiments," *J. Hydrol.* 45:169–184 (1980).

Knopman, D. S. and C. I. Voss "Behavior of Sensitivities in the One Dimensional Advection-Dispersion Equation: Implications for Parameter Estimation and Sampling Design," *Water Resour. Res.* 23(2 February):253–272 (1987).

Lee, S. T. and M. Okuyiga "Analysis of Dispersion in a Layered Porous Medium with Micro-Heterogeneity of Arbitrary Shapes and Size Distribution," *Soc. Pet. Eng.* SPE 15387:1–11 (1986).

Legatski, M. W. and D. L. Katz "Dispersion Coefficients for Gases Flowing in Consolidated Porous Media," *Soc. Pet. Eng. J.* 7:43–50 (1967).

Leverett, M. C. and W. B. Lewis "Steady Flow of Gas-Oil-Water Mixtures Through Unconsolidated Sands," (1941).

Mackay, D. M., P. V. Roberts, and J. A. Cherry "Transport of Organic Contaminants in Groundwater," *Environ. Sci. Technol.* 19(5):384–392 (1985).

Mackay, D. M. "Characterization of the Distribution and Behavior of Contaminants in the Subsurface," presented at NRC-WSTB Colloquium on *Ground Water and Soil Contamination Remediation: Are Science, Policy, and Public Perception Compatible?* (1989).

Matheron, G. and G. de Marsily "Is Transport in Porous Media Always Diffusive? A Counterexample," *Water Resourc. Res.* 16:901–907 (1980).

McCarthy, J. F. and J. M. Zachara "Subsurface Transport of Contaminants," *Environ. Sci. Technol.* 23(5):496–502 (1989).

Menzie, D. E., S. Dutta, R. Shadizadeh, and N. Malik "A New Metalic Method of Coating Oilfield Cores for Laboratory Studies," *J. Pet. Technol.* May (1988) pp. 35–45.

Mercer, J. W., and R. M. Cohen "A Review of Immiscible Fluids in the Subsurface: Properties, Models, Characterization and Remediation," *J. Contam. Hydrol.* 6:107–163 (1990).

Molz, F. J., O. Guven and J. G. Melville "An Examination of Scale Dependent Dispersion Coefficients," *Ground Water* 21(6):715–725 (1983).

Muskat, M., R. D. Wyckoff, H. G. Botset, and M. W. Meres "Flow of Gas-Liquid Mixtures Through Sands," *Trans. Am. Inst. Min. Metall. Pet. Eng.* 123:69–96 (1937).

Neretnieks, I., T. Eriksen, and P. Tahtinen "Tracer Movement in a Single Fissure in Granitic Rock: Some Experimental Results and Their Interpretation," *Water Resour. Res.* 18(4 August):849–858 (1982).

Palmer, C. D. and R. L. Johnson "Physical Processes Controlling the Transport of Contaminants in the Aqueous Phase," *Transport and Fate of Contaminants in the Subsurface,* EPA/625/4-89/019, U.S. Environmental Protection Agency, Cincinnati, OH (September, 1989).

Parker, J. C., R. J. Lenhard, and T. Kuppusamy "Modeling Multiphase Contaminant Transport in Ground Water and Vadose Zones," in *Hydrocarbons in Ground Water: Prevention, Detection and Restoration* (Worthington, Ohio: National Water Well Association, 1986), pp. 189–200.

Perkins, T. W. and O. C. Johnston "A Review of Diffusion and Dispersion in Porous Media, " *Soc. Pet. Eng. J.* 3:70–84 (1963).

Peters, E. J., W. H. Broman, and J. A. Broman "A Stability Theory for Miscible Displacement," *Soc. Pet. Eng. AIME,* SPE 13167:1–12 (1984).

Philip, J. R. "Field Heterogeneity: Some Basic Issues," *Water Resour. Res.* 16(2 April):443–448 (1980).

Pickens, J. F. and G. E. Grisak "Scale Dependent Dispersion in a Stratified Granular Aquifer," *Water Resour. Res.* 17(4 August):1191–1211 (1981).

Reible, D. D., T. H. Illangasekare, D. V. Doshi, and A. F. Ayoub "Development and Experimental Verification of a Model for Transport of Concentrated Organics in the Unsaturated Zone," in *Hydrocarbons in Ground Water: Prevention, Detection and Restoration,* (National Water Well Association, 1986), pp. 107–126.

Robson, S. G. "Feasibility of Digital Water Quality Modeling Illustrated by Application at Barstow, California." U.S. Geological Survey Water Resources Investigation Report 46-73, U.S. Government Printing Office, Washington, D.C. (1974).

Rumer, R. R., Jr. "Longitudinal Dispersion in Steady and Unsteady Flow," *J. Hydraul. Div. Am. Soc. Civil Eng.* HY4, 162:147–172 (1962).

Saffman, P. G. "A Theory of Dispersion in a Porous Medium," *J. Fluid Mech.* 6(3):321–349 (1959).

Sahimi, M., B. D. Hughes, L. E. Scriven, and H. T. Davis "Dispersion in Flow Through Porous Media-I, One-Phase Flow," *Chem. Eng. Sci.* 41(8):2103–2122 (1986).

Salter, S. J. and K. K. Mohantz "Multiphase Flow in Porous Media, Macroscopic Observations and Modeling," *Soc. Pet. Eng.* SPE 11017, (1982).

Santz, J. P. "An Analysis of Hydrodispersive Transfer in Aquifers," *Water Resour. Res.* 16(1):145–158 (1980).

Scheidegger, A. E. "On the Theory of Flow of Miscible Phases in Porous Media," in *Proc. IUGG General Assembly, Toronto 2* (1957), pp. 236–242.

Schwartz, F. W. "Macroscopic Dispersion in Porous Media: The Controlling Factors," *Water Resour. Res.* 13(4 August):743–752 (1977).

Smith, L. "A Stochastic Analysis of Steady-State Groundwater Flow in a Bounded Domain," Ph.D. Dissertation, University of British Columbia, Vancouver (1978).

Smith, L. and F. W. Schwartz "Mass Transport 3. Role of Hydraulic Conductivity Data in Prediction," *Water Resour. Res.* 17(5 October):1463–1479 (1981).

Sposito, G., W. A. Jury, and V. K. Gupta "Fundamental Problems in the Stochastic Convection-Dispersion Model of Solute Transport in Aquifers and Field Soils," *Water Resour. Res.* 22(1 January):77–88 (1986).

Sudicky, E. A., J. A. Cherry, and E. O. Frind "Migration of Contaminants in Groundwater in a Landfill: A Case Study. 4. A Natural Gradient Dispersion Test," *J. Hydrol.* 63:81–108 (1983).

Sudicky, E. A. and E. O. Frind "Contaminant Transport in Fractured Porous Media: Analytical Solutions for a System of Parallel Fractures," *Water Resour. Res.* 18(6 December):1634–1642 (1982).

Tang, D. H., E. O. Frind, and E. A. Sudicky "Contaminant Transport in Fractured Porous Media: Analytical Solution for A Single Fracture," *Water Resour. Res.* 17(3 June):555–564 (1981).

Taylor, G. I. "Dispersion of Solute Matter in Solvent Flowing Slowly Through a Tube," *Proc. R. Sci. Soc.* 219(1137):186–203 (1953).

Tyler, S. W., M. R. Whitbeck, M. W. Kirk, J. W. Hess, L. G. Everett, D. K. Kreamer, B. H. Wilson, and J. van Ee "Processes Affecting Subsurface Transport of Leaking Underground Tank Fluids," U.S. Environmental Protection Agency, Las Vegas, NV, EPA/600/6-87/005 (June 1987).

van Dam, J. "The Migration of Hydrocarbons in a Water Bearing Stratum," in *The Joint Problems of the Oil and Water Industries* (1967).

Wang, H. F. and M. P. Anderson "Advective-Dispersive Transport," in *Introduction to Groundwater Modeling: Finite Difference and Finite Element Methods* (San Francisco: W.H. Freeman, 1982), pp. 173–199.

Wheatcraft, S. W. and S. W. Tyler "An Explanation of Scale-Dependent Dispersivity in Heterogeneous Aquifers Using Concepts of Fractal Geometry," *Water Resour. Res.* 24(4 April):566–578 (1988).

Witherspoon, P. A., J. S. Y. Wang, K. Iwai, and J. E. Gale "Validity of Cubic Law for Fluid Flow in a Deformable Rock Fracture," *Water Resour. Res.* 16(16 December):1016–1024 (1980).

3

ABIOTIC PROCESSES

3.1 INTRODUCTION

In the previous chapter, hydrodynamic processes were considered. Contaminant properties and subsurface conditions that result in interactive processes often cause the rate of contaminant transport to differ from the rate of ground water flow. In this chapter abiotic (nonbiological) processes active in the subsurface are considered. Abiotic processes affect contaminant transport by causing interactions between the contaminant and the stationary subsurface material (e.g., sorption, ion exchange) or by changing the form of the contaminant (e.g., hydrolysis, redox reactions) which may subsequently interact with the subsurface material. In the case of sorption, the sorptive capacity of the soil for the contaminant must be satisfied before the contaminant will appear downgradient. The accumulation (sorption) of the contaminant on the subsurface media will slow down (retard) the rate of spread of the contaminant. During remediation of a contamination episode, sorbed contaminants will necessitate additional flushings of the pore spaces (ground water) to effect cleanup of the contaminated media (to cause complete desorption). Thus, to respond to contamination episodes and to predict the movement of contaminants in the subsurface, it is necessary to have a fundamental understanding of, and ability to quantify, the interactions of contaminants in the subsurface environment. Abiotic processes to be discussed in this chapter are sorption/ desorption, ion exchange, oxidation/reduction, precipitation/ dissolution, hydrolysis, cosolvation and ionization. Additional abiotic processes could be identified, but are beyond the scope of this discussion.

3.2 SORPTION/DESORPTION

   Sorption and desorption are two major mechanisms affecting the transport
of contaminants in the subsurface. The sorption of a variety of pesticides in
agricultural soils has been widely studied (Bailey and White, 1970; Hamaker
and Thompson, 1972; Rao and Davidson, 1980; Karickhoff, 1984; Sabatini
and Austin, 1990; to mention a few of the review articles). Based on this
experience, sorption is probably the most studied of the abiotic processes and
will thus be discussed more extensively than the other abiotic processes.

3.2.1 Sorption Fundamentals

   Adsorption is defined as the accumulation of a chemical at an interface
(Adamson, 1982). The interfaces of most interest in subsurface fate and
transport are liquid/solid interfaces and gas/solid interfaces. Adsorption,
absorption and sorption are three terms that refer to similar phenomena. Weber
(Weber, 1972; Weber et al., 1991) defines adsorption as the accumulation
occurring at an interface, absorption as the partitioning between two phases
(accumulation from ground water into organic carbon) and sorption as including
both adsorption and absorption. Often, the terms adsorption and sorption are
used interchangeably; such will be the case for this document. For contaminated
ground water, the solute would refer to the contaminant dissolved in the
solvent (ground water). When discussing sorption/desorption phenomena, the
sorbate is the contaminant (solute) and the sorbent is the phase or interface
where the accumulation occurs (solid phase). Some of the earliest studies of
adsorption investigated the accumulation of gases at solid surfaces (Langmuir,
1918; Freundlich, 1926).
   In general, sorption reactions may be classified as either sorbent or solvent
motivated (Weber, 1972). Sorbent motivated sorption occurs when an attraction
occurs between the sorbent (subsurface material) and the solute (contaminant)
and the contaminant accumulates at the surface due to the affinity of the
surface for the contaminant. An example of sorbent motivated sorption would
be a highly polar or ionizable contaminant interacting with the cation exchange
sites of clay minerals. Solvent motivated sorption occurs when the contaminant
is hydrophobic (water disliking, low water solubility). Examples of hydrophobic
contaminants are nonpolar organics which prefer nonpolar phases to the polar
water phase (as illustrated in the adage "likes dissolve likes"). Hydrophobic
contaminants will find it energetically favorable to accumulate at an interface
or partition into a nonpolar phase (e.g., associate with the organic content of

the subsurface medium) rather than remain in the water phase. Thus, for solvent motivated sorption, the accumulation of the contaminant is motivated by the "dislike" of the ground water phase (solvent) for the contaminant.

The adsorption attachment may be the result of one or a combination of electrostatic forces. Hamaker and Thompson (1972) include the following as electrostatic forces between chemicals and soils: van der Waals/London forces, hydrogen bonding, charge transfer, ligand exchange, ion exchange, direct and induced ion-dipole and dipole-dipole interactions, and chemisorption. The adsorption attachment can be classified into three categories: exchange, physical or chemical (Weber, 1972). Exchange adsorption (ion exchange) refers to the accumulation of the chemical at the sorption site due to the electrostatic attraction between the charged sites of the soil and the charged sites or polar moieties of the chemical. Physical adsorption (physisorption) is the result of van der Waals/London attraction or similar forces. Chemical adsorption (chemisorption) is the result of a chemical reaction between the solid surface and the chemical. Binding energies for these categories of attachment are: exchange, typically less than 50 kcal/mol; physisorption, 1 to 2 kcal/mol; chemisorption, typically greater than 50 kcal/mol (Hamaker and Thompson, 1972). While it is fundamentally expedient to separate the sorption attachment into these three categories, in actuality the sorption attachment is the result of a combination of forces.

The sorption of many neutral organic contaminants falls into the category of hydrophobic (solvent motivated) sorption with physisorption as the attachment mechanism. In a later section, sorbent motivated sorption with exchange attachment (ion exchange) will be discussed.

3.2.2 Equilibrium Sorption

3.2.2.1 Equilibrium Expressions

Equilibrium sorption assumes that the rate of sorption is relatively fast (compared to the rate of movement of ground water); hence, solution residence times sufficient to establish equilibrium exist. As in all cases, equilibrium assumptions are simplifications of kinetic expressions where the reactions are sufficiently fast so that they can be considered to be instantaneous. This simplification is illustrated below. Sorption can be thought of as a reaction between a soil and a chemical (C) resulting in sorbed chemical (q). Equation 3.1 shows such a reaction with forward (sorption, k_s) and reverse (desorption, k_d) kinetic constants. An example of a kinetic adsorption expression to describe this reaction is shown in Equation 3.2 (Davidson and McDougal, 1973). The term on the left hand side of Equation 3.2 defines the time rate of change of q (the mass of chemical adsorbed in the solid phase normalized by the mass of soil). The forward rate of this reaction (sorption) is taken to be first order with respect to the solute concentration in the ground water phase (C) and the reverse rate of this reaction (desorption) is assumed to be first order with respect to mass of chemical associated with the solid phase (q).

$$Soil + C \; \underset{k_d}{\overset{k_s}{\rightleftharpoons}} \; q \tag{3.1}$$

$$\frac{\partial q}{\partial t} = k_s \frac{\eta}{\rho_s(1-\eta)} C - k_d q \tag{3.2}$$

where
q = mass of chemical sorbed normalized by mass of soil (M/M)
k_s = first order adsorption rate constant (1/t)
η = pore water fraction (L³/L³)
ρ_s = solid phase particle density (M/L³)
k_d = first order desorption rate constant (1/t)
C = liquid phase chemical concentration (M/L³)
K_p = linear equilibrium partition coefficient (L³/M)

In the case of equilibrium conditions, the rate of the forward reaction is equal to the rate of the reverse reaction such that a steady state condition is achieved (although molecules are continuously sorbing and desorbing under equilibrium conditions, there is no net change in q and C). Thus, as q is no longer a function of time under equilibrium conditions, the left hand side of Equation 3.2 is zero and the expression simplifies to Equation 3.3. Valocchi (1985), Parker and Valocchi (1986), and Bahr and Rubin (1987) have discussed criteria for determining the validity of the local equilibrium assumption for a given set of

$$q = \left[\frac{k_s}{k_d} \frac{\eta}{\rho_s(1-\eta)} \right] C_e = K_p C_e \tag{3.3}$$

where
C_e = liquid phase equilibrium concentration (M/L³)
K_p = linear equilibrium partition coefficient (L³/M)

conditions. Valocchi (1985) evaluated the time moments for various equilibrium and nonequilibrium solute transport models. The temporal moments

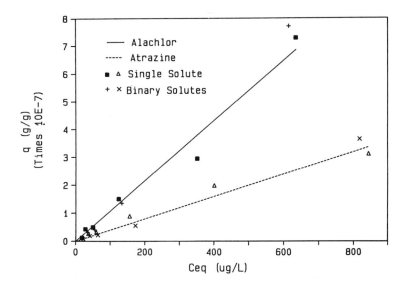

Figure 3.1. Linear sorption isotherm.

serve as indicators of the shape of the breakthrough curves. The first, second and third moments describe the mean breakthrough time (retardation), the degree of spreading (dispersion) and the degree of nonequilibrium (asymmetry), respectively. Parker and Valocchi (1986) further discussed the utilization of time moment analyses for solute transport predictions.

The simplest equilibrium sorption expression is the linear equilibrium expression shown in Equation 3.3. This expression assumes that a plot of q vs C_e (adsorption isotherm) will result in a straight line on arithmetic paper with a slope of K_p (see Figure 3.1). If this plot is curvilinear on arithmetic paper, the Freundlich or Langmuir expressions are suggested. If the data plotted on log-log paper are linear, then the Freundlich expression (Equation 3.4) may be utilized. Taking the log of the Freundlich expression results in Equation 3.5. The slope of the log-log plot

$$q = K_{fr} C_e^N \qquad (3.4)$$

$$\log q = \log K_{fr} + N \log C_e \qquad (3.5)$$

where K_{fr} = Freundlich partition coefficient $((L^3/M)^N)$
N = Freundlich exponent coefficient

of q vs C_e is N and the value of q at $C_e = 1.0$ (log $C_e = 0$) is K_{fr}. Another approach to determining the Freundlich parameters from the data would be

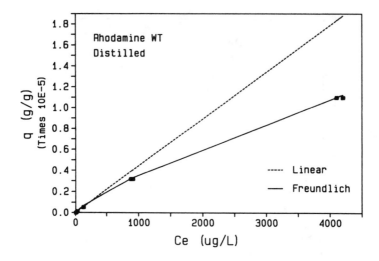

Figure 3.2.   Freundlich isotherm (Sabatini and Austin, 1991).

to solve a system of nonlinear equations produced by experimental values of q and C_e and the Freundlich expression (an approach greatly simplified by the advent of math software packages available for use on microcomputers). Figure 3.2 shows data described by the Freundlich relationship. It is observed that for a value of N = 1.0, the Freundlich relationship (Equation 3.4) simplifies to the linear adsorption expression (Equation 3.3). Depending on the specific subsurface material and chemical of interest and the concentration range of the chemical, some researchers have found the assumption of linear equilibrium adsorption to be valid (Karickhoff et al., 1979; Brown and Flagg, 1981), while others have found the Freundlich expression more appropriate and have observed N values in the range of 0.7 to 1.2 (Hamaker and Thompson, 1972; Rao and Davidson, 1980).

When nonlinear isotherms are evidenced, it should be noted that the isotherm may still be linear at low concentrations (µg/l). This agrees with Langmuir's conceptual adsorption model (Langmuir, 1918). This model assumes a finite number of adsorption sites each with equal affinity for the adsorbate (chemical). As more of the sites become occupied, the probability of the chemical mass still in solution finding one of the remaining adsorption sites decreases. This results in nonlinearity of the isotherm at higher concentrations (at higher concentrations, a smaller fraction of the chemical in solution ends up adsorbed to the soil). Another explanation for nonlinear isotherms is that the adsorption sites have a distribution of affinities for adsorbing the chemicals and that the most favorable sites are filled first. At higher equilibrium concentrations, the less favorable adsorption sites are utilized. The resulting reduction in the incremental adsorption with increasing concentration would cause the isotherm to be nonlinear at higher equilibrium concentrations. When the isotherm is nonlinear, Rao and Davidson (1980) state that the error of assuming linear adsorption may be negligible relative to the errors introduced by other assumptions inherit in the analysis. For nonpolar organics, the linear isotherm has seen the greatest utilization with the Freundlich isotherm being the second most utilized expression.

The Langmuir model is a nonlinear model that was derived from fundamental concepts with simplifying assumptions. Some of these assumptions are as follows: (1) the molecules are adsorbed on definite sites, (2) each sorption site accommodates only one molecule (monolayer), (3) no interactions exist between adjacent sites, and (4) each site has an equal affinity for the adsorbate (Benefield et al., 1982). The Langmuir model is shown in Equation 3.6.

$$q = \frac{aK_L C_e}{1 + K_L C_e} \qquad (3.6)$$

where   a = mass of chemical required to saturate a unit mass of soil (M/M)

   K_L = Langmuir constant (L^3/M)

The Langmuir model is basically a limited growth model; it is similar in form to those utilized in describing microbial population dynamics. For sorption, the limiting capacity is the discrete number of sorption sites available and the restriction that only monolayer sorption will occur (the "a" term is a capacity term). The Langmuir model differs from the Freundlich model by virtue of the limited growth aspect of the nonlinearity; no maximum value is dictated in the Freundlich model. At high enough liquid phase concentrations to saturate the surface, the plot of q vs C_e will become horizontal.

One method of solving for the Langmuir parameters would be to linearize the Langmuir equation. This can be accomplished by inverting Equation 3.6 and simplifying (Equation 3.7). In this form, the Langmuir constants can be determined from the slope and intercept of the data plotted as $1/q$ vs $1/C_e$ (see Figure 3.3). Alternately,

$$\frac{1}{q} = \frac{1}{aK_L} \frac{1}{C_e} + \frac{1}{a} \qquad (3.7)$$

the Langmuir parameters can be determined by solving the system of nonlinear equations generated by the data and the Langmuir expression. The Langmuir model is widely utilized when considering the transport of electrolytes, e.g., nutrients and metals, in soils (Harter and Baker, 1977; Veith and Sposito,

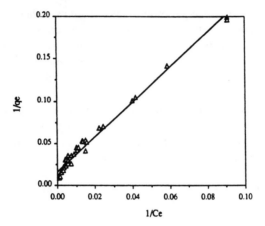

Figure 3.3.   **Langmuir isotherm for sorption of trichlorobenzene on Ann Arbor soil (Reprinted with permission from *Water Research*, Vol. 25, Weber, W. J., Jr., P. M. McGinley, and L. E. Katz, "Sorption Phenomena in Subsurface Systems: Concepts, Models and Effects on Contaminant Fate and Transport". Copyright 1991, Pergamon Press).**

1977; Brown and Combs, 1985). It can also be utilized over large concentration ranges for neutral organic contaminants (Weber et al., 1991 — see Figure 3.3).

The BET (Brunauer, Emmett, and Teller; 1938) model was developed to allow for the occurrence of multilayer adsorption (Benefield et al., 1982). This model is an extension of the Langmuir model (which only allows for the occurrence of a monolayer). Upon complete coverage of a surface with a monolayer (or possibly sooner), it may become thermodynamically favorable for the sorbate to begin "stacking", adsorbing to previously adsorbed molecules. This will increase the sorptive capacity of the surface and will result in increasing values of q as C_e increases.

Giles et al. (1960) devised a classification scheme to characterize the sorption mechanisms for a variety of sorbents and solutes based on their isotherms. Figure 3.4 shows the major types of isotherms observed for sorption. This figure divides the isotherms into four classes (S, L, H, and C) and three subgroups for each class (1, 2, or 3). The S-shaped isotherm is nonlinear and convex with respect to the abscissa, suggesting that the solute initially has difficulty in competing with the solvent for sorption sites. The L-type isotherm (Langmuir) is nonlinear and concave with respect to the abscissa, suggesting that the solute is preferentially sorbed with respect to the solvent. As the sorption progresses fewer sorption sites remain, making it more difficult for the solute to find an available site, thus decreasing the incremental level of

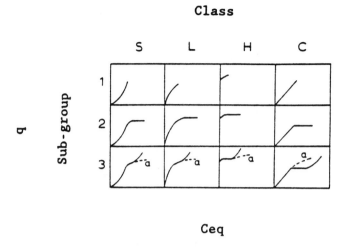

Figure 3.4.   Isotherm classification system (Adapted from Giles et al., 1960).

sorption. The H-type isotherm (high affinity), a special case of the L-type isotherm, occurs when the solute has such a great affinity for the solid that high levels of surface accumulation can occur with little solute remaining in solution. Beyond a threshold, the isotherm becomes similar in shape to the L-type isotherm. The C-type isotherm (constant-partition) is linear, suggesting that the availability of sites is constant as sorption progresses. For further discussion of these categories and the sorbent and solute properties that dictate their occurrence, and for examples of soil/chemical systems that satisfy each category, the reader is referred to Weber and Miller (1989).

For the case of multiple solutes, it is possible that the solutes will compete for sorption sites (possibly due to differences in sorption energies — sorbent or solute heterogeneity — or due to site limitations). Weber et al. (1991) discussed the application of the ideal adsorbed solution theory (IAST) model for characterizing multicomponent equilibrium sorption in the subsurface environment. The IAST model requires isotherms for each solute and sorbent, from which the spreading pressure is computed (the spreading pressure is the reduction in surface tension produced by sorption of a solute). IAST requires the spreading pressures of all components of the system to be the same at equilibrium. Weber et al. (1991) present single solute and bisolute sorption data in which competitive sorption was evidenced (sorption of tetrachloroethylene when dichlorobenzene was present versus when tetrachloroethylene was present as a single solute). The deviation of the bisolute isotherm from the single solute isotherm was predicted by the IAST model (as illustrated in Figure 3.5). The interested reader is referred to Weber et al. (1991) for more details on the IAST model.

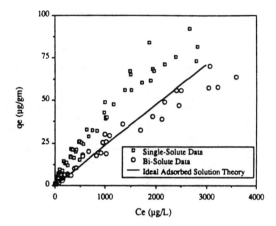

Figure 3.5.   Single and bisolute sorption (Reprinted with permission from *Water Research,* Vol. 25, Weber, W. J., Jr., P. M. McGinley, and L. E. Katz, "Sorption Phenomena in Subsurface Systems: Concepts, Models and Effects on Contaminant Fate and Transport". Copyright 1991, Pergamon Press).

3.2.2.2 One-Dimensional Modeling

Several modeling techniques are applicable to the prediction of chemical transport in ground water. This section will focus on mechanistic, deterministic modeling of solute transport. The reader is directed to Jury (1983) and Sposito et al. (1986) for treatment of stochastic and functional modeling approaches to solute transport. The fundamental governing equation for chemical transport under saturated ground water conditions is the advection and dispersion equation with sorption. This equation is derived from flux balance considerations about an elemental volume and, for the one-dimensional flow case, results in the partial differential equation shown in Equation 3.8 (Freeze and Cherry, 1979).

$$\frac{\partial C}{\partial t} = -v_x \frac{\partial C}{\partial x} + D_x \frac{\partial^2 C}{\partial x^2} - \frac{\rho_b}{\eta} \frac{\partial q}{\partial t} \qquad (3.8)$$

where D_x = hydrodynamic dispersion coefficient (L^2/t)
        x = dimension of solute transport (L)
       v_x = pore water (seepage) velocity (L/t)

Utilization of equilibrium sorption relationships in the governing partial differential equation requires that they be substituted into the $\partial q/\partial t$ term in

Equation 3.8. For linear equilibrium adsorption the $\partial q/\partial t$ term is as shown in Equation 3.9. Substituting Equation 3.9 into Equation 3.8 and simplifying yields Equation 3.10. The r_f term can be thought of as the ratio of the time of appearance downgradient of a sorbing chemical to that of a nonsorbing

$$\frac{\partial q}{\partial t} = \frac{\partial q}{\partial C}\frac{\partial C}{\partial t} = K_p\frac{\partial C}{\partial t} \tag{3.9}$$

$$\left[1 + \frac{\rho_b}{\eta}K_p\right]\frac{\partial C}{\partial t} = r_f\frac{\partial C}{\partial t} = -v_x\frac{\partial C}{\partial x} + D_x\frac{\partial^2 C}{\partial x^2} \tag{3.10}$$

where r_f = retardation factor (dimensionless).

(conservative) chemical. For the retardation factor to be dimensionless, it is necessary for K_p to have the inverse units of ρ_b; if ρ_b is in units of g/cm^3, K_p must be in units of cm^3/g. Dividing both sides of Equation 3.10 by r_f results in Equation 3.11. In this equation, D_x/r_f is referred to as the effective

$$\frac{\partial C}{\partial t} = -\frac{v_x}{r_f}\frac{\partial C}{\partial x} + \frac{D_x}{r_f}\frac{\partial^2 C}{\partial x^2} \tag{3.11}$$

dispersion coefficient and v_x/r_f is referred to as the effective velocity for the sorbing chemicals. As r_f increases (as sorption increases), the effective velocity of the chemical decreases (the chemical is more significantly retarded relative to the ground water). Exploring the physical significance of the retardation factor, the $(\rho_b/\eta)K_p$ term is [(mass solid/unit volume)/(volume voids /unit volume)] $\times$ [(mass solute adsorbed/mass solid)/(mass solute solution/volume solution)] or, simplifying, [mass solute adsorbed/mass solute solution]. Thus, this ratio indicates the number of volumes of chemical solution necessary to satisfy the sorption capacity of the material (assuming plug flow hydrodynamics and complete sorption of the chemical). Adding a value of one to this term (one pore volume) considers that the pore volume is initially devoid of chemical. Therefore, the retardation factor indicates the number of pore volumes required for the centroid of the chemical breakthrough to appear at a specified location downgradient.

Analytical solutions to the governing partial differential equation for solute transport with equilibrium adsorption (Equation 3.10) are available for simple boundary conditions. One solution to this equation is a simple modification of the analytical solution to the advection-dispersion equation (no adsorption). In the case when sorption is negligible (K_p = 0 and r_f = 1.0), Equations 3.10 and 3.11 simplify to the advection-dispersion equation. Given that the pore water velocity (v_x) is taken to be a constant in the solution to the simple advection-dispersion equation, and assuming that the retardation factor is a

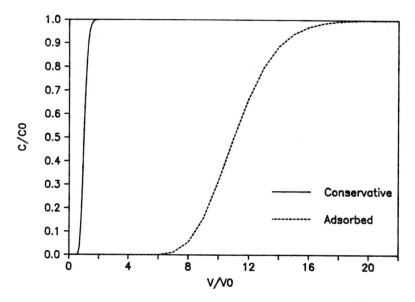

Figure 3.6.   Conservative and sorbing breakthrough curves.

constant, one can simply substitute the effective pore water velocity (v_x/r_f, also a constant) for v_x in the solution of the advection-dispersion equation. Equation 3.12 is an example of such a solution for the given initial and boundary conditions.

$$\frac{C}{C_o} = 0.5\left[erfc\left(\frac{xr_f - v_x t}{2\left(D_x r_f t\right)^{1/2}}\right) + \exp\left(\frac{v_x x}{D_x}\right) erfc\left(\frac{xr_f + v_x t}{2\left(D_x r_f t\right)^{1/2}}\right)\right] \qquad (3.12)$$

for $C\,(x,0) = 0.0 \qquad x \geq 0$
        $C\,(0,t) = C_o \qquad t \geq 0$
        $C\,(\infty,t) = 0.0 \qquad t \geq 0$

where erfc = complimentary error function
      exp = exponent

Figure 3.6 demonstrates breakthrough curves of conservative and sorptive compounds, demonstrating model output from Equation 3.12. The abscissae for Figure 3.6 (V/V_o) is relative pore volumes. This represents the amount of flow which has passed through the media relative to the volume of pores. This axis can also be thought of as relative time; the time of ground water flow relative to the time necessary for one pore volume of flow. van Genuchten and Alves (1982) have compiled analytical solutions for various boundary conditions for linear equilibrium adsorption. Parker and van Genuchten (1984) have published a manual describing a program (CXTFIT) that will determine transport parameters from laboratory and field data allowing for linear, equilibrium sorption, and for zero and first order reactions, allowing for several different initial conditions.

For the Freundlich expression, substitution of the appropriate $\delta q/\delta t$ term into Equation 3.8 produces the expression shown in Equation 3.13.

$$\left[1 + \frac{\rho_b}{\eta} NK_{fr} C^{N-1}\right] \frac{\partial C}{\partial t} = D_x \frac{\partial^2 C}{\partial x^2} - v_x \frac{\partial C}{\partial x} \tag{3.13}$$

The expression in the parentheses on the left-hand side of Equation 3.13 is not a constant as it was with linear sorption, but is instead a function of the chemical concentration. When the Freundlich exponent is one (the Freundlich expression simplifies to the linear expression), the expression within the parentheses on the left-hand side of Equation 3.13 is independent of C, is a constant and is the same as r_f (when $N = 1.0$, K_p and K_{fr} are the same). In the general case, the use of the Freundlich expression complicates the solution of the governing differential equation and may necessitate the use of numerical approximation techniques (finite difference, finite element, etc).

3.2.2.3 Parameter Determination

Equilibrium sorption parameters may be determined in the laboratory by conducting batch or column studies. The relative advantage of conducting batch studies is that a variety of factors (chemical concentration, pH, background ions, etc.) can be analyzed with limited effort. However, column studies more closely mimic the field scale by virtue of their continuous flow nature, the solid to liquid ratio in the column, the potential surface access limitations introduced by the first two factors, etc. At the same time, column studies are much more labor intensive. The use of batch studies as preliminary screening tools and column studies to confirm observations made in the batch studies is suggested as a reasonable approach.

In batch studies, a series of reactors with varying ratios of soil mass to chemical concentration are shaken until equilibrium sorption is known to exist. Typically, a constant mass of soil is utilized and the chemical concentration is varied to eliminate the impacts of varying soil to liquid ratio on the level of sorption ("solids effect", see O'Connor and Connolly, 1980; Voice et al., 1983; Mackay and Powers, 1987). The equilibrium chemical

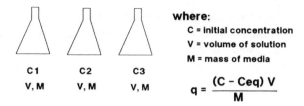

where:

C = initial concentration
V = volume of solution
M = mass of media

$$q = \frac{(C - Ceq) \, V}{M}$$

* Add soil and solute, equilibrate
* Measure equilibrium concentration (Ceq)
* Calculate q (mass sorbed/mass media) as above
* Plot q versus Ceq, slope is Kp

Figure 3.7. Overview of batch isotherm test and determination of K_p.

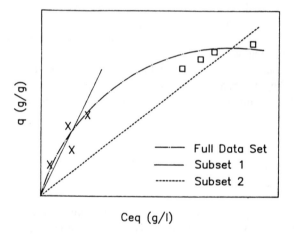

Figure 3.8. **Linear vs nonlinear sorption isotherm (Sabatini and Austin, 1990).**

concentration in the liquid phase is determined for each reactor and the chemical sorbed to the soil is calculated by mass balance. Each reactor produces a point on the isotherm (plot of q vs C_e). The batch method is overviewed in Figure 3.7, including the method of calculating q from the data.

Care must be taken in analyzing the results from batch studies or the use of data collected by others. Figure 3.8 demonstrates isotherms plotted for one set of data and two subsets of the same data. This figure points out the danger of extrapolating data beyond the experimental range from which the data were collected. Individual plots for the two subsets of the data can appear to be linear, while when the two subsets are combined it becomes apparent that the data are curvilinear. Extrapolating the linear results from the lower range to the higher concentrations would have resulted in higher predicted values of q than would actually occur, leading to the prediction that the soil would have a greater adsorptive capacity for the chemical than it actually does. Thus the chemical would appear downgradient sooner than predicted. Rao and Davidson (1979) observed this when attempting to extrapolate information from pesticide adsorption at the μg/l range to the mg/l range. Equilibrium studies at the higher concentrations resulted in N values in the range of 0.75 to 0.92, while previous work at the lower level had indicated linear adsorption. Modeling efforts using linear equilibrium parameters from the lower concentrations predicted a greater lag in the appearance of the pesticide than observed. Utilizing the Freundlich expression in the modeling resulted in improved predictive capabilities.

Sorptive parameters can also be determined from column studies in the laboratory. The results of the column study will be a breakthrough curve (a plot of chemical concentration vs time or, in normalized terms, relative concentration vs pore volumes passed). While conducting the column breakthrough study, the breakthrough of a conservative (nonadsorbing) chemical (e.g., chloride, tritiated water, etc.) should be evaluated (see Figure 3.6). This will allow evaluation of the hydrodynamics of flow through the porous medium and will allow determination of the dispersion coefficient (D_x).

Several analyses can be utilized in determining the r_f (or K_p) from the column data. The first approach would be to utilize the definition of the retardation factor. Under ideal conditions, the retardation factor corresponds to the centroid of the breakthrough curve, i.e., the value of the pore volume, where the relative concentration of the chemical breakthrough curve is 0.5, would correspond to the retardation factor. However, if nonlinear or nonequilibrium conditions exist, this methodology could result in erroneous estimates (the relative significance of the error is a function of the level of adsorption, etc.).

A second approach for determining r_f values from column data would be to utilize mass balance considerations of the chemical adsorbed during the column study. Knowing the influent concentration of the chemical and monitoring the concentration of the chemical exiting the column, it is possible to calculate the mass of chemical adsorbed (from the area above a C vs V plot). At such time as the concentration of the chemical exiting the column is equal to the concentration entering the column ($C/C_0 = 1.0$), it is assumed that the concentration throughout the column is equal to the inlet concentration. At this point, the column study can be considered to be at a steady state condition; any additional sorption occurring in the column can be considered to be relatively insignificant (undetectable). The mass of chemical adsorbed to the soil can be determined from mass balance considerations (total mass entering column minus mass exiting column minus mass in the pore water in the column). Given the mass of chemical sorbed and the mass of soil (as determined when packing the column), a value of q (mass of chemical adsorbed to the soil normalized by the mass of soil) can be determined. The chemical concentration causing this level of adsorption is the initial concentration (which is constant throughout the column). Thus, the column can be treated as a single point isotherm and a value of K_p can be determined. Utilizing this approach, along with additional parameters determined during the packing of the column (bulk density (ρ_b) and the porosity (η), using either a value of ρ_s determined for the material or using a default value of 2.65 g/cm³), a value of r_f can be determined. This mass balance approach is the same as determining the mass of sorption from the area above the breakthrough curve.

A third approach for determining r_f values from column data would be to utilize a model to describe the column data by fitting the r_f (such as CXTFIT, Parker and van Genuchten, 1984). In this approach, the dispersion coefficient (D_x) determined from the conservative tracer will be used. If the model assumes linear sorption and nonlinearity is in fact evidenced, then the accuracy of the fitted parameters must be evaluated. A fourth approach is to determine the ratio of the areas above the breakthrough curves for the conservative and adsorbing chemicals (this can be assumed to be equal to the retardation factor).

Parameter determination utilizing laboratory techniques may fail to account for field conditions not captured in the laboratory, e.g., background aqueous conditions, field heterogeneities in f_{oc}, and hydraulic conductivity. Field methods are advantageous from this standpoint. Several levels of field efforts can be used for these purposes, varying from single hole analyses to full scale field tracer tests. Mehran et al. (1987) discuss the analysis of subsurface conditions to determine *in situ* sorption constants. In coring a site for determining subsurface conditions, the soil samples can be analyzed for mass of chemical adsorbed normalized by mass of soil, or q (be sure to subtract chemicals associated with the moisture content of the soil when determining this parameter). Collecting a ground water sample at this location will provide a value of C_e which correlates to the q values determined above. Given values for q and C_e, a value for K_p can be determined. Mackay (1989) discusses several other devices that can be utilized for determining transport parameters from single hole analyses.

Field scale tracer studies can also be conducted to determine transport parameters. The scope of tracer studies can range from several to many wells and can use natural or induced gradients. The migration of a plume past a point (which would be similar to a breakthrough curve) can be studied, or the migration of the entire plume with time can be monitored (moment analyses can be utilized to determine the transport parameters). This technology is predominantly at the research scale, with probably the best known field scale study being the Borden, Ontario site. This study was conducted under natural gradient conditions, used 5000 sampling points, and analyzed 19,900 samples over a 3 year period (Mackay et al., 1986; Freyberg, 1986; Roberts et al., 1986; Curtis et al., 1986). Background information and results from this study are discussed in the final chapter. The time necessary to conduct this study could have been reduced by utilizing forced gradient conditions and/or conducting a smaller scale study. Another large scale study was conducted at the Otis Air Force Base on Cape Cod, Massachusetts. The site used 640 sampling points, with as many as 10,000 water samples per sampling episode (LeBlanc et al., 1989). These efforts are very laborious and costly, and the cost must be weighed against the improved definition of the subsurface system when considering whether to conduct a field scale tracer study. A number of studies have also been conducted at the Mobile, Alabama site using smaller scale methodologies (Guven et al., 1985; Guven et al., 1986; Molz et al., 1985; Molz et al., 1986). Several guidance documents exist to aid in the design of tracer studies (Davis et al., 1985; Molz et al., 1986).

Using organic contaminants in field tracer studies may be prohibitive from economic concerns (due to the cost of analyzing for organic contaminants at low levels) and public health/environmental concerns (due to exposure of the environment, and possibly the public, to contaminants of concern). Several researchers have proposed the use of surrogate organic chemicals that behave similarly to the contaminant of concern or whose properties can be used to characterize the sorptive processes active in the subsurface, but that are environmentally acceptable and that are more easily quantified (Palmer et al., 1990; Sabatini and Austin, 1991). Examples of chemicals evaluated for this purpose are fluorescent dyes and surfactants. This technology is still in the developmental stages, but preliminary results are encouraging.

3.2.2.4 Estimation Methods

As can be realized from the previous section, much time and expense would be necessary to determine equilibrium adsorption parameters for all possible combinations of soils and chemicals. For this reason, much research has been conducted investigating relationships capable of predicting the linear sorption coefficients based on readily available or easily obtainable parameters. For nonionic organic chemicals and aquifer materials, it has been observed that the fraction organic content (f_{oc}) of the subsurface material is the dominant soil characteristic affecting sorption (Bailey and White, 1970; Karickhoff et al., 1979; Chiou et al., 1983, etc.). While the partition coefficient (K_p) for a chemical has been observed to vary significantly from soil to soil (with varying f_{oc}), it was observed that normalization of the K_p values by the respective values of f_{oc} resulted in a parameter (K_{oc}) that was much less variable (independent of the soil and a function only of the chemical). Another way of expressing this is that K_p was observed to be proportional to f_{oc} with K_{oc} being the proportionality constant. The definition of K_{oc} is shown in Equation 3.14. The merit of this

$$K_{oc} = K_p / f_{oc} \qquad (3.14)$$

normalization has been supported by research in which the organic content of the soil was removed, resulting in significant decreases in sorption (Miller and Weber, 1986). Bailey and White (1970) attribute the significance of the f_{oc} of the soil with respect to the sorption of nonionic organic chemicals to the fact that the organic matter of the soil has the highest combined cation exchange capacity and surface area of the soil size separates. Hamaker and Thompson (1972) list typical values of f_{oc} for surface soils to be in the range of 1 to 8% while alluvial sand aquifer materials have been reported with f_{oc} values in the range of 0.02 to 1.0% (Schwarzenbach and Westall, 1981; Abdul et al., 1986; etc.).

Given the K_{oc} of a chemical and determining a value of the f_{oc} of the subsurface material, a value for K_p can be calculated. Determination of K_{oc} values for all chemicals is still a great undertaking. Researchers have thus evaluated a number of parameters for estimating the K_{oc} value of a chemical based on fundamental properties of the chemical. Remembering that for nonionic organic contaminants the sorption is solvent motivated, the solubility of the chemical is a logical parameter to be considered for estimating the K_{oc} of the chemical. Another parameter which has been proposed for estimation purposes is the octanol-water partition coefficient (K_{ow}) of the chemical. The K_{ow} of a chemical describes the partitioning of an organic chemical between a polar phase (water) and a relatively nonpolar phase (1-octanol). The octanol-water partitioning is likened to the partitioning (sorption) in an aquifer of a chemical between the water and the porous media organic content. Values of K_{ow} are available for many compounds (see Appendix A).

In the event that measured K_{ow} values are not available, Lyman et al. (1982) have summarized estimation methods for predicting the K_{ow} value for a chemical based on either fragment constants or other solvent/water partition coefficients for the chemical. Weber et al. (1986) and Chin et al. (1986) have discussed the use of high performance reverse phase liquid chromatography (HPRPLC) to indirectly estimate values for water solubility and K_{ow} for organic compounds. Chromatographic retention times from HPRPLC were observed to successfully predict the water solubility and K_{ow} of a variety of organic compounds. Two other parameters that have been evaluated (to a lesser degree) for predicting values of K_{oc} are parachor (P), where $P = (V_m)^{1/4}$ and V_m is the molar volume of the chemical (Hamaker and Thompson, 1972), and the first-order molecular connectivity index (^1X, a topology parameter that is a measure of the projected size of a molecule; Sabljic, 1987).

Table 3.1 Expressions Estimating K_{oc} based on Aqueous Solubility (S) and K_{ow}

Aqueous Solubility	Reference
$\log K_{oc} = -0.55 \log S$ (mg/L) + 3.64	Kenaga and Goring, 1980
$\log K_{oc} = -0.621 \log S$ (mg/L) + 3.95	Hassett et al., 1983
$\log K_{om} = -0.729 \log S$ (mol/L) + 0.001[a]	Chiou et al., 1983

K_{ow}	
$\log K_{oc} = \log K_{ow} - 0.21$	Karickhoff et al., 1979
$\log K_{oc} = 0.999 \log K_{ow} - 0.202$	Hassett et al., 1980
$\log K_{oc} = 0.544 \log K_{ow} + 1.377$	Kenaga and Goring, 1980
$\log K_{oc} = 0.937 \log K_{ow} - 0.006$	Brown and Flagg, 1981
$\log K_{oc} = 0.72 \log K_{ow} + 0.49$	Schwarzenbach and Westall, 1981

[a] $K_{om} = K_p / f_{om}$, where f_{om} = fraction organic matter
$f_{om} = 1.72 * f_{oc}$ (SSSA, 1982)

A number of empirical expressions have been proposed to relate K_{oc} to either the water solubility or the K_{ow} of a chemical. Table 3.1 summarizes a number of these expressions. These expressions were developed by a number of researchers, using a variety of chemicals and solids (including soils, river and lake sediments, and subsurface materials). While these expressions are intended to be generally applicable, their empirical nature makes their use somewhat limited to the conditions for which they were developed. As an example of this fact, the development of two of these expressions will be discussed. Karickhoff et al. (1979) investigated the adsorption of chemicals from two chemical classes (polycyclic aromatic hydrocarbons (PAHs) and chlorinated hydrocarbons) and three natural river and lake sediments. Linear adsorption was indicated and the corresponding expression shown in Table 3.1 was proposed. Brown and Flagg (1981) investigated the adsorption of chemicals from the triazine and dinitroaniline families with natural lake sediments. Using the combined data of their study and the data from Karickhoff et al. (1979), the expression in Table 3.1 was developed. The difference in the two relationships is an indication of the variability in the data collected by the two research efforts. While the relationships listed in Table 3.1 do differ, these differences may be acceptable, especially for preliminary analyses. In general, the relationships based on K_{ow} have been more widely utilized and more successful. Karickhoff et al. (1979) state that the octanol-water partitioning more closely parallels the chemical adsorption in the soil system than does aqueous solubility and thus, expressions based on K_{ow}

prove to be much better estimators than expressions based on aqueous solubility.

Given the K_{ow} value for a chemical and the f_{oc} value of the soil, an estimate of the K_p value, and thus the adsorption of the chemical on the soil, can be made. While the ease of obtaining K_{ow} and f_{oc} values makes the use of these expressions attractive, caution must be taken in applying them. Banerjee et al. (1985) found that at f_{oc} levels less than 0.2% or clay content to f_{oc} ratios greater than 60, ineractions with the mineral surfaces may become significant. One approach to evaluate this effect is to assess the relative surface area (adsorption sites) from the soil size separates. McCarty et al. (1981) have proposed a simplified expression (Equation 3.15) to determine the critical f_{oc} (f_{oc}^*) where the mineral surface area

$$f_{oc}^* = \frac{S}{200} \frac{1}{\left(K_{ow}\right)^{0.84}}$$

(3.15)

becomes significant. In Equation 3.15, S is the specific surface area of the mineral fraction in m^2/gm. The critical f_{oc} value is observed to decrease as the value of K_{ow} increases (the chemical is more sorptive with respect to the f_{oc}) and as S decreases (lower mineral surface area). As the f_{oc} of the formation approaches the critical value of f_{oc}, the validity of the relationships in Table 3.1 becomes questionable.

When considering ionic chemicals (such as the organic pesticides diquat and paraquat; Bailey and White, 1970), the cation exchange capacity becomes a significant indicator of the level of sorption. Schwarzenbach and Westall (1985) observed that K_p values for ionizable hydrophobic organics (e.g., chlorinated phenols) were a function of the pH and the ionic strength of the aqueous solvent. The sorption of ionizable organics was thus observed to be in contrast to the partitioning model for neutral organics wherein sorption is independent of pH and ionic strength (Sabatini and Austin, 1991). The empirical relationships discussed above have assumed the porous medium to be saturated with aqueous solvent. Chiou et al. (1985) demonstrated the deviations in sorption that occur when the soil is dehydrated and when an organic solvent is present in place of the aqueous solvent. The researchers demonstrated that, in aqueous systems, the water molecules compete with the solute for sorption sites on mineral surfaces. As the medium is dehydrated, more of the organics are able to sorb onto the mineral surfaces. Also, the researchers demonstrated that in the presence of organic solvents, the solute is highly soluble; thus, the sorption of the solute significantly decreases. These exceptions to the simple partitioning model indicate the importance of understanding the fundamentals of sorption; thus, having the knowledge to decide when the use of various empirical relationships for estimating K_p is justified.

As a final comment, it should be mentioned that the role of absorption (partitioning) and adsorption in the association of nonionic organic compounds with the fraction organic content of subsurface materials has been debated. Some researchers have suggested that the linear nature of adsorption of organic compounds over the solubility range of the compounds and the lack of competitive sorption serve to indicate partitioning (absorption) while other researchers have countered these arguments. While significant from a fundamental standpoint, a detailed presentation of this discussion is beyond the scope of this document. The interested reader is referred to the literature for discussions of information relative to and implications of this topic (Chiou, 1989; Mingelgrin and Gerstl, 1983; Chiou et al., 1985; Chiou et al., 1983; Karickhoff, 1984; Rao and Davidson, 1979; Weber, et al., 1991 — to list a few).

3.2.3 Equilibrium Desorption

A common misconception is that once a chemical is adsorbed it no longer poses a threat to the ground water; however, pesticide adsorption to soil organic matter has been observed to be reversible. When the chemical concentration in the soil pore water decreases (as the chemical front passes), desorption of the pesticide from the solid phase to the pore water phase occurs (the free energy gradient is from the soil to the ground water). Modeling efforts often assume that desorption is completely reversible (the desorption curve is symmetrical to the adsorption curve). Several researchers have observed asymmetry (hysteresis) in the desorption curve (Swanson and Dutt, 1973; Miller and Chang, 1989). Swanson and Dutt (1973) found that the desorption data could be described by the Freundlich relationship with the value of N_{ads}/N_{des} being 2.3.

Hysteresis of desorption can be evaluated in laboratory batch studies in conjunction with adsorption isotherm studies. Upon attainment of equilibrium adsorption in a reactor, the supernatant of each reactor is removed and re-placed with contaminant free water and shaken until equilibrium desorption occurs. This process is repeated for each reactor resulting in a series of desorption data points for each original adsorption data point. Thus, each reactor from the adsorption study results in a desorption isotherm. Figure 3.9 shows an example of an adsorption isotherm and three desorption isotherms (indicating hysteresis of desorption). The triangles in Figure 3.9 correspond to the data points (reactors) that are used to establish the adsorption isotherm. Each triangle is the starting point of a desorption isotherm. The squares along each desorption isotherm correspond to desorption data points which are determined as outlined above. If the desorption was completely reversible (no

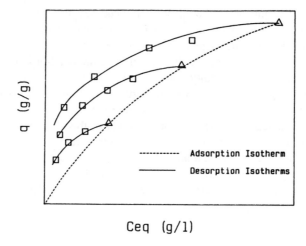

Figure 3.9.   Hysteresis of desorption-isotherm (Sabatini and Austin, 1990).

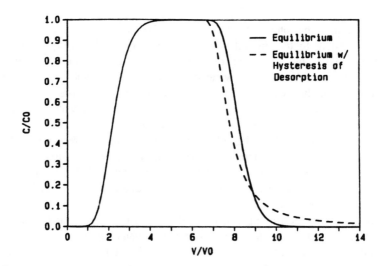

Figure 3.10.   Desorption hysteresis-breakthrough curve (Sabatini and Austin, 1990).

hysteresis of desorption), the desorption data points would fall on the adsorption isotherm.

The effect of hysteresis of desorption on a breakthrough curve is shown in Figure 3.10. The ordinate of Figure 3.10 is relative concentration (C coming out the column relative to C going in) and the abscissa is relative pore volumes or time (one pore volume is the amount of flow necessary to replace one volume of pore water). Figure 3.10 shows one graph in which hysteresis

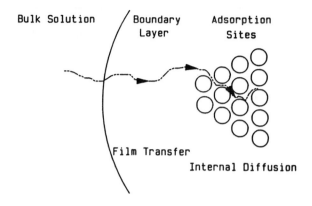

Figure 3.11. Dual resistance conceptualization (Sabatini and Austin, 1990).

of desorption is modeled and one graph in which desorption is considered as symmetrical. The graphs demonstrate the loss of symmetry when including hysteresis of desorption. Much remains to be learned about the nature, kinetics and modeling of desorption and its influence on the transport of chemicals in ground water. It is apparent that this phenomenon can greatly affect the time and volume of ground water necessary for "pump and treat" technologies and for determining the time necessary for a slug of contaminants to pass a given point, e.g., a well. It should be mentioned that this phenomenon has been observed in other fields for systems other than soil and ground water (Harwell, 1990).

3.2.4 Nonequilibrium Sorption/Desorption

Early research showed that equilibrium adsorption expressions were not always able to predict accurately the results observed in column studies with the greatest deviations occurring at higher pore water velocities (Kay and Elrick, 1967; van Genuchten et al., 1974; Valocchi, 1985). In attempting to predict the nonequilibrium adsorption experimentally observed, it is necessary to have a conceptual framework of the adsorption process and to isolate the rate limiting step(s). The adsorption process is considered to consist of three basic steps (Weber, 1972; Benefield et al., 1982; Weber et al., 1990). First, the adsorbate must diffuse from the aqueous phase (bulk liquid) to the soil or aggregate surface (film transport). Second, the adsorbate must diffuse through the intra-aggregate or intraparticle pores to the adsorption site (intraparticle diffusion). Third, the adsorbate undergoes the actual adsorption step (adsorption). This conceptualization is demonstrated in Figure 3.11. Rate

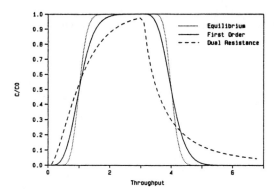

Figure 3.12.   Equilibrium vs nonequilibrium breakthrough and elution curves (Sabatini and Austin, 1990).

limitations by the film transport and/or the intraparticle diffusion steps would be classified as physical nonequilibrium while rate limitations of the adsorption step would be classified as chemical nonequilibrium. Figure 3.12 illustrates the difference in the shapes of breakthrough curves for equilibrium and nonequilibrium conditions.

The first attempts by researchers to describe nonequilibrium adsorption during solute transport assumed the adsorption step was rate limiting. It was felt that the higher pore water velocities did not allow sufficient contact time (residence time) for the adsorption to reach equilibrium. A later hypothesis was that the actual chemical adsorption step was not limiting but that diffusion of the pesticide from the aqueous phase (bulk liquid) to the final adsorption site becomes limiting at higher pore water velocities (physical nonequilibrium). While physical and chemical nonequilibrium approaches are fundamentally different, some researchers have argued that it is difficult to distinguish between the two processes by evaluating experimental data or by modeling attempts (van Genuchten, 1981; Nkedi-Kizza et al., 1984; Skopp, 1986). Nkedi-Kizza et al. (1984) have discussed the equivalence of two conceptual models (one physical and one chemical nonequilibrium). Introduction of dimensionless variables into each model resulted in the same dimensionless form of the equations.

A complete review of the chemical and physical nonequilibrium expressions that have been proposed is beyond the scope of this effort. The interested reader is directed to other references for presentation of additional nonequilibrium adsorption expressions not covered (Travis and Etnier, 1981; Rao and Jessup, 1983; Brusseau and Rao, 1989; Weber et al., 1991). The goal here will be to briefly review some of the nonequilibrium adsorption expressions proposed in an effort to establish the types of modeling approaches that have been used.

3.2.4.1 Chemical Nonequilibrium Expressions

Davidson and McDougal (1973) proposed the use of a first order kinetic adsorption expression as shown in Equation 3.2. This expression is first order with respect to both C and q with rate coefficients for adsorption and desorption (k_s and k_d, respectively). This first order expression reduces to the linear adsorption expression at equilibrium ($\partial q/\partial t = 0.0$). Hornsby and Davidson (1973) used a kinetic adsorption expression that was first order with respect to q but Nth order with respect to C. This expression, as shown in Equation 3.16, simplifies to the Freundlich expression at equilibrium.

$$\frac{\partial q}{\partial t} = k_s \frac{\eta}{\rho_s(1-\eta)} C^N - k_d q \qquad (3.16)$$

Cameron and Klute (1977) discussed the use of a combined equilibrium and kinetic adsorption (two site or bicontinuum) model. This approach was justified by the authors based on heterogeneities present in the soil. The authors state that quite different processes, such as rapid sorption with the soil organic matter (equilibrium) and slow adsorption on mineral surfaces (kinetic), may be involved in the overall adsorption process. The authors state that diffusion limited sites could fall into the kinetic (physical) site category. Rao et al. (1979) investigated the use of two site (bicontinuum) models to describe nonequilibrium breakthrough curves. Some fraction of the adsorption sites was assumed to participate in equilibrium adsorption while the remaining portion of the sites was assumed to participate in kinetic (physical or chemical) adsorption.

3.2.4.2 Physical Nonequilibrium Expressions

As researchers began to realize that chemical nonequilibrium expressions were not able to describe and/or predict the experimentally observed data, the utility of physical nonequilibrium expressions received increased attention. Davidson and McDougal (1973), Hornsby and Davidson (1973), and van Genuchten et al. (1974) are examples of early research efforts that found chemical nonequilibrium expressions to be inadequate and suggested the investigation of physical nonequilibrium expressions. The fact that researchers often observed pesticide adsorption to be virtually complete in batch kinetic studies in less than 1 hr (Leenher and Alrichs, 1971) also supported the conclusion that the adsorption step was not rate limiting.

Skopp and Warrick (1974) discussed a two-phase model for describing solute transport of sorptive solutes in soils. The two phases considered by the authors were the mobile phase (bulk liquid) and the stationary phase (boundary layer of pores). Transport in the mobile phase was assumed to occur by advection and dispersion while transport through the stationary phase was assumed to be due to diffusion (advection was assumed to be negligible in this phase).

van Genuchten and Wierenga (1976) developed a model to describe the mass transfer of solute in sorbing porous media. The model development involved dividing the soil matrix into five regions: (1) air spaces, (2) mobile (dynamic) water located in the larger (inter-aggregate) pores, (3) immobile (stagnant) water located inside aggregates and at the contact points of aggregates and/or particles, (4) dynamic soil, located sufficiently close to the mobile water phase for assumed equilibrium to exist between solute in the mobile phase and the soil phase, and (5) stagnant soil region, where adsorption by soil is diffusion limited through immobile liquid. Equation 3.17 was developed from these considerations. In Equation 3.17, the subscript "m" refers to mobile phase and "im" refers to immobile phase. The mass transfer between the mobile and immobile phases was considered to be first order, as shown in Equation 3.18. Sensitivity analyses were conducted for this model with a range in shapes of predicted breakthrough curves observed. These predicted

$$\Theta_m \frac{\partial C_m}{\partial t} + \Theta_{im} \frac{\partial C_{im}}{\partial t} + f\rho_b \frac{\partial q_m}{\partial t} + (1-f)\rho_b \frac{\partial q_{im}}{\partial t}$$

$$= \Theta_m D_x \frac{\partial^2 C_m}{\partial x^2} - \Theta_m v_{x,m} \frac{\partial C_m}{\partial x} \tag{3.17}$$

$$\Theta_{im} \frac{\partial C_{im}}{\partial t} + (1-f)\rho_b \frac{\partial q_{im}}{\partial t} = \alpha'(C_m - C_{im}) \tag{3.18}$$

where θ_m = mobile phase water content (L^3/L^3)
θ_{im} = immobile phase water content (L^3/L^3)
f = fraction adsorption sites in dynamic region
ρ_b = bulk soil density (M/L^3)
α' = first order mass transfer coefficient ($1/t$)

breakthrough curves include shapes similar to those experimentally observed but which previous modeling attempts had been unsuccessful in predicting. Parker and van Genuchten (1984) produced a bulletin discussing the use of this model (including later additions).

Rao et al. (1979) investigated the use of two-site (bicontinuum) models to describe nonequilibrium breakthrough curves. One of the sites was considered to experience equilibrium adsorption and the other site was considered to experience physical nonequilibrium adsorption. The physical nonequilibrium condition suggests that diffusion limited sites exist. Diffusion from the mobile phase to the immobile phase was assumed to be first order (see Equation 3.18). In this case, the bicontinuum model is conceptually the same as the two phase model of van Genuchten and Wierenga (1976).

Miller (1984) proposed a physical nonequilibrium model that incorporated film transport and intraparticle diffusion (see Figure 3.11 for conceptualization) as the source of the nonequilibrium breakthrough (Miller and Weber, 1986). This effort differed from ones previously discussed in that diffusion was considered to be Fickian rather than first order. This approach has been labeled the Fickian physical nonequilibrium model. This model was developed using mass transfer and mass balance concepts and resulted in the relationships shown in Equations 3.19 and 3.20.

$$\frac{\partial q}{\partial t} = D_s \frac{1}{r^2} \frac{\partial}{\partial r} \left[r^2 \frac{\partial q}{\partial r} \right] \tag{3.19}$$

$$k_f \left(C - C_s \right) = D_s \rho_s \frac{\partial q}{\partial r} \qquad \text{when } r = R \tag{3.20}$$

where D_s = intraparticle diffusion coefficient (L^2/t)
   r = radial dimension for particle (L)
   k_f = external film transfer coefficient (L/t)
   C_s = equilibrium pesticide concentration at exterior of particle (M/L^3)
   R = radius of particle (L)

Good predictive capabilities for the laboratory column studies were observed using parameters determined in completely mixed batch reactors.

Crittenden et al. (1986) developed a Fickian physical nonequilibrium model similar to that of Miller (1984). The model of Crittenden et al. (1986) was based on the presence of aggregates (or diffusion limited regions in the absence of physical aggregates) in the soil which caused the nonequilibrium breakthrough curves (see Figure 3.11 for conceptualization). This model included intra-aggregate diffusion both in the pore space and along the pore surfaces. The authors referred to the model as the dispersed flow, pore, and surface diffusion (DFPSDM) model. The authors conducted sensitivity analyses on the model to determine the relative significance of dispersion, film transport and intraparticle (intra-aggregate) diffusion on the shape of the breakthrough curve. The authors concluded that under most conditions the intraparticle diffusion would be the limiting case. The authors also discussed techniques for determining the model parameters separate from the column data by estimation techniques. Hutzler et al. (1986) concluded that, while the DFPSDM appeared to be an improved mechanistic model, their work suggested that an additional kinetic mechanism should be included in the model. Roberts et al. (1987) utilized the model of Crittenden et al. (1986) in an attempt to predict the data of Nkedi-Kizza et al. (1982). Estimation techniques were used to predict the necessary parameters separate from the column data with good results. The authors concluded that hydrodynamic dispersion governed the nonequilibrium breakthrough curves at low velocities and internal pore diffusion dominated at higher pore water velocities (3 to 143 cm/hr). The external mass transfer was concluded to play a minor role under all the experimental conditions investigated.

It is not necessary for a soil to be aggregated for diffusion limited physical nonequilibrium to be experienced. Bouchard et al. (1988b) designed their column experiments to prevent the existence of diffusion limited regions for the transport of atrazine, diuron, and hexazinone in low organic carbon aquifer materials. The authors found the level of nonequilibrium to increase with increasing organic carbon content of the soil. The adsorption was determined to be linear for the solutes studied in the concentration ranges investigated, eliminating nonlinear isotherms as the cause for the nonequilibrium breakthrough curves observed. It was thus concluded that the nonequilibrium was caused by diffusion limitations into the organic carbon matrix of the soils investigated. Lee et al. (1988) arrived at similar conclusions concerning nonequilibrium breakthrough curves observed while investigating the movement of TCE and p-xylene in two sand aquifer materials.

Nkedi-Kizza et al. (1989) evaluated nonequilibrium transport of atrazine and diuron by plotting relative concentration vs relative pore volume normalized by the retardation factor of the compound, $(V/V_o)/r_f$, also referred to as throughput (Miller, 1984). This consolidates all the breakthrough curves to a common range of normalized time. When equilibrium conditions are satisfied, the spreading forces should be common for conservative and sorbing chemicals and the curves should overlay one another when plotted on these axes. Nkedi-Kizza et al. (1989) found that atrazine and diuron breakthrough curves did not overlay with the curve for tritiated water when plotted on thenormalized scale; however, when the organic content of the soil was reduced from 0.2 % to <0.01% (by H_2O_2 treatment), the curves did overlay. This suggests that the source of the nonequilibrium transport was intraorganic matter diffusion (IOMD). Brusseau and Rao (1989) discussed IOMD and intramineral diffusion (IMD) as possible sources of physical nonequilibrium. The authors analyzed data from the literature for a wide range of chemicals and soils and concluded that IOMD appeared to be the source of nonequilibrium sorption. Brusseau et al. (1989) proposed a model to account for multiple sources of nonequilibrium (e.g., intra-aggregate diffusion and IOMD limitations). For further discussion on the source of nonequilbrium transport, the reader is referred to the following references: Szecsody and Bales, 1989; Wu and Gschwend, 1988; Wu and Gschwend, 1986; Karickhoff and Morris, 1985.

3.2.5 Summary

A rather detailed discussion of sorption has been presented; ranging from rather sophisticated efforts to understand the fundamental mechanisms of nonequilibrium transport to rather simple empirical expressions for estimating linear equilibrium sorption coefficients. While the more sophisticated efforts are necessary to aid in the development of fundamental models, the simpler expressions are more easily utilized by the practicing professional addressing fate and transport processes. In evaluating sorption models, it must be remembered that the subsurface is a complex system and that other factors may significantly affect the movement of contaminants (distributions of hydraulic conductivity, flow velocity, organic content, etc.). The professional must determine what level of sophistication is merited in the description of the sorption processes in light of other variabilities present in the system.

Subsequent sections will discuss other abiotic processes. Space prohibits these processes from being discussed in the detail that sorption has been; however, this should not be interpreted to mean that sorption is the most significant process in all cases. Depending on the chemical and the environmental conditions, other abiotic processes may in fact be dominant (e.g., precipitation for metals). Many of the approaches and limitations which have been discussed for sorption are equally applicable to the other abiotic processes.

3.3 COSOLVATION/IONIZATION

Solvent motivated sorption has been presented as one of the two motivations of sorption. Solvent motivated sorption suggests that the sorption is due to a "dislike" of the solvent for the solute (that the solute is more thermodynamically favorable in the sorbed phase). In this scenario, anything that acts to increase the presence of the solute in the solvent (increased solubility) will also act to decrease the sorption. Cosolvation and ionization are two examples of processes that act to increase the chemical solubility, the former by altering the solvent phase and the latter by altering the chemical form.

3.3.1 Cosolvation

Cosolvation refers to a mobile phase consisting of multiple solvents that are miscible in one another. For purposes of this discussion, the cosolvents are ground water and miscible organic solvents. When considering the migration of contaminants from land disposal sites for hazardous organic contaminants, it is possible that organic solvents miscible with ground water may be present in the landfill. The organic solvents may migrate to the ground water, resulting in cosolvency. The presence of the organic cosolvent in the solvent phase will act to decrease the solvophobicity of hydrophobic organic contaminants that may also be leaching from the landfill, resulting in increasing solubility of the organic solute(s) and thus decreasing sorption (increasing mobility of the organic solute). This is in keeping with the basic axiom that likes dissolve likes; the less polar the mobile phase, the more soluble are nonpolar organics. Based on this conceptual model, it would be expected that as the fraction of cosolvent increases (as the polarity of the mobile phase decreases), the level of sorption will decrease.

Rao et al. (1985) developed a theoretical model in an attempt to account for the impact of cosolvation (water and organic solvents) on the sorption and migration of hydrophobic organic contaminants (referred to as the solvophobic model). The model was based upon the assumption that sorption was the result of hydrophobic (or, more generally, solvophobic) interactions, i.e., that sorption was solvent motivated. The authors considered that solute-sorbent interactions (sorption) are influenced by the following interactions (listed in decreasing order of significance): solute-solvent, solvent-solvent, solvent-sorbent, and sorbent-sorbent. The solute-sorbent and solute-solvent interactions were incorporated into the modeling approach. The authors used existing theories regarding the solubility of chemicals in aqueous and binary mixed solvents; these suggested that the solubility of a hydrophobic solute in a mixed solvent system is a function of the fraction of cosolvency (f^c) and the nature of the solute (defined by a parameter σ^c, discussed below). The solute was generally considered to be comprised of a nonpolar moiety (hydrocarbonaceous surface area, HSA, Å^2) and a polar moiety (polar surface area, PSA, Å^2), with the total surface area (TSA) being the sum of HSA and PSA. These properties of the solute, and the interfacial free energies of solute-solvent interactions, are integrated into the parameter σ^c.

Rao et al. (1985) proposed Equation 3.21 to account for deviations in K_p from water (K_p^w) due to the mixed solvents (K_p^m). Equation 3.22 is the definition of σ^c. From Equation 3.21, it is evident that the

$$\ln\left(K_p^m / K_p^w\right) = -\alpha\sigma^c f^c$$

or (3.21)

$$\ln\left(K_p^m\right) = \ln\left(K_p^w\right) - \alpha\sigma^c f^c$$

$$\sigma^c = \Delta\gamma^c HSA / (kT)$$ (3.22)

where K_p^m = partitioning coefficient from mixed solvent
K_p^w = partitioning coefficient from water
α = empirical constant
f^c = fraction of cosolvent
$\Delta\gamma^c$ = difference in the interfacial free energy (ergs/$\mathring{A}^2$) at the aqueous interface and the organic cosolvent interface, respectively, for HSA
HSA = hydrocarbonaceous surface area, $\mathring{A}^2$
k = Boltzman constant, ergs/°K
T = temperature, °K

relative ratio of partitioning coefficients (K_p^m/K_p^w) decreases exponentially as the fraction cosolvency (f^c) increases. Equation 3.21 suggests that a plot of ln (K_p^m/K_p^w) vs f^c should be linear with a negative slope. Equation 3.22 indicates that the value of σ^c, at a given concentration, is a function only of the solute and the solvents (and does not depend on the sorbent). These observations with respect to the model are consistent with the conceptual model described above (decreasing sorption as solute becomes more soluble in the solvent mixture, etc.). The theoretical model can handle multiple cosolvents. The researchers conducted preliminary verification of the model by analyzing published data for various systems.

Nkedi-Kizza et al. (1985) evaluated the theoretical solvophobic model of Rao et al. (1985) by evaluating the sorption of anthracene and two herbicides (atrazine and diuron) by soils from aqueous solutions and binary solvent mixtures. The cosolvent mixtures evaluated were methanol-water and acetone-water. Methanol and acetone were selected by the researchers based on their high miscibility with water, their variation in protonation (methanol is a proton donor while acetone is a proton acceptor), and their likelihood of occurrence in landfill leachates. Figure 3.13 shows the sorption of anthracene with four soils (two with very similar f_{oc}) for varying f^c values of methanol-water cosolvents (plotted as ln K_p^m vs f^c). The linear nature of the data when plotted in this manner agrees with the solvophobic model of Rao et al. (1985). Also, it is observed that the slopes are similar for all soils, again agreeing with the solvophobic model. A similar plot for acetone-water cosolvents

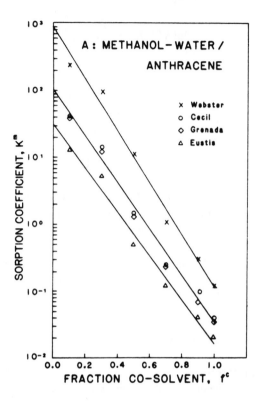

Figure 3.13.   Effect of fraction cosolvent on sorption (K^m) of anthracene from methanol-water for four soils (Reprinted with permission from Nkedi-Kizza, P., P. S. C. Rao, and A. G. Hornsby. "Influence of Organic Cosolvents on Sorption of Hydrophobic Organic Chemicals by Soils," *Enviromental Science and Technology* **19(10):975–979. Copyright 1985, American Chemical Society).**

(utilizing two soils) indicated deviations from linearity at f^c values greater than 0.25. It was hypothesized that these deviations were due to solvent-sorbent interactions (potential changes in the organic content of the porous medium due to the acetone). Figure 3.14 shows data for the sorption of anthracene from both solvent systems plotted as relative sorption [ln (K_p^m/ K_p^w)] vs f^c. The sorption data, for a range of soils and for a given solvent system, is observed to be described by a single line when plotted in this manner. This indicates that the data, in this form, are independent of the soil (in keeping with the solvophobic theory, as discussed above). Thus, it is possible to account for the organic cosolvent effects on sorption by a single parameter (σ^c) which incorporates solvent and solute properties. The researchers were also able to verify the dependence of σ^c on HSA and $\Delta\gamma^c$

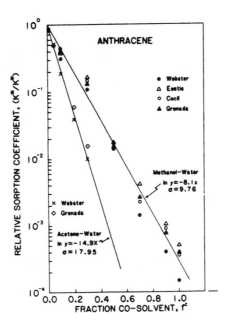

Figure 3.14. Relative sorption coefficient (K^m/K^w) for anthracene sorption from methanol-water and acetone-water (Reprinted with permission from Nkedi-Kizza, P., P. S. C. Rao, and A. G. Hornsby. "Influence of Organic Cosolvents on Sorption of Hydrophobic Organic Chemicals by Soils," *Enviromental Science and Technology* 19(10):975–979. Copyright 1985, American Chemical Society).

(see Equation 3.21) by virtue of data obtained in this research and from previous research. The observations made for anthracene were verified by data collected for atrazine and diuron. The data collected in this research thus served to verify the theoretical solvophobic model proposed by Rao et al. (1985). Nkedi-Kizza et al. (1985) suggested that these results could also prove beneficial when conducting sorption studies for highly hydrophobic chemicals (when sorption to reactor vessels, etc., are difficult problems). The authors suggested that data could be collected at several levels of cosolvents (where the problems of sorption on vessel walls is negligible) with subsequent extrapolation to $f^c = 0$ (for determination of K_p^w, see Figure 3.13).

The use of reversed-phase liquid chromatography (RPLC) for investigating the thermodynamics and mechanisms of sorption in cosolvent systems has been evaluated for methanol/water systems (Woodburn et al., 1989). The researchers used the enthalpy-entropy compensation model and RPLC data to infer variations in sorption mechanisms for PAHs and monohalobenzenes as compared with alkylbenzenes for methanol/water RPLC systems (the reader is referred to Woodburn et al. (1989) for additional details of this method). Batch tests with a variety of soils confirmed observations made from the RPLC studies, suggesting that the sorption mechanisms for a given chemical were similar for RPLC and soil systems, and for binary solvent systems. This suggests that evaluation of thermodynamics and mechanisms of sorption for hydrophobic organic chemicals on soils in cosolvent systems can be successfully investigated using RPLC studies, thus simplifying the analyses.

Solvophobic theory indicates that as the fraction of cosolvent decreases, the partitioning increases (as fraction cosolvent approaches zero, the partitioning approaches that for water). Thus, it is necessary to determine the fraction of cosolvency to be expected from the contamination episode. At relatively low fractions of cosolvency, the impact may be relatively insignificant. Also, the fraction of cosolvency can be expected to change as the plume disperses (due to dilution of the plume with cosolvent free ground water). Cosolvency may be expected to be significant primarily near the source of the ground water contamination (Johnson et al., 1989). The temporal (spatial) dependency of the fraction of cosolvency (f^c) with plume migration serves to further complicate this phenomenon. Research is underway to study this aspect of cosolvent transport (Wood and Rao, 1991).

3.3.2 Ionization

Certain compounds can either gain or lose a proton as a function of pH and thus go from a neutral form to an ionic form. For organic compounds, this ionization will serve to greatly increase the solubility of the chemical in the ground water (a polar solvent). Thus, as the solubility increases the sorption (due to solvent motivated sorption) will decrease. The loss of a proton produces anionic compounds. For anionic organic compounds, the adsorption to clay minerals (sorbent motivated sorption) will be negligible. Thus, the net result will be a significant decrease in the sorption of the chemical (and thus a significant increase in the chemical mobility). The gain of a proton (or loss of a hydroxyl) will result in the formation of a positive ion. In this case, the ionic compound may associate to a greater degree with the cation exchange (CEC) capacity of the clay minerals (ion exchange adsorption). The overall impact on sorption (mobility) will depend on the relative sorption of the neutral and ionic forms of the compound.

The ionization of an organic compound as a function of pH can be illustrated by 2,4,5-trichlorophenol. As a function of pH, 2,4,5-trichlorophenol may ionize to form 2,4,5-trichlorophenolate (an anion). This is illustrated in Figure 3.15. The ionization of 2,4,5-trichlorophenol results in a loss of a proton (H^+). The loss of protons will increase as the proton concentration in solution decreases

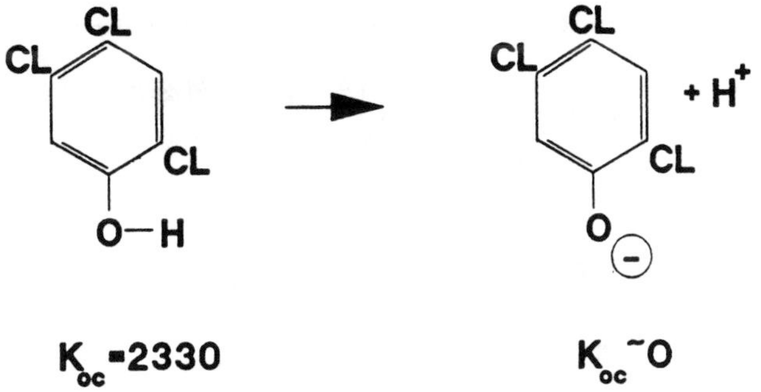

Figure 3.15. Ionization and K_{oc} values for 2,4,5-trichlorophenol (Johnson et al., 1989).

(or as the hydroxyl concentration increases, i.e., as the pH increases). As illustrated in Figure 3.15, the K_{oc} value of the neutral compound is 2330 while for the anion the K_{oc} value is approximately zero, greatly increasing the mobility of this compound. Figure 3.16 shows the K_{oc} values as a function of pH.

Ionization can be represented by a chemical reaction as shown in Equation 3.23 with the expression describing the equilibrium position of this reaction shown in Equation 3.24. In the example discussed above, RH would correspond to 2,4,5-trichlorophenol and R⁻ would refer to 2,4,5-trichlorophenolate. The [ ] refers to concentration (activity) and K is the equilibrium constant of the reaction. The form of this reaction is the same

$$RH \rightleftharpoons R^- + H^+ \qquad (3.23)$$

$$K = \frac{\left[R^-\right]\left[H^+\right]}{\left[RH\right]} \qquad (3.24)$$

as that for a weak acid in that the reaction is not complete (not all of the reactants go to products). If the hydrogen ion concentration (activity) in solution is initially low (high pH), the reaction will tend to proceed from left to right to a greater extent. If the system is at equilibrium and the hydrogen ion concentration increases (decreasing pH), the reaction will proceed from right to left (Le Chatelier's principle of equilibrium chemistry). Thus, the degree of ionization is a function of the pH of the system, as discussed above (and demonstrated in Figure 3.16). This relationship can be quantified as shown below. Equation 3.25 results from taking the log of Equation 3.24 and Equations 3.26 and 3.27 are simplifications of Equation 3.25 (remember that by convention, $pH = -\log [H^+]$, $pK = -\log K$, etc.).

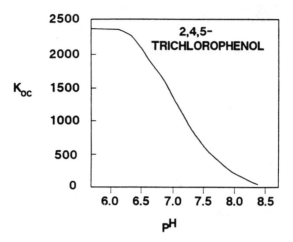

Figure 3.16. K_{oc} vs pH for 2,4,5-trichlorophenol (Johnson et al., 1989).

$$\log K = \left(\log\left[R^-\right] + \log\left[H^+\right]\right) - \log[RH] \qquad (3.25)$$

$$-\log K = -\log\left[H^+\right] + \log[RH] - \log\left[R^-\right] \qquad (3.26)$$

$$pK - pH = \log[RH] - \log\left[R^-\right] \qquad (3.27)$$

It can be observed from Equation 3.27 that when the value of pH equals the value of pK, the nonionic and ionic concentrations of the compound will be equal. As the pH increases above the pK, Equation 3.27 indicates that the log [RH] < log [R⁻]; i.e., the concentration of the ionic form is greater than the nonionic form. Conversely, as the pH decreases below the pK, the relative concentration of the nonionic form is greater. This is in keeping with the conceptual model discussed above. The value of pK thus is valuable in assessing the relative form of ionic compounds. For acids (compounds releasing protons) the pK is referred to as pK_a, and for bases it is referred to as pK_b. Thus, knowing the pK_a and / or pK_b values for a chemical and the pH of the subsurface environment, the relative form of ionizable organic compounds in this subsurface environment can be determined (and the concomitant relative mobility of the compounds assessed).

As previously discussed, the sorption of neutral organic contaminants is independent of the ionic strength of the ground water. However, for ionizable compounds sorption is dependent on ionic strength conditions. Westall et al. (1985) showed the partitioning of chlorinated phenols to be affected by the pH and ionic strength of the aqueous phase. For the sorption of ionic organic compounds, a more complete knowledge of the subsurface environment is required.

When cationic organic chemicals sorb (ion exchange) to soil surfaces, the level of sorption of nonionic hydrophobic contaminants can be affected. A soil may have very low values of f_{oc}, and thus exhibit little sorption of hydrophobic organic contaminants. However, if this soil has an appreciable clay fraction, the exchange of organic cations onto the clay fraction will convert the hydrophilic clay surface to a hydrophobic surface. The hydrophobic surface present in the altered medium will sorb hydrophobic organic compounds to a much greater extent than did the unaltered medium surface, significantly altering the mobility of the hydrophobic chemical in the subsurface. Several examples of laboratory studies demonstrating this phenomenon are presented by Bouchard et al. (1988a), Lee et al. (1989), and Bronawell et al. (1990).

3.4 ION EXCHANGE

Ion exchange is a specific category of adsorption. As discussed previously, sorption can be considered as either adsorbent or solvent motivated. Solvent motivated sorption (partitioning) typically occurs for neutral organic chemicals in ground water with the accumulation occurring in the fraction organic content of the media. This section will consider adsorbent motivated sorption; the accumulation occurs due to an affinity of the solid surface for the chemical. Typically, the adsorbent surface contains a charge deficiency and requires the accumulation of ions near the solid/liquid interface to neutralize the surface charge. In subsurface media, the mineral fraction most commonly involved in ion exchange is the clay fraction.

3.4.1 Ion Exchange Fundamentals

Ion exchange occurs when the adsorbent charge deficiency can be neutralized more efficiently by ions in solution than by those ions currently adsorbed. For example, if sodium ions (monovalent) have accumulated at the interface and suddenly calcium ions (divalent) appear, the surface excess can be more efficiently neutralized by the calcium ions than by the sodium ions. Thus, the sodium ions will desorb and the calcium ions will adsorb; an exchange of ions will occur.

An example of ion exchange that may be familiar to the reader is a home water softener. A home water softener functions by accumulating the divalent cations from the ground water onto the adsorbent (ion exchange resin) while releasing monovalent ions, thus reducing the water hardness. Equation 3.28 shows a general equation for the exchange of a multivalent cation (B^{n+}) for a monovalent cation (A^+) on a resin (R^-) and Equation 3.29 shows this equation for exchange of calcium for sodium ions. In the subsurface, the resin would

$$nR^-A^+ + B^{n+} \rightleftharpoons R_n^- B^{n+} + nA^+ \qquad (3.28)$$

$$2R^-Na^+ + Ca^{2+} \rightleftharpoons R_2^- Ca^{2+} + 2Na^+ \qquad (3.29)$$

be natural mineral surfaces, e.g., clay minerals, and the potential mono- and multivalent cations are numerous. The most common ions occurring naturally in soil are as follows (listed in decreasing order of occurrence for nonmarine deposits): cations, Ca^{2+}, Mg^{2+}, Na^+, and K^+; anions, SO_4^{2-}, Cl^-, PO_4^{3-}, and NO_3^- (Mitchell, 1976). The phenomenon of ion exchange is reviewed by Thomas (1977).

A solution of a high concentration of monovalent ions (brine, NaCl) can, by mass action, cause sodium to displace calcium (monovalent to displace divalent) ions from the adsorption sites (even though the reverse process would occur under low ionic strength conditions). The end result will be a media with sodium cations associated with the exchange sites with the calcium displaced in the mobile phase. This process will be of concern in brine contamination episodes, e.g., oil drilling brines. This is also how home water softeners are recharged upon exhaustion (Benefield et al., 1982).

Since clay minerals are frequently the dominant source of ion exchange in the subsurface, it is important to have a fundamental understanding of clay mineralogy and the source(s) of charge deficiencies in clay minerals. Clays are commonly layered aluminosilicates. Two basic structural units include a tetrahedral structure about Si^{4+} and an octahedral structure about Al^{3+} (Hillel, 1982). Clay minerals are made up of these basic structural units. For example, kaolinite is made up of one tetrahedral and one octahedral layer (1:1). Montmorillonite is comprised of two tetrahedral layers and one octahedral layer (2:1). Table 3.2 summarizes several clay minerals, the type of layering for the mineral (1:1 vs 2:1) and properties of the mineral.

Three sources responsible for the charge deficiencies (and concomitant exchange capacity) for clay minerals can be identified (Mitchell, 1976). One source of exchange capacity is isomorphous substitution. An example of isomorphous substitution is a tetrahedral structure with Al^{3+} at the center rather than Si^{4+}. The term isomorphous substitution may be misleading, for example, it may suggest that the tetrahedral structure initially had Si^{4+} at the

Table 3.2 Summary of Clay Mineral Characteristics

Type	Mineral	Octahedral Layer Cations	Tetrahedral Layer Cations	Size (μm)	CEC (meq/100 gm)	Specific Gravity	Specific Surface Area[a] (m²/gm)
1:1	Kaolinite	Al_4	Si_4	0.05 – 4.0	3 – 15	2.60 – 2.68	10 – 20
2:1	Illite	$(Al, Mg, Fe)_{4\text{-}6}$	$(Al, Si)_8$	0.003 – 10.0	10 – 40	2.6 – 3.0	65 – 100
2:1	Montmor- Illonite	$Al_{3.34}Mg_{0.64}$	Si_8	0.001 – 10.0	80 – 150	2.35 – 2.70	50 – 120 700 – 840
2:1	Vermiculite	$(Mg, Fe)_6$	$(Al, Si)_8$	0.003 – 10.0	100 – 150	Not given	40 – 80 870

Source: Adapted from Mitchell, 1976.

[a] When two ranges of specific surface area (SSA) are listed, the lower range of numbers corresponds to primary SSA and the higher range corresponds to secondary SSA.

Table 3.3 Surface Charge of Common Minerals

Mineral	PZC
"Al(OH)$_3$" (amorph)	7.5–8.5
Al$_2$O$_3$	9.1
CuO$_3$	9.5
"Fe(OH)$_3$" (amorph)	8.5
MgO	12.4
MnO$_2$	2–4.5
SiO$_2$	2–3.5
Clays	
Kaolinite	3.3–4.6
Montmorillonite	2.5
Asbestos	
Chrysotile	10–12
Crocidolite	5–6
CaCO$_3$	8–9
Ca$_5$(PO$_4$)$_3$OH	6–7
FePO$_4$	3
AlPO$_4$	4

Source: Adapted from Montgomery, 1985.

center which at some time was substituted by Al^{3+}. In actuality, isomorphous substitution refers to the fact that during the formation of the clay minerals, Al^{3+} was present rather than Si^{4+}. Thus, no physical substitution (or displacement) occurred (Mitchell, 1976). However, isomorphous substitution results in a net decrease in the charge of the structure and thus results in a charge deficiency for the resulting mineral. Another common isomorphous substitution is the formation of the octahedral structure with Mg^{2+} rather than Al^{3+}. A second source of exchange capacity is the occurrence of broken bonds at particle edges and noncleavage surfaces of the minerals. A third source of exchange capacity is the replacement of the hydrogen of an exposed hydroxyl group.

The surface charge of clays and many other mineral surfaces is a function of pH. At low pH (excess hydrogen ions), the surface can exhibit a positive net charge while at higher pH (excess hydroxyl ions) the surface can exhibit a negative charge. The nature of the surface, and of the sorption, is thus impacted by the pH of the ground water and/or of the contamination. At an intermediate pH the surface exhibits a neutral charge; this pH is classified as the point of zero charge (pzc). The pzc differs for the various mineral surfaces, as observed in Table 3.3. At a pH above the pzc of the mineral, the surface exhibits a net negative charge while at a pH below the pzc the surface exhibits a net positive charge. Thus, the surface charge exhibited by a mineral surface can be determined by evaluating the pH of the solution relative to the pzc of the mineral. A neutral pH is above the pzc of clays but below the pzc of aluminum oxide (see Table 3.3). Thus, at a neutral pH clays will exhibit a negative surface charge (as discussed above), while aluminum oxide will exhibit a positive surface charge. It should be noted that at very low values of pH the clay surfaces may also exhibit a positive surface charge.

An important consideration in addressing the potential for ion exchange is to know the hierarchy of preferred ions on adsorption sites. It has already been stated that higher valency ions will displace lower valency ions. However, the question remains as to what sequence will occur for ions of the same valency (univalent). For univalent ions, it has been observed that smaller ions are preferentially exchanged for larger ions. The sequence in which ions are preferentially exchanged is referred to as the selectivity of the mineral surface for the ions. Following is an example of ion selectivity (from least to greatest — Mitchell, 1976): $Na^+ < Li^+ < K^+ < Rb^+ < Cs^+ < Mg^{2+} < Ca^{2+} < Ba^{2+} < Cu^{2+} < Al^{3+} < Fe^{3+} < Th^{4+}$. However, this selectivity sequence can be negated by mass action effects at high concentrations of individual ions. As mentioned previously, brine solutions can displace multivalent cations with monovalent cations due to the high concentration of the monovalent cations and mass action effects.

Two important considerations with respect to the transport of exchanging chemicals are the level of adsorption of the exchanging ions and the rate at which the exchange takes place. The level of adsorption for a subsurface media is greatly influenced by the surface area containing exchange sites. This parameter is typically discussed in terms of the CEC of the medium and is expressed in units of milliequivalents of exchanging cations per 100 grams of solids (meq/100 g). For example, a typical range of CEC values for kaolinite is 3 to 15 meq/100 g, for illite is 10 to 40 meq/100 g and for montmorillonite is 80 to 150 meq/100 g (Mitchell, 1976). Table 3.2 summarizes CEC values for several clay minerals. The relative ranges of CEC for the various clay minerals indicate that montmorillonite has the greatest charge deficiency of the clay minerals mentioned. The CEC ranges also indicate that montmorillonite and vermiculite are the clay minerals that will exchange cations to the greatest extent (and thus decrease the mobility of the cations to the greatest extent).

Clay fractions are also high in specific surface area (SSA, surface area per unit mass of mineral, m^2/gm) as compared to other size fractions. Clay minerals have values of SSA in the range of 10 to 900 m^2/gm (Mitchell, 1976), depending on the specific mineral. Table 3.2 summarizes SSA values for several clay minerals. It is observed that, in general, increasing values of SSA correlate with increasing values of CEC. Montmorillonite and vermiculite are listed as having primary and secondary SSA. The primary SSA is due to particle surfaces independent of interlayer zones while secondary SSA are those surfaces exposed by expanding the lattice so that polar fluids can penetrate between layers. As a basis of comparison, the sand size fraction with the highest surface area (corresponding to the smallest size sand particles) has a SSA value of approximately 10^{-2} m^2/gm (three orders of magnitude less than that for the clay minerals with the smallest SSA). The SSA of sand particles (assuming spherical particles) can be approximated by SSA = 3/$(\rho_s r)$, where ρ_s is the particle density, and r is the radius of the sphere. In comparing the relative surface areas of sand and clay fractions, it is observed that clay minerals can act as the dominant soil fraction with respect to surface interactions even when present as a minor fraction with respect to mass (this was also observed previously for organic matter).

The rate of ion exchange may become an issue, especially during artificially high ground water velocities (e.g., during pump and treat remediation). The rate of exchange varies with the clay mineralogy. With kaolinite, the exchange rate is virtually instantaneous. For illite, the exchange rate is slower, with units of hours being reasonable. For montmorillonite, the exchange rate requires even greater time (Mitchell, 1976). In general, the rate of exchange decreases with increasing location of the exchange sites in interlayer regions (secondary exchange). The exchange can be conceptualized as taking place in three steps: film transfer, intraparticle (interlayer) diffusion, and the exchange step. This is similar to the conceptualization described previously for sorption (refer to Figure 3.11). For the case of exchange adsorption, it may be more appropriate to replace the sphere in Figure 3.11 with plates (to represent clay minerals rather than sand particles); however, in the case of clay aggregates (clay balls) the sphere may be more reasonable.

The cations most commonly thought of as participating in cation exchange processes are inorganic elements or compounds. However, organic cations are equally susceptible to ion exchange. The sorption of organic cations, and the influence of ground water conditions (e.g., pH, ionic strength, etc.) on sorption of organic cations, have been discussed in a previous section ("Ionization").

Valocchi (1980) and Valocchi et al. (1981) evaluated the transport of ion exchanging solutes during ground water recharge. The study evaluated both homovalent and heterovalent exchange (exchange of ions of similar valency and differing valency, respectively) in binary and ternary systems (two and three ions in solution, respectively). In the case of binary systems, the concentration profiles corresponded to a single chromatographic front (one breakthrough curve). For ternary systems, two chromatographic fronts were evidenced which were separated by an intermediate plateau zone. A model was developed for analyzing these systems with the resulting model applied to a field project involving direct injection of municipal effluents subject to advanced (tertiary) treatment. Model parameters were determined using well logs, core samples, and tracer breakthrough studies. Chemical parameters were evaluated in the laboratory using batch studies. With this information, the model was used to predict the transport of sodium, ammonia, potassium, magnesium, and calcium at various observation wells at the field site. The predicted results agreed very closely with the field data, helping to substantiate the fundamental modeling approach. Valocchi (1984) also discussed the use of an "effective" K_p approach (similar to that discussed for solvent motivated sorption of hydrophobic chemicals on the organic content of the soil) for describing the transport of ion exchanging contaminants. Persaud and Wierenga (1982) and Persaud et al. (1983) conducted modeling efforts, along with laboratory batch and column studies, to evaluate the sorption and miscible displacement of inorganic cations (calcium, sodium, lithium, and cesium).

Ceazan et al. (1989) conducted field studies with respect to the exchange and transport of ammonium and potassium in a sand and gravel aquifer. The field site is located on a glacial outwash on Cape Cod, Massachusetts. The unconfined aquifer below the Otis Air Base has been contaminated by the discharge of secondary treated wastewater onto sand infiltration beds. The water table of the unconfined aquifer is approximately 6.5 m below the infiltration beds. These practices have resulted in contamination of the ground water with ground water plumes delineated as shown in Figure 3.17. Figure 3.17 shows cross-sectional views of the contaminant plumes with the sand infiltration beds shown as the left ordinate (y axis). In Figure 3.17, contaminant plumes for the following chemicals are shown: (A) ammonium, (B) calcium, (C) magnesium, and (D) and potassium. The concentration of these chemicals in the infiltration beds and the concentration of the chemicals in the natural ground water are also shown. It is observed that the calcium and magnesium plumes have migrated further than the ammonium or potassium plumes. The researchers discounted the occurrence of other processes (such as nitrification of the ammonium) as being responsible for the relative extent of the plume migration, and thus attribute the retardation of ammonium and potassium to adsorption (cation exchange). Laboratory batch studies demonstrated a linear adsorption isotherm for ammonium, confirming exchange as a retardation mechanism. A logical question based on observations from Figure 3.17 is what did ammonium and potassium exchange with on the adsorption sites.

To further analyze the results discussed above, Ceazan et al. (1989) conducted two forced gradient field tracer studies at the Otis Air Base site (one of which will be discussed here). For the forced gradient studies, chemicals were injected at a depth of 10 m below the land surface and the appearance of the ions was monitored at an observation well 1.5 m away. For the tracer test of interest, 20 meq/L of NH_4^+ and Br^- (as NH_4Br) was injected as a pulse over a 2-min period. Figure 3.18 shows the appearance of the NH_4^+ and Br^- at the observation well and suggests that the NH_4^+ was retarded with respect to Br^-. As the researchers observed, the relative duration of this study (less than 12 hr) makes it highly unlikely that bioactivity was significant. This served to reinforce the previous observation that adsorption (exchange) was responsible for the retardation of ammonium. For ammonium to adsorb, it

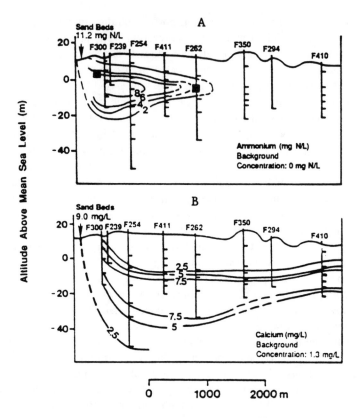

Figure 3.17.   Cross section of ground water plume (Reprinted with per-
                mission from Ceazan, M. L., E. M. Thurman, and R. L.
                Smith. "Retardation of Ammonium and Potassium Trans-
                port through a Contaminated Sand and Gravel Aquifer:
                The Role of Cation Exchange," *Enviromental Science and
                Technology* 23(11):1402–1408. Copyright 1989, American
                Chemical Society).

must exchange for cations (displace cations previously adsorbed). Monitoring
of pH indicated that the ammonium was not exchanging for hydrogen ions.
Monitoring for calcium and magnesium resulted in the data shown in Figure
3.19 (note that the ordinate values in Figure 3.19 have units of µeq/L). The
increase in calcium and magnesium corresponds to the period of time in
which the ammonium was being adsorbed, indicating that ammonium (a
monovalent cation) displaced calcium and magnesium (divalent cations) in
the transport process. Ion selectivity suggests that divalent cations would be
prefentially adsorbed over monovalent cations (the opposite of observations
made in this research). The authors attributed the exchange of monovalent

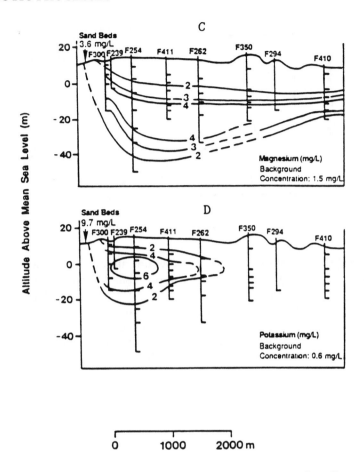

Figure 3.17.   Cross section of ground water plume (continued).

ammonium for divalent cations to mass action effects caused by the high concentration of ammonium in the pulse (see discussion above for mass action effects). Note that the ammonium was present in the pulse at the meq/L level while the divalent cations were present in the natural ground water at the µeq/L level. Comparison of Figures 3.18 and 3.19 during the period of time in which ammonium begins to appear at the observation well further confirms this observation. It is observed that as the ammonium concentrations increase (as the pulse of ammonium reaches the observation well), the calcium and magnesium concentrations decrease below the background levels (the initial portion of the data). This is explained by the fact that as the ammonium desorbs and thus appears in solution downgradient (in this case, the observation well), the desorption necessarily requires that it was preferentially displaced by other cations. The decrease in divalent cations observed in Figure 3.19

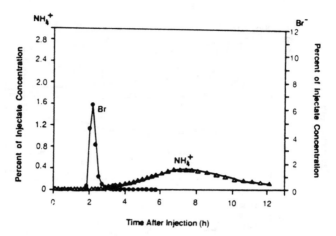

Figure 3.18.   Breakthrough of bromide and ammonium in field study
                (Reprinted with permission from Ceazan, M. L., E. M.
                Thurman, and R. L. Smith. "Retardation of Ammonium
                and Potassium Transport through a Contaminated Sand
                and Gravel Aquifer: The Role of Cation Exchange,"
                Enviromental Science and Technology 23(11):1402–1408.
                Copyright 1989, American Chemical Society).

would correspond to those cations which displaced the ammonium and caused
the ammonium to migrate downgradient (the divalent cations exchanging for
the ammonium were in the natural ground water displacing behind the tracer
pulse). The researchers suggested that the mass of ammonium ions desorbed
corresponds to the loss of mass of divalent cations due to exchange from the
ground water flushing behind the pulse (the area below the ammonium elution
curve approximately equals the depressed area of the divalent cation curve).
Thus, it is observed that the divalent cations present in the natural ground
water are preferentially exchanged for the monovalent cation ammonium,
except when mass action reverses this process. These observations are
consistent with the conceptual model presented above and predictive
models proposed by others (Valocchi et al., 1981; Persaud and Wierenga,1982).

3.5 OXIDATION/REDUCTION REACTIONS

Oxidation and reduction refer to the transfer of electrons and concomitant
species change of ions or compounds. Oxidation refers to the loss of electrons

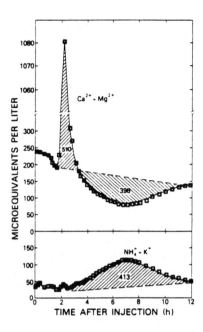

Figure 3.19. Breakthrough of calcium and magnesium in field study (Reprinted with permission from Ceazan, M. L., E. M. Thurman, and R. L. Smith. "Retardation of Ammonium and Potassium Transport through a Contaminated Sand and Gravel Aquifer: The Role of Cation Exchange," *Enviromental Science and Technology* **23**(11):1402–1408, Copyright 1989, American Chemical Society).

and reduction refers to the gain of electrons. Oxidation and reduction occur in tandem; electrons given up by one compound must be gained by another compound. By convention, oxidation and reduction reactions are considered as half reactions. An example of a reduction half reaction is given in Equation 3.30.

$$Fe(OH)_3(s) + 3H^+(aq) + e^-(aq) \rightarrow Fe^{2+}(aq) + 3H_2O(\ell) \qquad (3.30)$$

In this reaction, iron goes from the +3 to +2 state, gaining an electron and thus being reduced. Notice that this reaction causes iron to go from a solid state to a dissolved state. The reduction of iron in the subsurface results in the solubilization of iron in the ground water. Domestic users of ground water with high levels of iron will observe iron precipitating in bathtubs and staining clothes (assuming the ground water has not been treated to remove the soluble iron). This occurs because the ground water, which reaches the surface in a highly reduced state, is exposed to the atmosphere (oxygen), resulting in oxidation of the iron. The oxidation of the iron is expressed by the reverse of Equation 3.30. The reverse process (oxidation) causes iron to go from its soluble form to its insoluble complex.

As mentioned above, the reduction half reaction shown in Equation 3.30 occurs in tandem with an oxidation half reaction. An example of an oxidation half reaction that might occur is shown in Equation 3.31. In this reaction, glucose is oxidized to carbon dioxide. Notice that

$$\frac{1}{24}C_6H_{12}O_6(aq)+\frac{1}{4}H_2O \rightarrow \frac{1}{4}CO_2(g)+H^+(aq)+e^-(aq) \qquad (3.31)$$

this reaction is written such that only one electron is transferred. This convention makes it easy to combine oxidation and reduction half reactions because the electrons will drop out as the half reactions are added. This is convenient in balancing the overall reaction since electrons do not exist as a stable entity in aqueous solutions. In Equation 3.31, the carbon is oxidized from the neutral state to the +4 state, a net loss of four electrons per carbon atom. Thus, one electron is lost for each 1/4 of a carbon atom (or 1/24 of a glucose molecule). By combining the reduction and oxidation half reactions presented above, a complete oxidation/reduction reaction can be written. The complete oxidation/reduction reaction is shown in Equation 3.32. Notice that the electrons do not appear in this equation.

$$Fe(OH)_3(s)+\frac{1}{24}C_6H_{12}O_6(aq)+2H^+(aq) \rightarrow$$
$$Fe^{2+}(aq)+\frac{11}{4}H_2O(\ell)+\frac{1}{4}CO_2(g) \qquad (3.32)$$

The ability of an oxidation-reduction reaction to occur is a function of the redox (shorthand for reduction/oxidation) potential. Just as the concentration (activity) of free protons (hydrogen ions, H^+) can be utilized to assess the acid-base status of the soil environment, so also the activity of the free electron can be utilized to assess the oxidation-reduction potential of the subsurface. The redox potential is often defined in terms of the negative logarithm of the free-electron activity as shown in Equation 3.33, where (e^-) corresponds to the free electron activity. As with

$$pE = -\log(e^-) \qquad (3.33)$$

pH, large values of pE indicate low values of electron activity. Low values of electron activity favor the existence of electron poor (oxidized) species. This is analogous to the fact that at high values of pH the hydrogen ion

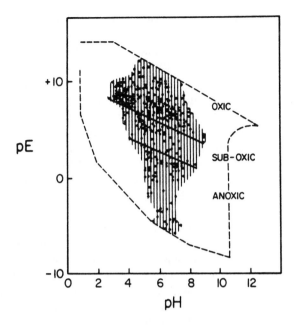

Figure 3.20. pE versus pH diagram showing soil regions (From *The Chemistry of Soils* by Garrison Sposito. Copyright 1989 by Oxford University Press, Inc. Reprinted with permission).

concentration or activity is low and proton poor species are favored. Small values of pE indicate high electron activity and thus correspond to electron rich species (reduced), just as small values of pH correspond to high proton activities and thus proton rich species — acids. One way in which pH and pE differ is that pH values typically do not go below zero while pE values below zero are common. Values of pE can range from as high as 13 (highly oxidized, electron poor) to as low as –6 (highly reduced, electron rich) in subsurface environments (Sposito, 1989).

The subsurface environment can be divided up into redox regions as follows (for a pH of 7.0): pE > 7, oxic; 2 < pE < 7, suboxic; and pE < 2, anoxic. Suboxic soils differ from oxic soils in having pE values low enough to deplete $O_2(g)$ but not low enough to deplete sulfate ions (Sposito, 1989). Figure 3.20 shows a plot of typical pE values as a function of pH for subsurface systems and shows the oxic, suboxic and anoxic regions. The shaded portion of Figure 3.20 corresponds to regions commonly observed in soil environments. The dashed segment corresponds to that region at which microorganisms are able to function. This is important as many of the redox reactions are microbially mediated (principally bacteria; Sposito, 1989). The types of substrates available also impact the microbial activity and the related redox reactions. Typically, the subsurface is a closed system (limiting the replenishment of oxygen, etc.). In this situation, the oxygen is readily depleted and reducing environments (low values of pE, anoxic conditions) result.

Naturally occurring chemical elements that are most commonly impacted by redox reactions in the subsurface are C, N, O, S, Mn, and Fe. Other chemicals (that may be present in the subsurface due to contamination) that are susceptible to redox reactions include As, Se, Cr, Hg, and Pb (Johnson et al., 1989). In a closed system, the chemicals may be expected to be successively reduced in the order shown in Table 3.4 (Sposito, 1989). As the pE of the soil environment drops below a value of 11, sufficient electron activity exists to reduce $O_2(g)$ to $H_2O(l)$. Below a pE value of 5, free oxygen is not stable in neutral soils. At pE values above 5 but below 11, the oxygen is consumed in the aerobic respiration of microorganisms. With decreasing values of pE, reduction of compounds occurs in the following order: N, Mn, Fe, and S. In general, aerobic microorganisms do not function below pE values of 5, denitrifying bacteria thrive in the pE range of 0 to 10, and sulfate reducing bacteria do not function well at pE values above 2. Observation of the reduction half reactions in Table 3.4 indicates that protons (hydrogen ions) are frequently consumed as part of the reduction half reaction, thus increasing the value of pH.

Determination of the equilibrium position of redox reactions is similar to that for other reactions. By convention, the reduction of $H^+(aq)$ to $H_2(g)$ is assigned a value of log K equal to zero (where K is the equilibrium constant). Table 3.5 lists some important reduction half reactions and corresponding values of log K. When adding half reactions, the values of log K are simply added. If the reverse of the half reaction is used (oxidation half reaction instead of the reduction half reaction as listed in the table), the sign of the log K value is reversed. For example, the log K value of the redox reaction shown in Equation 3.32 would be the sum of the log K value for Equation 3.30 (16.4) and the negative of the log K value for Equation 3.31 (0.2), or 16.6. Thus, the redox equilibrium constant could be written as shown in Equation 3.34. Given this equation and all but one of the activities, it would be possible to determine the remaining parameter. One can also use this type of expression as a

$$K = \frac{\left(Fe^{2+}\right)\left(H_2O\right)^{11/4}\left(CO_2\right)^{1/4}}{\left(Fe(OH)_3\right)\left(C_6H_{12}O_6\right)^{1/24}\left(H^+\right)^2} = 10^{16.6} \tag{3.34}$$

plotting equation.

The redox potential can also be expressed in terms of volts (E_h). Equation 3.35 can be utilized to convert from pE to E_h (Sposito, 1989). At standard

Table 3.4 Reduction Reactions Initiated in Neutral Soils as a Function of pE

Reduction Half-Reaction	Range of pE_{init}
$1/4\ O_2(g) + H^+(aq) + e^-(aq) = 1/2\ H_2O(\ell)$	5.0–11.0
$1/2\ NO_2^-(aq) + H^+(aq) + e^-(aq) = 1/2\ NO_2^-(aq) + 1/2\ H_2O(\ell)$	3.4–8.5
$1/2\ NO_3^-(aq) + 6/5\ H^+(aq) + e^-(aq) = 1/10\ N_2(g) + 3/5\ H_2O(\ell)$	
$1/8\ NO_3^-(aq) + 5/4\ H^+(aq) + e^-(aq) = 1/8\ NH_4^+(aq) + 3/8\ H_2O(\ell)$	
$1/2\ MnO_2(s) + 2H^+(aq) + e^-(aq) = 1/2\ Mn^{2+}(aq) + H_2O(\ell)$	3.4–6.8
$Fe(OH)_3(s) + 2H^+(aq) + e^-(aq) = Fe^{2+}(aq) + 3H_2O(\ell)$	1.7–5.0
$FeOOH(s) + 2H^+(aq) + e^-(aq) = Fe^{2+}(aq) + 2H_2O(\ell)$	
$1/8\ SO_4^{2-}(aq) + 9/8\ H^+(aq) + e^-(aq) = 1/8\ HS^-(aq) + 1/2\ H_2O(\ell)$	–2.5–0.0
$1/4\ SO_4^{2-}(aq) + 5/4\ H^+(aq) + e^-(aq) = 1/8\ S_2O_3^-(aq) + 5/8\ H_2O(\ell)$	
$1/8\ SO_4^{2-}(aq) + 5/4\ H^+(aq) + e^-(aq) = 1/8\ H_2S(aq) + 1/2\ H_2O(\ell)$	

From *The Chemistry of Soils* by Garrison Sposito. Copyright 1989 by Oxford University Press. Reprinted with permission.

Table 3.5 Reduction Half Reactions and Equilibrium Constants

Reduction Half-Reaction	log K
$1/4\ O_2(g) + H^+(aq) + e^-(aq) = 1/2\ H_2O(\ell)$	20.8
$H^+(aq) + e^-(aq) = 1/2\ H_2(g)$	0.0
$1/2\ NO_3^-(aq) + H^+ + e^-(aq) = 1/2\ NO_2^-(aq) + 1/2\ H_2O(\ell)$	14.1
$1/4\ NO_3^-(aq) + 5/4\ H^+(aq) + e^-(aq) = 1/8\ N_2O(aq) + 5/8\ H_2O(\ell)$	18.9
$1/5\ NO_3^-(aq) + 6/5\ H^+(aq) + e^-(aq) = 1/10\ NO_2(aq) + 3/5\ H_2O(\ell)$	21.1
$1/8\ NO_3^-(aq) + 5/4\ H^+(aq) + e^-(aq) = 1/8\ NH_4^+(aq) + 3/8\ H_2O(\ell)$	14.9
$1/2\ MnO_2(s) + 2H^+(aq) + e^-(aq) = 1/2\ Mn^{2+}(aq) + H_2O(\ell)$	20.7
$1/2\ MnO_2(s) + 1/2\ HCO_3^-(aq) + 3/2\ H^+(aq) + e^-(aq) = 1/2\ MnCO_3(s) + H_2O(\ell)$	20.2
$Fe(OH)_3(s) + 3H^+(aq) + e^-(aq) = Fe^{2+}(aq) + 3H_2O(\ell)$	16.4
$FeOOH(s) + 3H^+(aq) + e^-(aq) = Fe^{2+}(aq) + 2H_2O(\ell)$	11.3
$1/2\ Fe_3O_4(s) + 4H^+(aq) + e^-(aq) = 3/2\ Fe^{2+}(aq) + 2H_2O(\ell)$	14.9
$1/2\ Fe_2O_3(s) + 3H^+(aq) + e^-(aq) = Fe^{2+}(aq) + 3/2\ H_2O(\ell)$	11.1
$1/4\ SO_4^{2-}(aq) + 5/4\ H^+(aq) + e^-(aq) = 1/8\ S_2O_3^{2-}(aq) + 5/8\ H_2O(\ell)$	4.9
$1/8\ SO_4^{2-}(aq) + 9/8\ H^+(aq) + e^-(aq) = 1/8\ HS^-(aq) + 1/2\ H_2O(\ell)$	4.3
$1/8\ SO_4^{2-}(aq) + 5/4\ H^+(aq) + e^-(aq) = 1/8\ H_2S(aq) + 1/2\ H_2O(\ell)$	5.1
$1/2\ CO_2(g) + 1/2\ H^+(aq) + e^-(aq) = 1/2\ CHO_2^-(aq)$	-3.8
$1/4\ CO_2(g) + 7/8\ H^+(aq) + e^-(aq) = 1/8\ C_2H_3O_2^-(aq) + 1/4\ H_2O(\ell)$	1.2
$1/4\ CO_2(g) + 1/12\ NH_4^+(aq) + 11/12\ H^+(aq) + e^-(aq) = 1/12\ C_3H_4O_2NH_3(aq) + 1/3\ H_2O(\ell)$	0.8
$1/4\ CO_2(g) + e^-(aq) = 1/24\ C_6H_{12}O_6(aq) + 1/4\ H_2O(\ell)$	-0.2
$1/8\ CO_2(g) + H^+(aq) + e^-(aq) = 1/8\ CH_4(g) + 1/4\ H_2O(\ell)$	2.9

From *The Chemistry of Soils* by Garrison Sposito. Copyright 1989 by Oxford University Press. Reprinted with permission.

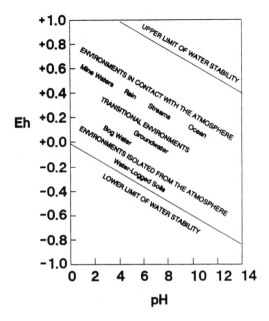

Figure 3.21.   pE vs pH diagram showing soil regions (Johnson et al., 1989).

temperature (298°K), the equation simplifies to the latter term in Equation 3.35. Figure 3.21 is similar to Figure 3.20, but is expressed in terms of

$$E_h(volts) = \frac{RT\ln 10}{F} pE = 0.0592\,pE \qquad (3.35)$$

where E_h =   electrical potential
        R   =   universal gas constant
        T   =   temperature
        F   =   Faraday's constant

E_h vs pH (instead of pE vs pH) and describes the types of environments that correspond to specific regions of the graph. The pE values of Figure 3.21 agree well with the E_h values of Figure 3.20 using the conversion presented in Equation 3.35.

As an example of redox reactions occurring in the subsurface, consider the accidental spill of organic contaminants in a saturated zone that is initially oxic (aerobic). Assume that the ground water is a closed system, i.e., virtually no opportunity exists for oxygen to be replenished. Initially the oxygen will be reduced (act as the electron acceptor in the oxidation of the organic con-

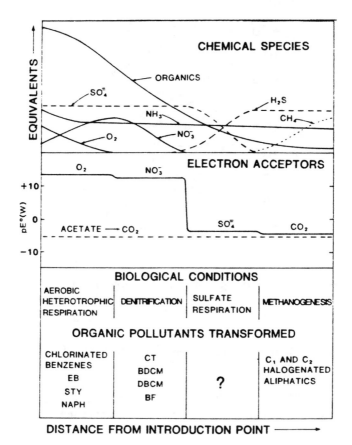

Figure 3.22. Redox reactions in the subsurface (EB =ethylbenzene, STY = styrene, NAPH = napthalene, CT = carbon tetrachloride, BDCM = bromodichloromethane, DBCM = dibromochloro- methane, BF = bromoform, Bouwer and McCarty, 1984).

taminants). As the oxygen is depleted, an alternate electron acceptor must be identified for the contaminant oxidation to continue. Examples of alternate electron acceptors, in order of preference (as indicated in Table 3.3), include NO_3^-, MnO_2, $Fe(OH)_3$, and SO_4^{2-}. As each subsequent electron acceptor is reduced, the next most desirable electron acceptor will be utilized (and the value of pE will concomitantly decline). Figure 3.22 is an example of such a scenario after Bouwer and McCarty (1984). Figure 3.22 shows loss of organics (due to microbial oxidation) and the resulting impacts on the elec- tron acceptors (in terms of equivalents of electron acceptors and the pE of the soil). For example, initially oxygen is present (with a concomitant high value of pE) and is subsequently depleted (under aerobic biological conditions). Figure 3.22 demonstrates the complete scenario of organic oxidation through various redox regions and concomitant microbial systems. The example discussed above was for the movement of the contaminant plume through ground water. A similar scenario could be described for the flooding of a normally unsaturated soil environment.

3.6 HYDROLYSIS

Hydrolysis is the reaction of a compound with water. Hydrolysis of organic compounds frequently results in the formation of alcohols and alkenes (Johnson et al., 1989). Examples of these types of reactions for chlorinated compounds are shown in Equations 3.36 and 3.37. Hydrolysis may function when biodegradation cannot and produces a

$$RX + HOH \rightarrow ROH + HX \tag{3.36}$$

$$H_3C - CH_2X \rightarrow H_2C = CH_2 + HX \tag{3.37}$$

product that is susceptible to biodegradation; in this manner, hydrolysis can be a significant process affecting fate and transport. Certain functional groups are potentially susceptible to hydrolysis (Mabey and Mill, 1978): amides, carbamates, epoxides, aliphatic and aromatic esters, alkyl and aryl halides, nitriles, and phosphorous esters.

Hydrolysis may be biologically mediated or may occur independent of the biosystem (through abiotic processes). For biotic hydrolysis, the biomass population and/or the specific enzyme concentration (activity) will significantly impact the level of hydrolysis. Under abiotic conditions, environmental factors such as pH, dissolved organic matter, dissolved metal ions, etc., may impact the level of hydrolysis. Biotic and abiotic hydrolysis are not mutually exclusive processes; they may jointly contribute to the breakdown of a compound (Wolfe et al., 1989). For purposes of this discussion, abiotic hydrolysis will be emphasized.

In the situation where a compound is susceptible to competing reactions, it may be difficult to isolate the hydrolysis reaction. For this reason, hydrolysis reactions are most frequently measured in the laboratory, and even then it may be necessary to adjust the pH, temperature, etc., to isolate the hydrolysis reaction. For example, the hydrolysis of 1,2,4-trichlorobenzene was studied in the laboratory at 70°C to speed up the hydrolysis reaction and to prevent loss of the chemical due to biodegradation (Ellington et al., 1986).

Hydrolysis data are commonly summarized as first order decay coefficients. Although hydrolysis reactions do not necessarily always follow first order kinetics, the relative ease of incorporating a first order reaction into the transport models commonly dictates its use. This reaction can be represented by the expression shown in Equation 3.38. This expression assumes that the time rate of loss of the chemical due to hydrolysis is proportional to the liquid phase concentration of the chemical (raised to the first power, thus first order) with K_h being the proportionality constant. Rearranging and integrating from initial concentration (C_0) at time equal to zero to final concentration (C) at time t results in Equation 3.39.

$$\frac{dC}{dt} = -K_h C \tag{3.38}$$

$$\ln\left(\frac{C}{C_0}\right) = -K_h t \tag{3.39}$$

The value of K_h can be determined by plotting values of ln (C/C_0) vs time and determining the slope of the plot. The rate constant (K_h) can include contributions from acid- or base-catalyzed hydrolysis, nucleophilic attack by water or catalysis by buffers in the reaction medium (Wolfe et al., 1989). An alternate approach uses the hydrolysis half-life (the time till the final concentration is one half of the initial concentration). Substituting C/C_0 equal to 0.5 and $t_{1/2}$ into Equation 3.39 and solving for K_h results in Equation 3.40. It is best to

$$K_h = 0.69 / t_{1/2} \tag{3.40}$$

utilize the entire data set whichever approach is utilized. Figure 3.23 shows a plot of hydrolysis data collected for 1,2,4-trichlorobenzene (as mentioned above, this data was collected at 70°C to prevent biodegradation as a competing reaction). The resulting first order hydrolysis coefficient is 4.3×10^{-3} hr^{-1}. As a point of interest, the first order reaction, described here for hydrolysis, is commonly used for many types of reactions. Radioactive decay is typically modeled in this fashion and half-lives are commonly used to describe radioactive decay. Biodegradation may also be described using a first order reaction. First order reaction rate constants or half-lives are thus common input values for fate and transport models and can be used to account for a variety of processes.

Several environmental factors have been observed to affect the rate of hydrolysis. Aqueous phase factors affecting the rate of hydrolysis of organic compounds in the subsurface include pH, temperature, and dissolved metal catalysis (Wolfe et al., 1989). The effect of pH can be attributed to either specific acid-base effects or to a change in the speciation of the compound (see next section for a discussion of speciation). Acid-catalyzed hydrolysis

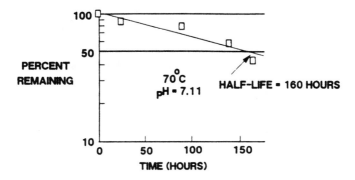

Figure 3.23.   Hydrolysis data for 1,2,4-trichlorobenzene (Johnson et al., 1989).

reaction rates increase with decreasing pH, while base-catalyzed rates increase with increasing pH. For neutral hydrolysis reactions, the rates are observed to be independent of system pH (Wolfe et al., 1989). Temperature effects on the rate of hydrolysis of chemicals can be pronounced. The Arrhenius equation is commonly used to adjust hydrolysis rate constants for temperature (Benefield et al., 1982; Wolfe et al., 1989). For a 10°C change in temperature, the hydrolysis rate constant could change by a factor of 2.5 (increasing temperature corresponding to increasing rate constant). Metal ions, such as calcium, copper, magnesium, iron, cobalt, and nickel, have been observed to catalyze the hydrolysis of certain chemicals when the metal ions are present at high concentrations. However, when the metals ions are present at concentrations more common to natural ground water, research suggests that their impact on hydrolysis rates is insignificant (Wolfe et al., 1989).

Several factors with respect to the soil medium that impact the hydrolysis rate constants include soil type, sorption potential, and soil water content (Wolfe et al., 1989). The proportion of sand/silt/clay has been observed to affect the hydrolysis of organic compounds. The greater the fraction of clays, the greater the hydrolysis rate. Clays with greater surface area/CEC (2:1 clays) have higher hydrolysis rates than clays with less surface area/CEC (1:1 clays). It has been suggested that the metal cations associated with clay minerals serve to catalyze the hydrolysis and thus are responsible for the increased hydrolysis rate. Research studying the effect of sorption of organic compounds on the rate of hydrolysis of the compound is inconclusive (Wolfe et al., 1989). Some researchers have observed increases in the sorption of organic compounds to decrease the hydrolysis rate of the organic compound and increase the persistence of the chemical. Other researchers have reported higher rates of hydrolysis of chemicals sorbed onto sterile (abiotic) soils than when the chemical was present in aqueous solution, suggesting surface catalysis of the hydrolysis reaction (Wolfe et al., 1989). Additional research is necessary to further analyze the effect of chemical sorption on the abiotic hydrolysis of the chemical. The water content of the soil serves as a reaction medium for biotic and abiotic processes. Thus, increasing water contents have been shown to increase the rate of hydrolysis (Wolfe et al., 1989).

3.7 SPECIATION: DISSOLUTION AND PRECIPITATION

Inorganic chemicals can occur in many forms (species) depending on the environmental conditions (e.g., pH, pE, organic, and inorganic ligands). As the environmental conditions change, the speciation of the chemical may also change. Thus, it is important to have a fundamental understanding of the chemical species possible for a given chemical in a range of environmental conditions and to be able to determine the species that will exist under a specified set of environmental conditions. Knowledge of the species present is vital as the various species of a chemical will have differing transport and fate potentials.

In ground water, six categories have been suggested in which an element or compound may exist (Johnson et al., 1989): (1) "free" ions (surrounded only by water molecules), (2) insoluble species, e.g., Ag_2S, $BaSO_4$, (3) metal/ligand complexes, e.g., $Al(OH)^{2+}$, Cu-humate, (4) adsorbed species, e.g., lead adsorbed onto a ferric hydroxide surface, (5) species held on a surface by ion exchange, e.g., calcium ions on clay mineral surfaces, and (6) species that differ by oxidation state, e.g., manganese [II] and [IV], iron [II] and [III], and chromium [III] and [VI]. It is important to know the form (speciation, location) of the element (compound) as this will greatly impact the mobility, reactivity, toxicity, etc. of the chemical. Knowledge of the total concentration of a chemical (irrespective of speciation) is of little value when assessing the fate and transport of the chemical in the subsurface.

The natural composition of ground water is significantly impacted by the dissolution and weathering of minerals. Dissolution refers to the complete solubilizing of all components within a mineral. For example, the dissolution of halite (NaCl) results in sodium and chloride in the ground water, the dissolution of gypsum ($CaSO_4 \cdot 2H_2O$) results in calcium and sulfate in the ground water, etc. Weathering is a partial solubilizing of certain elements within the mineral, leaving behind other elements. For example, the weathering of aluminosilicates (such as feldspars) contributes cations (e.g., calcium, magnesium, potassium, sodium, and silica) to the ground water. Secondary weathering products from the weathering of aluminosilicates may result, such as kaolinite and montmorillonite (Johnson et al., 1989). The opposite of dissolution is precipitation. A change in environmental conditions (pH, redox potential, temperature, mass of chemical, etc.) in the subsurface

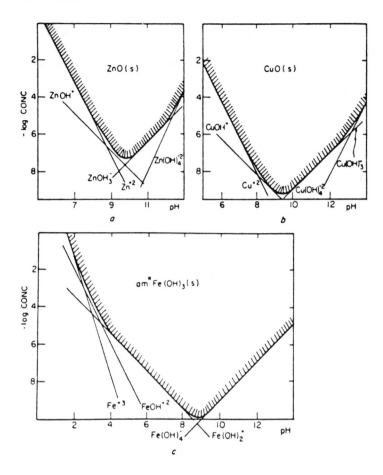

Figure 3.24. Speciation of Zn, Cu, and Fe (*Aquatic Chemistry*, W. Stumm and J. J. Morgan. Copyright 1970, John Wiley & Sons. Reprinted by permission of John Wiley & Sons).

may cause the saturation limit of a chemical to be exceeded, resulting in precipitation of the chemical.

As mentioned previously, the speciation of a chemical is a function of the pH of the system. Figure 3.24 shows an example of the speciation of zinc, copper and iron as a function of pH. The ordinate of these plots is the negative of the log of the species concentration (thus, increasing numbers correspond to decreasing concentrations) and the abscissa are pH. Each line on the plot corresponds to an equilibrium relationship between soluble and solid forms of the metals (determined by the solubility product, a thermodynamic constant). The cumulative solubility for all species is the region denoted by the dashed lines (see Stumm and Morgan, 1981 or Benefield et al., 1982 for a detailed discussion of the mechanics of producing this type of

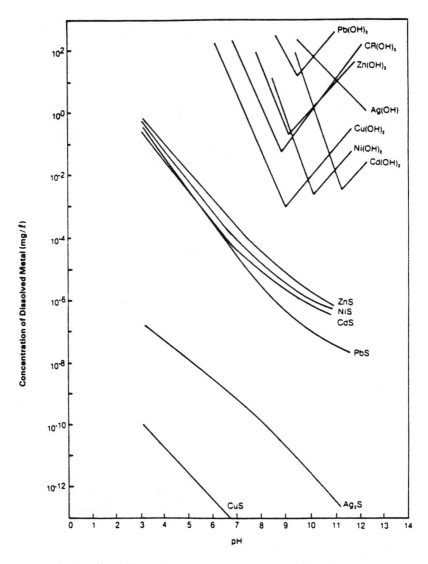

Figure 3.25.   Solubility of metal hydroxides and sulfides (U.S. Environmental Protection Agency, 1985).

plot). In Figure 3.24, as the pH increases the solubility is observed to decrease for Zn, Cu, and Fe and reach a solubility minimum at a pH between 8 and 10. As the pH further increases, the solubility of the metals increases, this time in the anionic form. The solubility of metal species as a function of pH (and the occurrence of a solubility minimum as a function of pH) depends on the specific metal and the form of the precipitate. Figure 3.25 shows the solubility of metal hydroxides and metal sulfides as a function of pH. It is observed that metal hydroxides exhibit a solubility minimum at a pH of 8 to 12 while the solubility of the metal sulfides continually decreases with increasing pH.

The above discussion has centered on ground water systems where the dissolution/precipitation of discrete minerals is of concern. It is not uncommon in ground water systems for multiple minerals to be present and for common ions to exist amongst the minerals. The contribution of ions from one mineral will affect the solubility of other minerals containing the same ion (referred to as the "common ion effect"). This serves to further complicate the system and increase the difficulty in determining the solubility limits of various species. Algorithms have been developed to aid in this process. One example of such an algorithm is MINTEQA1, an equilibrium metal speciation model developed for the EPA (Brown and Allison, 1987). MINTEQA1 utilizes the thermodynamic expressions and constants necessary to assess the equilibrium speciation of metals in a variety of subsurface environments based on geochemical principles. A manual is available that explains the fundamentals of the model and the mechanics of using the model (Brown and Allison, 1987).

The point of dissolution/precipitation of a system can be altered by instigation of remediation. Pump and treat remediation will remove the pore water and either will reinject treated water or cause the migration of upgradient ground water. Either scenario could result in decreases in the aqueous phase concentration of the metals. Thus, the equilibrium or steady state position of the system prior to remediation will be altered and dissolution of metal precipitate may occur. If the metal is the target of the remediation, significant time may be necessary to sufficiently dissolve the metal to reach acceptable levels in the ground water. Also, in the pump and treat scenario, the rate of pore water migration may act to limit the ability of the dissolution process to take place (it may be kinetically limited).

REFERENCES

Abdul, A. S., T. L. Gibson, and D. N. Rai "The Effect of Organic Carbon on the Adsorption of Fluorene by Aquifer Materials," *Haz. Wastes Haz. Mat.* 3(4):429–440 (1986).

Adamson, A. W. *Physical Chemistry of Surfaces,* 4th ed. (New York: John Wiley & Sons, Inc., 1982).

Bahr, J. M. and J. Rubin "Direct Comparison of Kinetic and Local Equilibrium Formulations for Solute Transport Affected by Surface Reactions," *Water Resour. Res.* 23(3):438–452 (1987).

Bailey, G. W. and J.L. White "Factors Influencing the Adsorption, Desorption and Movement of Pesticides in Soil," in *Residue Reviews,* Vol. 32, F. A. Gunther and J. D. Gunther, Eds. (New York: Springer-Verlag, 1970), pp. 29–92.

Banerjee, P., M. D. Piwoni, and K. Ebeid "Sorption of Organic Contaminants to a Low Carbon Subsurface Core," *Chemosphere* 14(8):1057–1067 (1985).

Benefield, L. D., J. F. Judkins, and B. L. Weand *Process Chemistry for Water and Wastewater Treatment* (Englewood Cliffs, NJ: Prentice Hall, 1982).

Bouchard, D. C., R. M. Powell, and D. A. Clark "Organic Cation Effects on the Sorption of Metals and Neutral Organic Compounds on Aquifer Material," *J. Environ. Sci. Health* A23(6):585–601 (1988a).

Bouchard, D. C., A. L. Wood, M. L. Campbell, P. Nkedi-Kizza, and P. S. C. Rao "Sorption Nonequilibrium During Solute Transport," *J. Contam. Hydrol.* 2:209–223 (1988b).

Bouwer, E. J. and P. L. McCarty "Modeling of Trace Organics Biotransformation in the Subsurface," *Ground Water* 22(4):433–440 (1984).

Brown, D. S. and J. D. Allison "MINTEQAI, An Equilibrium Metal Speciation Model: User's Manual," EPA/600/3–87/012, U.S. Environmental Protection Agency, Athens, GA (1987).

Brown, D. S. and G. Combs "A Modified Langmuir Equation for Predicting Sorption of Methyl-acridinium Ion in Soils and Sediments," *J. Environ. Qual.* 14:195–199 (1985).

Brown, D. S. and E. W. Flagg "Empirical Prediction of Organic Pollutant Sorption in Natural Sediments," *J. Environ. Qual.* 10(3):382–386 (1981).

Brownawell, B. J., H. Chen, J. M. Collier, and J. C. Westall "Adsorption of Organic Cations to Natural Materials," *Environ. Sci. Technol.* 24(8):1234–1241 (1990).

Brunauer, S., P. H. Emmett, and E. Teller "Adsorption of Gases in Multimolecular Layers," *J. Am. Chem. Soc.* 60:309 (1938).

Brusseau, M. L. and P. S. C. Rao "Sorption Nonideality During Organic Contaminant Transport in Porous Media," *CRC Crit. Rev. Environ. Control* 19:33–99 (1989).

Brusseau, M. L., R. E. Jessup, and P. S. C. Rao "Modeling the Transport of Solutes Influenced by Multiprocess Nonequilibrium," *Water Resour. Res.* 25(9):1971–1988 (1989).

Cameron, D. R. and A. Klute "Convective-Dispersive Solute Transport With a Combined Equilibrium and Kinetic Adsorption Model," *Water Resour. Res.* 13(1):183–188 (1977).

Ceazan, M. L., E. M. Thurman, and R. L. Smith "Retardation of Ammonium and Potassium Transport through a Contaminated Sand and Gravel Aquifer: The Role of Cation Exchange," *Environ. Sci. Technol.* 23(11):1402–1408 (1989).

Chin, Y., W. J. Weber, Jr., and T. C. Voice "Determination of Partition Coefficients and Aqueous Solubilities by Reverse Phase Chromatography — II, Evaluation of Partitioning and Solubility Models," *Water Res.* 20(11):1443–1450 (1986).

Chiou, C. T. "Theoretical Considerations of the Partition Uptake of Nonionic Organic Compounds by Soil Organic Matter," in *Reaction and Movement of Organic Chemicals in Soils*, B. L. Sawhney and K. Brown, Eds. (Madison, WI: Soil Science Society of America, SSSA Special Publication No. 22, 1989), pp. 1–29.

Chiou, C. T., P. E. Porter, and D. W. Schmedding "Partition Equilibria of Nonionic Organic Compounds between Soil Organic Matter and Water," *Environ. Sci. Technol.* 17(4):227–231 (1983).

Chiou, C. T., T. D. Shoup, and P. E. Porter "Mechanistic Roles of Soil Humas and Minerals in the Sorption of Nonionic Organic Compounds from Aqueous and Organic Solutions," *Org. Geochem.* 8(1):9–14 (1985).

Crittenden, J. C., N. J. Hutzler, D. G. Geyer, J. L. Oravitz, and G. Friedman "Transport of Organic Compounds with Saturated Groundwater Flow: Model Development and Parameter Sensitivity," *Water Resour. Res.* 22(3):271–284 (1986).

Curtis, G. P., P. V. Roberts, and M. Reinhard "A Natural Gradient Experiment on Solute Transport in a Sand Aquifer: 4. Sorption of Organic Solutes and its Influence on Mobility," *Water Resour. Res.* 22(13):2059–2067 (1986).

Davidson, J. M. and J. R. McDougal "Experimental and Predicted Movement of Three Herbicides in a Water-Saturated Soil," *J. Environ. Qual.* 2(4):428–433 (1973).

Davis, S. N., D. J. Campbell, H. W. Bentley, and T. J. Flynn "Ground Water Tracers," Cooperative Agreement CR-810036, U.S. Environmental Protection Agency, Ada, OK (1985).

Ellington, J. J., F. E. Stancil, and W. D. Payne "Measurement of Hydrolysis Rate Constants for Evaluation of Hazardous Waste Land Disposal: Volume 1. Data on 32 Chemicals," EPA/600/3–86/043, U.S. Environmental Protection Agency, Athens, GA (1986).

Freeze, R. A. and J. A. Cherry *Groundwater* (New York: John Wiley & Sons, Inc., 1979).

Freundlich, H. *Colloid and Capillary Chemistry* (New York: E.P. Dutton and Company Publishers, 1926).

Freyberg, D. L. "A Natural Gradient Experiment on Solute Transport in a Sand Aquifer: 2. Spatial Moments and the Advection and Dispersion of Nonreactive Tracers," *Water Resour. Res.* 22(13):2031–2046 (1986).

Giles, C. H., T. H. MacEwan, S. N. Nakhwa, and D. Smith "Studies in Sorption, Part XI: A System of Classification of Solution Adsorption Isotherms, and Its Use in Diagnosis of Adsorption Mechanisms and in Measurement of Specific Surface Areas of Solids," *J. Chem. Soc.* 163:3973–3993 (1960).

Guven, O. R., W. Falta, J. Molz, and J. G. Melville "Analysis and Interpretation of Single-Well Tracer Tests in Stratified Aquifers," *Water Resour. Res.* 21:676–684 (1985).

Guven, O. R., W. Falta, J. Molz, and J. G. Melville "A Simplified Analysis of Two-Well Tracer Tests in Stratified Aquifers," *Ground Water* 24:68–82 (1986).

Hamaker, J. W. and J. M. Thompson "Adsorption." in *Organic Chemicals in the Soil Environment,* Vol. 6, C. A. Goring and J. W. Hamaker, Eds. (New York: Marcel Dekker, Inc., 1972), pp.49–143.

Harter, R. D. and D. E. Baker "Applications and Misapplications of the Langmuir Equation to Soil Adsorption Phenomena," *Soil Sci. Soc. Am. J.* 41:1077–1080 (1977).

Harwell, J. Professor of Chemical Engineering, University of Oklahoma, Norman, OK, personal communication (1990).

Hassett, J. J., W. L. Banwart, and R. A. Griffin "Correlation of Compound Properties with Sorption Characteristics of Nonpolar Compounds by Soils and Sediments: Concepts and Limitations," in *Environment and Solid Wastes,* C. W. Francis and S. I. Auerbach, Eds. (Boston: Butterworths, 1983), pp. 161–178.

Hassett, J. J., J. C. Means, W. L. Banwart, and S. G. Wood "Sorption Properties of Sediments and Energy-Related Pollutants," EPA/600/3-80-041, U.S. Environmental Protection Agency, Washington, D.C. (1980).

Hillel, D. *Introduction to Soil Physics,* (Orlando, FL: Academic Press, 1982).

Hornsby, A. G. and J. M. Davidson "Solution and Adsorbed Fluometuron Concentration Distribution in a Water-Saturated Soil: Experimental and Predicted Evaluation," in *Proc. Soil Sci. Soc. Am.* 37:823–828 (1973).

Hutzler, N. J., J. C. Crittenden, J. S. Gierke, and A. S. Johnson "Transport of Organic Compounds with Saturated Groundwater Flow: Experimental Results," *Water Resour. Res.* 22(3):285–295 (1986).

Johnson, R. L., C. D. Palmer, and W. Fish "Subsurface Chemical Processes," in *Transport and Fate of Contaminants in the Subsurface.* EPA/625/4-89/019, U.S. Environmental Protection Agency, Cincinnati, OH and Ada, OK, 41–56 (1989).

Jury, W. A. "Chemical Transport Modeling: Current Approaches and Unresolved Problems," in *Chemical Mobility and Reactivity in Soil Systems,* (Madison, WI: Soil Science Society of America, Special Publication No.11, 1983), pp. 49–64.

Karickhoff, S. W. "Organic Pollutant Sorption in Aquatic Systems." *J. Hydraul. Eng.* 110(6):707–735 (1984).

Karickhoff, S. W. and K. R. Morris "Sorption Dynamics of Hydrophobic Pollutants in Sediment Suspensions," *Environ. Toxicol. Chem.* 4:469–479 (1985).

Karickhoff, S. W., D. S. Brown, and T. A. Scott "Sorption of Hydrophobic Pollutants on Natural Sediments," *Water Res.* 13:241–248 (1979).

Kay, B. D. and D. E. Elrick "Adsorption and Movement of Lindane in Soils," *Soil Sci.* 104(5):314–322 (1967).

Kenaga, E. E. and C. A. I. Goring *ASTM Special Technical Publication 707*, American Society for Testing Materials, Washington, D.C. (1980).

Langmuir, I. "The Adsorption of Gases on Plane Surfaces of Glass, Mica and Platinum," *J. Am. Chem. Soc.* 40:1361 (1918).

Le Blanc, D. R., S. P. Garabedian, W. W. Wood, K. M. Hess, and R. D. Quadri "Natural Gradient Tracer Test in Sand and Gravel: Objective, Approach, and Overview of Tracer Movement," in *Toxic Waste-Ground Water Contamination Program Third Technical Meeting*, U.S. Geological Survey (1989).

Lee, J., J. R. Crum, and S. A. Boyd "Enhanced Retention of Organic Contaminants by Soils Exchanged with Organic Cations," *Environ. Sci. Technol.* 23:1365–1372 (1989).

Lee, L. S., P. S. C. Rao, M. L. Brusseau, and R. A. Ogwada "Nonequilibrium Sorption of Organic Contaminants During Flow through Columns of Aquifer Materials," *Environ. Toxicol. Chem.* 7:779–793 (1988).

Leenher, J. A. and J. L. Alrichs "A Kinetic and Equilibrium Study of the Adsorption of Carbaryl and Parathion upon Soil Organic Matter Surfaces," *Proc. Soil Sci. Soc. Am.* 35:700-705 (1971).

Mabey, W. R. and T. Mill "Critical Review of Hydrolysis of Organic Compounds in Water Under Environmental Conditions," *J. Phys. Chem. Ref. Data* 7:383–415 (1978).

Mackay, D. M. "Characterization of the Distribution and Behavior of Contaminants in the Subsurface," presented at NRC-WSTB colloquium entitled *Ground Water and Soil Contamination Remediation: Are Science, Policy and Public Perception Compatible?* (1989).

Mackay, D. M. and B. Powers "Sorption of Hydrophobic Chemicals from Water: A Hypothesis for the Particle Concentration Effect," *Chemosphere* 16:745–757 (1987).

Mackay, D. M., D. L. Freyberg, P. V. Roberts, and J. A. Cherry "A Natural Gradient Experiment on Solute Transport in a Sand Aquifer: 1. Approach and Overview of Plume Movement," *Water Resour. Res.* 22(13):2017–2029 (1986).

McCarty, P. L., M. Reinhard, and B. E. Rittmann "Trace Organics in Groundwater," *Environ. Sci. Technol.* 15(1):40–51 (1981).

Mehran, M., R. L. Olsen, and B. M. Rector "Distribution Coefficient of Trichloroethylene in Soil-Water Systems," *Ground Water* 25(3):275–282 (1987).

Miller, C. T. "Modeling of Sorption and Desorption Phenomena for Hydrophobic Organic Contaminants in Saturated Soil Environments," PhD Dissertation, University of Michigan, Ann Arbor, MI (1984).

Miller, C. T. and W. J. Weber, Jr. "Sorption of Hydrophobic Organic Pollutants in Saturated Soil Systems," *J. Contam. Hydrol.* 1:243–261 (1986).

Miller, C. T. and S. L. Chang "An Investigation of Desorption Hysteresis in Subsurface Systems," published abstract EOS, *Am. Geophys. Union,* October 24, 1989, p. 1093.

Mingelgrin, U. and Z. Gerstl "Reevaluation of Partitioning as a Mechanism of Nonionic Chemicals Adsorption in Soils," *J. Environ. Qual.* 12(1):1–11 (1983).

Mitchell, J. K. *Fundamentals of Soil Behavior* (New York: John Wiley & Sons, Inc., 1976).

Molz, F. J., O. Guven, J. G. Melville, and J. F. Keely "Performance and Analysis of Aquifer Tracer Tests With Implications for Contaminant Transport Modeling," EPA/600/2–86/062, U.S. Environmental Protection Agency, Ada, OK (1986).

Molz, F. J., J. G. Melville, O. Guven, R. D. Crocker, and K. T. Matteson "Design and Performance of Single-Well Tracer Test at the Mobile Site," *Water Resour. Res.* 22:1497–1502 (1985).

Montgomery, J. M. Consulting Engineers, Inc., *Water Treatment Principles and Design* (New York: John Wiley & Sons, Inc., 1985).

Nkedi-Kizza, P., M. L. Brusseau, P. S. C. Rao, and A. G. Hornsby "Nonequilibrium Sorption during Displacement of Hydrophobic Organic Chemicals and ^{45}Ca through Soil Columns with Aqueous and Mixed Solvents," *Environ. Sci. Technol.* 23:814–820 (1989).

Nkedi-Kizza, P., P. S. C. Rao, R. E. Jessup, and J. M. Davidson "Ion Exchange and Diffusive Mass Transfer During Miscible Displacement Through an Aggregated Oxisol," *Soil Sci. Soc. Am. J.* 46:471–476 (1982).

Nkedi-Kizza, P., J. W. Biggar, H. M. Selim, M. T. van Genuchten, P. J. Wierenga, P. J. Davidson, and D. R. Nielsen "On the Equivalence of Two Conceptual Models for Describing Ion Exchange During Transport Through an Aggregated Oxisol," *Water Resour. Res.* 20(8):1123–1130 (1984).

Nkedi-Kizza, P., P. S. C. Rao, and A. G. Hornsby "Influence of Organic Cosolvents on Sorption of Hydrophobic Organic Chemicals by Soils," *Environ. Sci. Technol.* 19(10):975–979 (1985).

O'Connor, D. J. and J.P. Connolly "The Effect of Concentration of Adsorbing Solids on the Partition Coefficient," *Water Res.* 14:1517–1523 (1980).

Palmer, C., D. A. Sabatini, and J. H. Harwell "Nonionic Surfactants as Adsorbing Ground Water Tracers," presented at *Ground Water Flow Systems and Land Use: Relation to Quality of Shallow Ground Water. Annual Meeting of the Association of Ground Water Scientists and Engineers,* Anaheim, CA, September 25–27, 1990.

Parker, J. C. and A. J. Valocchi "Constraints on the Validity of Equilibrium and First-Order Kinetic Transport Models in Structured Soils," *Water Resour. Res.* 22(3):399–407 (1986).

Parker, J. C. and M. T. van Genuchten "Determining Transport Parameters from Laboratory and Field Tracer Experiments," Virginia Agricultural Experiment Station, Bulletin 84-3 (1984).

Persaud, N. and P. J. Wierenga "A Differential Model for One-Dimensional Cation Transport in Discrete Homoionic Ion Exchange Media," *Soil Sci. Soc. Am. J.* 46:482–490 (1982).

Persaud, N., J. M. Davidson, and P. S. C. Rao "Miscible Displacement of Inorganic Cations in a Discrete Homoionic Exchange Medium," *Soil Sci.* 136(5):269–278 (1983).

Rao, P. S. C. and J. M. Davidson "Adsorption and Movement of Selected Pesticides at High Concentrations in Soil," *Water Res.* 13:375–380 (1979).

Rao, P. S. C. and J. M. Davidson "Estimation of Pesticide Retention and Transformation Parameters Required in Nonpoint Source Pollution Models," in *Environmental Impact of Nonpoint Source Pollution,* M. R. Overcash and J. M. Davidson, Eds. (Ann Arbor Science, Ann Arbor, MI, 1980), pp. 23–67.

Rao, P. S. C., R. E. Green, L. R. Ahuja. and J. M. Davidson "Evaluation of a Capillary Bundle Model for Describing Solute Dispersion in Aggregated Soils," *Soil Sci. Soc. Am. J.* 40:815–820 (1976).

Rao, P. S. C. and R. E. Jessup "Sorption and Movement of Pesticides and Other Toxic Organic Substance in Soils," in *Chemical Mobility and Reactivity in Soil Systems,* (Madison, WI: Soil Science Society of America, Special Publication 11:183–201 (1983)).

Rao, P. S. C., A. G. Hornsby, D. P. Kilcrease, and P. Nkedi-Kizza "Sorption and Transport of Hydrophobic Organic Chemicals in Aqueous and Mixed Solvent Systems: Model Development and Preliminary Evaluation," *J. Environ. Qual.* 14(3):376–383 (1985).

"Remedial Action at Waste Disposal Sites," EPA/625/6-85/006, U.S. Environmental Protection Agency (1985).

Roberts, P. V., M. N. Goltz, R. S. Summers, J. C. Crittenden, and P. Nkedi-Kizza "The Influence of Mass Transfer on Solute Transport in Column Experiments with an Aggregated Soil," *J. Contam. Hydrol.* 1:375–393 (1987).

Roberts, P. V., M. N. Goltz, and D. M. Mackay "A Natural Gradient Experiment on Solute Transport in a Sand Aquifer: 3. Retardation Estimates and Mass Balances for Organic Solutes," *Water Resour. Res.* 22(13):2047–2058 (1986).

Sabatini, D. A. and T. A. Austin "Adsorption, Desorption and Transport of Pesticides in Groundwater: A Critical Review," *J. Irrig. Drainage Div. — ASCE* 116(1):3–15 (1990).

Sabatini, D. A. and T. A. Austin "Characteristics of Rhodamine WT and Fluorescein as Adsorbing Groundwater Tracers," *Ground Water* 29(3): 341–349 (1991).

Sabljic, A. "On the Prediction of Soil Sorption Coefficients of Organic Pollutants from Molecular Structure: Application of Molecular Topology Model," *Environ. Sci. Technol.* 21(4):358–366 (1987).

Schwarzenbach, R. P. and J. Westall "Transport of Nonpolar Organic Compounds from Surface Water to Groundwater: Laboratory Sorption Studies," *Environ. Sci. Technol.* 15:1360–1367 (1981).

Schwarzenbach, R. P. and J. Westall "Sorption of Hydrophobic Trace Organic Compounds in Groundwater Systems," *Water Sci. Technol.* 17:39–55 (1985).

Skopp, J. "Analysis of Time-Dependent Chemical Processes in Soils," *J. Environ. Qual.* 15(3):205–213 (1986).

Skopp, J. and A. W. Warrick "A Two-Phase Model for the Miscible Displacement of Reactive Solutes in Soils," *Soil Sci. Soc. Am. J.* 38(4):545–550 (1974).

Sposito, G. *The Chemistry of Soils* (New York: Oxford Press, 1989).

Sposito, G., R. E. White, P. R. Darrah, and W. Jury "A Transfer Function Model of Solute Transport Through Soil - 3: The Convection-Dispersion Equation," *Water Resour. Res.* 22(2):255–262 (1986).

SSSA (Soil Science Society of America) *Methods of Soil Analysis: Part 2. Chemical and Microbiological Properties*, Page, A. L., R. H. Miller, and D. R. Keeney, Eds. (Madison, WI: American Society of Agronomy, Inc., 1982).

Stumm, W. and J. J. Morgan *Aquatic Chemistry*, 2nd ed. (New York: John Wiley & Sons, Inc., 1981).

Swanson, R. A. and G. R. Dutt "Chemical and Physical Processes that Affect Atrazine and Distribution in Soil Systems," *Proc. Soil Science Soc. Am.* 37:872–876 (1973).

Szecsody, J. E. and R. C. Bales "Sorption Kinetics of Low-Molecular-Weight Hydrophobic Organic Compounds on Surface-Modified Silica," *J. Contam. Hydrol.* 4:181–203 (1989).

Thomas, G. W. "Historical Developments in Soil Chemistry: Ion Exchange," *Soil Sci. Soc. Am. J.* 41:230–238 (1977).

Travis, C. C. and E. L. Etnier "A Survey of Sorption Relationships for Reactive Solutes in Soil," *J. Environ. Qual.* 10:8–17 (1981).

Valocchi, A. J. "Validity of the Local Equilibrium Assumption for Modeling Sorbing Solute Transport Through Homogeneous Soils," *Water Resour. Res.* 21(6):808–820 (1985).

Valocchi, A. J. "Describing the Transport of Ion-Exchanging Contaminants Using an Effective K_d Approach," *Water Resour. Res.* 20(4):499–503 (1984).

Valocchi, A. J. "Transport of Ion-Exchanging Solutes During Groundwater Recharge," PhD Dissertation, Stanford University, Palo Alto, CA (1980).

Valocchi, A. J., R. L. Street, and P. V. Roberts "Transport of Ion-Exchanging Solutes in Groundwater: Chromatographic Theory and Field Simulation," *Water Resourc. Res.* 17(5):1517–1527 (1981).

van Genuchten, M. T. and W. J. Alves "Analytical Solutions of the One-Dimensional Convective-Dispersive Solute Transport Equation," U.S. Department of Agriculture, Agricultural Research Service, Washington, D.C., Technical Bulletin No.1661 (1982).

van Genuchten, M. T., J. M. Davidson, and P. J. Wierenga "An Evaluation of Kinetic and Equilibrium Equations for the Prediction of Pesticide Movement through Porous Media," *Soil Sci. Soc. Am. Proc.* 38:29–35 (1974).

van Genuchten, M. T. and P. J. Wierenga "Mass Transfer Studies in Sorbing Porous Media I. Analytical Solutions," *Soil Sci. Soc. Am. J.* 40(4):473–480 (1976).

Veith, J. A. and G. Sposito "On the Use of the Langmuir Equation in the Interpretation of 'Adsorption' Phenomena," *Soil Sci. Soc. Am. J.* 41:697–702 (1977).

Voice, T. C., C. P. Rice, and W. J. Weber Jr. "Effect of Solids Concentration on the Sorptive Partitioning of Hydrophobic Pollutants in Aquatic Systems," *Environ. Sci. Technol.* 17:513-518 (1983).

Weber, J. B. and C. T. Miller "Organic Chemical Movement Over and Through Soil," in *Reaction and Movement of Organic Chemicals in Soils,* B. L. Sawhney and K. Brown, Eds. (Madison, WI: Soil Science Society of America, SSSA Special Publication No. 22, 1989), pp. 305–334.

Weber, W. J., Jr. *Physiochemical Process for Water Quality Control* (New York: John Wiley & Sons, Inc. 1972).

Weber, W. J., Jr., P. M. McGinley, and L. E. Katz "Sorption Phenomena in Subsurface Systems: Concepts, Models and Effects on Contaminant Fate and Transport," *Water Res.* (Vol. 25, 1991), pp. 499–528.

Weber, W. J., Jr., Y. Chin, and C. P. Rice "Determination of Partition Coefficients and Aqueous Solubilities By Reverse Phase Chromatography — I," *Water Res.* 20(11):1433-1442 (1986).

Westall, J. C., C. Leuenberger, and R. P. Schwarzenbach "Influence of pH and Ionic Strength on the Aqueous-Nonaqueous Distribution of Chlorinated Phenols," *Environ. Sc. Technol.* 19:193–198 (1985).

Wolfe, N. L., M. E. Metwally, and A. E. Moftah "Hydrolytic Transformations of Organic Chemicals in the Environment" in *Reaction and Movement of Organic Chemicals in Soils,* B. L. Sawhney and K. Brown, Eds. (Madison, WI: Soil Science Society of America, SSSA Special Publication No. 22, 1989), pp. 229–242.

Wood, L. and P. S. C. Rao "Application of Gradient Elution Techniques for Assessment of Organic Solute Mobility in Solvent Mixtures," abstract *65th Annual Colloid and Surface Science Symposium—American Chemical Society,* held at the University of Oklahoma, Norman, OK June 17–19, 1991.

Woodburn, K. B., L. S. Lee, P. S. C. Rao, and J. J. Delfino "Comparison of Sorption Energetics for Hydrophobic Organic Chemicals by Synthetic and Natural Sorbents from Methanol/Water Solvent Mixtures," *Environ. Sci. Technol.* 23:407–413 (1989).

Wu, S. and P. M. Gschwend "Sorption Kinetics of Hydrophobic Organic Compounds to Natural Sediments and Soils," *Environ. Sci. Technol.* 20:717–725 (1986).

Wu, S. and P. M. Gschwend "Numerical Modeling of Sorption Kinetics of Organic Compounds to Soil and Sediment Particles," *Water Resour. Res.* 24(8):1373–1383 (1988).

4

<u>BIOTIC PROCESSES</u>

4.1 INTRODUCTION

Previous chapters have discussed hydrodynamic and abiotic processes affecting subsurface contaminant migration. Subsurface biotic processes include those biological phenomena that affect the transport and/or fate of compounds (organics, inorganics, or the microbes themselves) in the subsurface.

One example of a subsurface biotic process is the biodegradation of an organic compound by subsurface microbiota. This process is especially significant due to the fact that if mineralization of the organic is achieved (end products of CO_2, H_2O, etc.), the biotic process effectively renders the contaminant innocuous (the aquifer is self cleansing). It has only been in the last decade that the variety and magnitude of microorganisms present in aquifers, and thus the ability of aquifers to support biodegradation, have been realized (due largely to improved sampling methods). In the words of Devinny (1990), "Far from being sterile sand filters, aquifer soils are culture vats teeming with life." Degradation of a compound may decrease or completely eliminate the mass of the compound in the subsurface. However, if mineralization is not realized, intermediate breakdown products (metabolites) are generated which may be of equal or greater concern than the original compound. A second example of a subsurface biotic process is the migration of microorganisms in the subsurface. For example, much research has been conducted on the migration of viruses in the subsurface. In this case, the microorganisms themselves are of concern, and the biotic processes affecting their migration must be understood.

This chapter will focus on gaining a fundamental understanding of the impacts of biotic processes on subsurface quality. Topics to be discussed in this chapter include fundamentals of microbiology, subsurface microbiology, subsurface metabolic processes, modeling of subsurface biodegradation, and transport of microorganisms. Additional biotic processes could be identified, but are beyond the scope of this discussion.

4.2 FUNDAMENTALS OF MICROBIOLOGY

Before discussing subsurface microbiology, a review of the fundamentals of microbiology will be presented. This review will include discussions of nutrient requirements/metabolism of microorganisms, environmental factors affecting microorganisms, and the kinetics of growth for microorganisms.

4.2.1 Nutrition Requirements/Metabolism

Microorganisms require the following elements to maintain existing cells and to produce new cells: an energy source, a carbon source (new cellular material), inorganic elements (e.g., nitrogen, phosphorous, sulfur, potassium, calcium, and magnesium), and organic nutrients (growth factors) (Metcalf and Eddy, 1991). Microorganisms are frequently classified based on the sources they utilize for energy and carbon; these classifications are discussed below.

The two most common sources of carbon utilized by microorganisms are organic matter and carbon dioxide. Microorganisms that utilize organic matter as a source of cell carbon are referred to as heterotrophs while microorganisms that utilize carbon dioxide (or other inorganic substances) as a source of cell carbon are referred to as autotrophs. Autotrophic conversion of carbon dioxide to microbial biomass is a reductive process that requires more energy than oxidative synthesis of heterotrophic cell biomass. Cell energy is typically obtained from chemical oxidation reactions (chemotrophs) or from light (phototrophs — not pertinent to subsurface ecosystems). Chemotrophs may be heterotrophic or autotrophic depending on their carbon source. Examples of chemoheterotrophs (also known as chemoorganotrophs) include protozoa, fungi and most bacteria; nitrifying bacteria are a common chemoautotroph. Chemoheterotrophs typically obtain their energy from the oxidation of organic compounds while chemoautotrophs obtain energy from the oxidation of reduced inorganic compounds, e.g., ammonia, nitrite, and sulfide (Metcalf and Eddy, 1991).

Microorganisms require inorganic elements and organic nutrients for life processes. The need for C, H, O, N, and P as the major building blocks for biomass is demonstrated by the empirical relationship for biomass used in wastewater treatment as shown in Equation 4.1 (Benefield and Randall, 1980):

$$C_{60}H_{87}O_{23}N_{12}P \qquad (4.1)$$

Additional components required in lesser proportions include S, K, Mg, Ca, Fe, Na, and Cl. Trace elements necessary for biogrowth include Zn, Mn, Mo, Se, Co, Cu, Ni, V, and W (Metcalf and Eddy, 1991). Certain organic micronutrients must be present for microorganisms to function; these typically cannot be synthesized by the microorganisms and include vitamins, essential amino acids, and precursors for the synthesis of essential amino acids or other required synthesized compounds (Reynolds, 1982). It should be noted that the requirements for the major and minor nutrients and growth factors vary for different microorganisms, and that the nutrient which is depleted first becomes the limiting factor in further microbial activity.

Microorganisms can be further classified according to the type of metabolic process they use. Microorganisms that generate energy by electron transfer from an electron donor to an external electron acceptor are classified as undergoing respiratory metabolism (humans utilize oxygen as an electron acceptor). If the microorganisms do not utilize an external electron acceptor, they are classified as undergoing fermentative metabolism. Fermentation is less energy efficient than respiration and is thus typically a slower process and has a lower cell yield (Metcalf and Eddy, 1991).

Respiration involves the oxidation of one compound (this compound loses electrons) and the reduction of another compound (the electron acceptor). This is commonly referred to as a redox reaction (in this case a microbially mediated redox reaction). If the electron acceptor is free oxygen (molecular oxygen, O_2), the metabolism is known as aerobic respiration. If the electron acceptor is something other than free oxygen, the metabolism is referred to as anoxic or anaerobic respiration. Facultative anaerobes can undergo aerobic respiration if molecular oxygen is available while molecular oxygen is toxic to obligate anaerobes. Table 4.1 shows a series of reduction half reactions and their reverse oxidation reactions (Devinny, 1990). In the first entry, molecular oxygen (O_2) is reduced to form water as the oxygen accepts electrons (is reduced); in the reverse case, water is oxidized to form molecular oxygen as the oxygen loses electrons (is oxidized). The amount of energy released in the reduction reaction is equal to the amount of energy consumed during the oxidation reaction (the energy relationship is reversible). The reductions are listed in Table 4.1 in decreasing order of the amount of energy generated; the reduction of molecular oxygen to water generates the greatest amount of energy whereas the reduction of carbon dioxide to organic matter generates the least amount of energy. In the second column, the oxidation half reactions

Table 4.1 Reduction and Oxidation Half Reactions with Subsurface Microbial Significance[a]

Reductions	↔	Oxidations
Oxygen to water		Water to oxygen
$O_2 \rightarrow H_2O$		$H_2O \rightarrow O_2$
Nitrate to nitrogen gas		Nitrogen gas to nitrate
$NO_3^- \rightarrow N_2$		$NO_2^- \rightarrow NO_3^-$
Nitrate to nitrite		Nitrite to nitrate
$NO_3^- \rightarrow NO_2^-$		$NO_2^- \rightarrow NO_3^-$
Nitrite to ammonia		Ammonia to nitrite
$NO_2^- \rightarrow NH_3$		$NH_3 \rightarrow NO_2^-$
Ferric iron to ferrous iron		Ferrous iron to ferric iron
$Fe^{3+} \rightarrow Fe^{2+}$		$Fe^{2+} \rightarrow Fe^{3+}$
Carbohydrate to alcohol		Alcohol to carbohydrate
$CH_2O \rightarrow CH_3OH$		$CH_3OH \rightarrow CH_2O$
Sulfate to sulfide		Sulfide to sulfate
$SO_4^= \rightarrow H_2S$		$H_2S \rightarrow SO_4^=$
Carbon dioxide to methane		Methane to carbon dioxide
$CO_2 \rightarrow CH_4$		$CH_4 \rightarrow CO_2$
Nitrogen gas to ammonia		Ammonia to nitrogen gas
$N_2 \rightarrow NH_3$		$NH_3 \rightarrow N_2$
Hydrogen ion to hydrogen gas		Hydrogen gas to hydrogen ion
$H^+ \rightarrow H_2$		$H_2 \rightarrow H^+$
Carbon dioxide to organic matter		Organic matter to carbon dioxide
$CO_2 \rightarrow CH_2O$		$CH_2O \rightarrow CO_2$

Source: Devinny, 1990.

[a] Reduction, which consume electrons, listed in order of the energy they produce, most energy production first can be reversed to make a list of oxidation, which generate electrons, listed in order of the energy they consume, most energy consumption first.

are listed in order of decreasing amount of energy required; the oxidation of water to molecular oxygen will require the greatest amount of energy input whereas the oxidation of organic matter to carbon dioxide will require the least amount of energy input (Devinny, 1990).

When reduction and oxidation half reactions are combined, the net energy of the reaction is the energy released during the reduction half reaction minus the energy consumed during the oxidation half reaction. From Table 4.1, since the reduction half reactions are listed in order of decreasing energy release, any reduction half reaction combined with an oxidation half reaction listed below it (which will thus consume less energy than is generated during the reduction half reaction) will result in a net generation of energy. Thus, combination of the top entry for the reduction half reactions (molecular oxygen to water) and the bottom entry in the oxidation half reactions (organic matter to carbon dioxide) results in the greatest net generation of energy. This corresponds to aerobic degradation of organic matter (organic matter is oxidized and oxygen is the electron acceptor). The organic matter will also serve as the carbon source. Thus, microorganisms promulgating these reactions would be classified as aerobic chemoheterotrophs. It should be noted that the aerobic oxidations of organics do in fact result in the greatest amount of energy generation. Also, from Table 4.1 it can be observed that the combination of a reduction half reaction with an oxidation half reaction above it will require a net input of energy to cause the redox reaction to occur. This generally occurs only when the microorganism has a need for the resulting product.

The majority of higher life forms (multicelled organisms, including humans) utilize oxygen as an electron acceptor during respiration while oxidizing organic matter (e.g., carbohydrates). The net energy released provides higher life forms with the energy necessary to maintain life. These higher life forms are generally obligate aerobes and thus cannot exist in the absence of molecular oxygen (we certainly know this to be true for humans). We also know that high levels of organic contamination in surface water can deplete the dissolved oxygen in the water and result in fish kills. Plant life converts carbon dioxide to organic matter (e.g., carbohydrates) as they convert water to oxygen; from Table 4.1 it is observed that this is the most costly redox reaction possible, requiring large quantities of energy input. This is only possible because of the ability of plant life to harness the vast amount of energy available in sunlight (Devinny, 1990).

From Table 4.1 it is possible to determine if redox reactions are energetically (thermodynamically) favorable. While kinetics cannot be inferred from thermodynamics, it is possible to make inferences with respect to the relative rates of redox reactions utilizing Table 4.1. Microorganisms that are utilizing the more energetic reactions are generally more efficient and will thus likely dominate over other microorganisms in competing for the available resources. However, when some environmental factor becomes limiting for these microorganisms, those with the second most efficient form of metabolism will generally dominate. Thus, the general rule of thumb is that the microbially mediated redox reactions will occur in the order of the amount of energy generated. This is evidenced in the fact that facultative anaerobes will choose the more efficient aerobic respiration in the presence of oxygen. However, in certain situations other factors may serve to override this rule of thumb. Also, several reactions may in fact occur at the same time (Devinny, 1990).

It is possible for microorganisms to consume all the available substrate (organic matter) while oxygen is still present in ample quantities; alternately, it is possible for oxygen to be depleted with excess substrate remaining. Microbial systems that may result from these two scenarios are discussed

Table 4.2 Microbially Mediated Redox Reactions When Oxygen Is Abundant

Reduction[a]	Oxidation[b]	Redox Reactions of aerobic respiration and chemoautotrophy
Oxygen to water $O_2 \rightarrow H_2O$	Nitrogen gas to nitrate $N_2 \rightarrow NO_3^-$	9. (Does not occur, kinetically hindered)
	Nitrite to nitrate $NO_2^- \rightarrow NO_3^-$	8. Nitrification
	Ammonia to nitrite $NH_3 \rightarrow NO_2^-$	7. Nitrification
	Ferrous iron to ferric iron $Fe^{2+} \rightarrow Fe^{3+}$	6. Iron oxidation
	Sulfide to sulfate $H_2S \rightarrow SO_4^=$	5. Sulfide oxidation
	Methane to carbon dioxide $CH_4 \rightarrow CO_2$	4. Methane oxidation
	Ammonia to nitrogen gas $NH_3 \rightarrow N_2$	3. (Does not occur, kinetically hindered)
	Hydrogen gas to hydrogen ion $H_2 \rightarrow H^+$	2. Hydrogen oxidation
	Organic matter to carbon dioxide $CH_2O \rightarrow CO_2$	1. Aerobic respiration

Source: Devinny, 1990.

[a] Reduction of oxygen, which provides the most energy, is combined with oxidation of various substrates.

[b] Oxidation of various substrates beginning with the last, which consumes the least energy, and proceeding upward to produce redox reactions.

below. In the first scenario, oxygen is abundant while organic matter is not present. Oxygen will thus serve as the electron acceptor (will be reduced); however, a compound other than organic matter will serve as the electron donor (be oxidized). At the same time, organic matter will not be available to serve as a carbon source; thus, aerobic chemoautotraphs will dominate. Table 4.2 shows reactions that are possible (from Table 4.1) under these conditions (when oxygen is reduced and a compound other than organic matter is oxidized). The last entry in Table 4.2 is included for comparative purposes (it is heterotrophic); it is the entry that would result in the greatest energy release. The energy release thus is greatest at the bottom of Table 4.2

and least at the top (Devinny, 1990). The autotrophic redox reactions most common in the subsurface will be discussed below.

In the absence of complex organic matter, oxidation of methane to carbon dioxide is the most energetic transformation commonly observed in the subsurface. Methane may be present as the end product of other microorganisms active in the subsurface, in the vicinity of natural gas deposits or as the result of leaks in natural gas delivery systems. In the absence of organic matter and methane, the next most energetic transformation would include the oxidation of sulfide. Sulfide is not a common component of ground water, but may be present due to natural or contamination phenomena. Another source of electrons is the oxidation of iron. Iron may be solubilized (ferrous iron, Fe[II]) in ground water under reducing conditions (anaerobic); upon oxidation the iron precipitates (ferric iron, Fe[III]). Finally, in the absence of organic matter, methane, sulfide, or iron as electron donors, oxidation of ammonia to nitrite/nitrate may occur (note that this is the least energetically favorable of the reactions considered).

The second scenario is the reverse of the first; molecular oxygen is not available and organic matter is abundant. Under these conditions, a compound other than molecular oxygen will serve as the electron acceptor (be reduced). Table 4.3 shows the reactions possible under these conditions with the redox reactions decreasing in the amount of energy released from top to bottom entries. The most energetically favorable reaction in the absence of oxygen is the reduction of nitrate to nitrogen gas (denitrification). Agriculturalists are familiar with this phenomenon as it can result in the loss of nitrogen as a nutrient for crop growth (Alexander, 1977). Wastewater engineers utilize this process to remove nitrate (and thus nitrogen) from the waste stream to protect receiving water bodies from eutrophication (e.g., nitrogen may be the limiting nutrient preventing the occurrence of algal blooms). Denitrification can occur in the presence of low levels of free oxygen; thus, this process has been labeled anoxic by some to distinguish it from pure anaerobic systems where free oxygen is toxic (Grady and Lim, 1980). It is possible that nitrate may be reduced to nitrite or ammonia during metabolic activities. It is also possible that ferric iron (Fe[III]) may be reduced to ferrous iron (Fe[II]), rendering the iron soluble. This is the process that results in soluble iron in ground water. The reduction of carbohydrates to alcohols, along with the oxidation of organic matter, is an example of the fermentation process. Sulfate may be reduced to sulfide, with the potential for foul odors and immobilization of metals as metal sulfides. Carbon dioxide may be reduced to methane in anaerobic processes. Nitrogen gas may be reduced to ammonia, a process referred to as nitrogen fixation. Agriculturalists are familiar with this process as it takes nitrogen gas, which cannot be utilized during crop growth, and transforms it into ammonia, which is an available nitrogen form (Alexander, 1977). Finally, hydrogen ions may be reduced to hydrogen gas and carbon dioxide may be reduced to organic matter. It should be noted that Tables 4.1, 4.2, and 4.3 are

Table 4.3 Microbially Mediated Redox Reactions When Organic Matter is Abundant

Reduction[a]	Oxidation	Redox Reactions of aerobic respiration and anaerobic respiration
Oxygen to water $O_2 \rightarrow H_2O$	Organic matter to carbon dioxide $CH_2O \rightarrow CO_2$	1. Aerobic respiration
Nitrate to nitrogen gas $NO_3^- \rightarrow N_2$		2. Denitrification
Nitrate to nitrite $NO_3^- \rightarrow NO_2^-$ or Nitrite to ammonia $NO_2^- \rightarrow NH_3$		3. Assimilatory ammonification of fermentative nitrate reduction
Ferric iron to ferrous iron $Fe^{3+} \rightarrow Fe^{2+}$		4. Iron reduction
Carbohydrate to alcohol $CH_2O \rightarrow CH_3OH$		5. Fermentation
Sulfate to sulfide $SO_4^= \rightarrow H_2S$		6. Sulfide generation
Carbon dioxide to methane $CO_2 \rightarrow CH_4$		7. Methane generation
Nitrogen gas to ammonia $N_2 \rightarrow NH_3$		8. Nitrogen fixation
Hydrogen ion to hydrogen gas $H^+ \rightarrow H_2$		9. Hydrogen generation
Carbon dioxide to organic matter $CO_2 \rightarrow CH_2O$		10. Fermentation

Source: Devinny, 1990.

[a] Reduction of various substrates beginning with the first, which provides the most energy, is combined with oxidation of organic matter.

far from complete; however, they do serve to illustrate the types of metabolism that may be encountered in a subsurface system (Devinny, 1990).

The biochemical reactions discussed above are carried out by microorganisms either for energy or synthesis (new cells). A third type of metabolism has been observed (cometabolism) in which the microorganisms neither use the energy generated nor the products in synthesis. In these reactions a primary substrate is utilized for energy and synthesis. Cometabolism occurs as a side effect of the main metabolism, often due to enzymes (organic catalysts) excreted by microorganisms (Edmonds, 1978). Cometabolism thus has implications for contaminants that cannot be utilized as a main substrate but are susceptible to degradation as result of metabolism of another contaminant.

4.2.2 Environmental Factors

A myriad of factors can serve to limit the occurrence or extent of microbial metabolism of a subsurface contaminant. Certainly, the compounds discussed in the previous sections will limit metabolism when present below certain levels. Common examples of additional limiting factors include: pH, temperature, osmotic pressure, toxics, substrate concentration, moisture content, missing members of a microbial consortia, and substrate structure. Each of these limiting factors will be further discussed below.

In general, most bacteria find the optimum pH range to be 6.5 to 7.5 and are not able to survive at pH values greater than 9.5 or below 4.0 (Metcalf and Eddy, 1991). Microbial activity generally increases with increasing temperature (approximately a twofold activity increase for a 10°C temperature increase). Microorganisms have different optimal and allowable temperature ranges in which they can function, and may be classified as follows: 0 to 10°C, psychrophilic; 10 to 45°C, mesophilic; and 45 to 75°C, thermophilic. Since microorganisms utilize osmosis in their metabolic processes, the salinity of the system (which affects the osmotic pressure) must be within certain limits. Most microorganisms are not affected by the aqueous salt content if it is between 500 to 35,000 mg/L (Reynolds, 1982).

The presence of certain compounds may be toxic to the microorganisms. For example, heavy metals are commonly toxic to microorganisms. The presence of acids and bases may also be toxic to the microorganisms (Reynolds, 1982). It is even possible for the substrate to be toxic to the microorganism at elevated concentrations; while at very low concentrations it may not be sufficient to allow microbial response (Suflita, 1989a). Moisture is a vital component for microorganisms. While ample moisture will be available in the saturated zones of the subsurface, portions of the vadose zone (unsaturated zones) may have insufficient moisture to maintain bioactivity (Alexander, 1977). Usually, several microorganisms (a microbial consortium) act jointly to metabolize a given organic contaminant. If a member of the requisite consortia is missing, this may limit the metabolism of the compound (Suflita, 1989a).

The structure of a contaminant can significantly affect the ability of microorganisms to metabolize the chemical. For example, compounds with large structures (such as polymers) that are unable to enter microbial cells will be poorly metabolized, whereas the monomers may be readily metabolized. The structure of the compound will also affect its physical state (solubility, volatility, sorption, etc.) and will thus influence its susceptibility to biodegradation. From the data of DeWeerd et al. (1986), it is observed that the type and position of the halogen substituent influenced the rate of aryl-reductive dehalogenation by an anaerobic consortium of bacteria. While the brominated and iodinated benzoates were transformed by these bacteria, no fluorinated benzoates and only one chlorinated benzoate were transformed. It was also observed that increasing halogen size resulted in slower degradation rates. The metasubstituted halobenzoates always exhibited the highest rate of biodegradation. The effect of chlorine positions on the biodegradability and bacterial metabolism of over 30 isomers of polychlorinated biphenyls has been researched by Furukawa et al. (1978, 1979), with the chlorine substitution again observed to affect the biodegradation. In general, xenobiotic compounds are complex and the microbes have had little time to adapt metabolic pathways to handle these compounds. Microorganisms have demonstrated an amazing ability to adapt to their environment. Thus, it is reasonable to expect that, given enough time, the microorganisms will adapt such that they may not only survive in the presence of these xenobiotic compounds, but may actually be capable of utilizing them (Suflita, 1989a).

4.2.3 Growth Kinetics

Growth of microorganisms is primarily a function of the presence of viable microbial cells and the requisite factors previously discussed (carbon source, energy source, electron acceptor, nutrients, etc.). The various growth phases that a biomass may undergo in response to an influx of substrate will be illustrated by considering a batch system into which excess substrate and a small quantity of viable bacterial cells are added (it is assumed that the additional substances and environmental conditions necessary for growth are present). The response of the biomass to this influx of substrate can be divided into six phases: (1) lag phase, (2) acceleration phase, (3) exponential phase, (4) declining growth phase, (5) stationary phase, and (6) endogenous phase (Benefield and Randall, 1980). These six phases are illustrated in Figure 4.1. The lag phase corresponds to a period when the microbes are becoming adjusted to their new environmental conditions. During the lag phase the number of viable cells does not increase, and the cell size and the rate of metabolic activity are at a maximum. The extent of the lag phase is a function

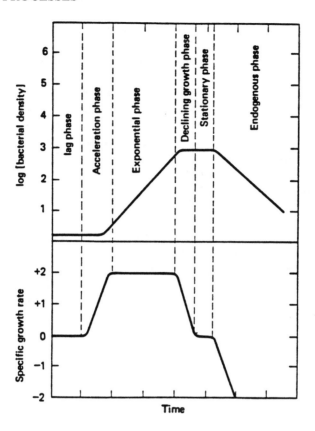

Figure 4.1. **Phases of microbial growth in closed system (Benefield and Randall, 1980).**

of the activity and magnitude of the initial biomass population and also the nature of the substrate. During the acceleration phase, decreasing generation times and increasing growth rates are evidenced for the microbial cells. During the exponential growth phase, the maximum rate of biomass growth and substrate conversion is experienced. The cell generation time is a minimum, the cell density is constant and the cell size is at a minimum. Towards the end of the exponential growth phase, some condition becomes limiting and the system is not able to maintain the rate of growth. The limiting condition may include insufficient substrate, depletion of electron acceptor, insufficient nutrients, buildup of toxic end products (especially in closed systems where these end products are not carried away), etc. As a result of these limiting conditions, the system will enter the declining growth phase. During the declining growth phase the cell generation time increases, and cell death occurs at a greater rate causing the rate of net biomass growth to decrease.

As the limiting conditions become more pronounced, the stationary phase is encountered. During this phase, no net accumulation in biomass is experienced. This can be the result of the fact that the rate of growth equals the rate of death (zero population growth) or could be due to the cells remaining in a state of suspended animation (Benefield and Randall, 1980). In the endogenous phase, the rate of cell death exceeds the rate of cell growth (population decline), endogenous metabolism is evidenced, and cell lysis occurs (Benefield and Randall, 1980; Reynolds, 1982; Metcalf and Eddy, 1991). It should be emphasized that the scenario described above has assumed a batch system (a closed system) which once established does not allow for subsequent inputs or outflows from the system. The subsurface microbial response to a contaminant release would be similar, although inflows and outflows will be realized.

The biomass growth rate has been observed to be a function of the biomass population and the concentration of a limiting nutrient (Benefield and Randall, 1980). Normalizing the biomass growth rate by the biomass concentration yields the specific biomass growth rate. Equation 4.2 is one expression for relating the specific biomass growth rate to the limiting nutrient concentration.

$$\mu = \mu_m \frac{S}{K_s + S} \tag{4.2}$$

where μ = specific growth rate (t^{-1})
μ_m = maximum value of μ at saturation concentrations of growth-limiting substrate (t^{-1})
S = residual growth-limiting substrate concentration (m/t)
K_s = saturation constant numerically equal to the substrate concentration at which $\mu = \mu_m/2$ (M/L^3)

Equation 4.2 is illustrated in Figure 4.2 over a range of substrate (limiting nutrient) concentrations. At low values of substrate (S), the growth rate is limited by S and is low. As the level of S increases, it no longer acts as the limiting factor. At high values of S, some other factor is limiting and changes in the level of S do not impact the biomass growth rate (it is constant). Since the rate of biomass increase is related to the rate of substrate use, this expression (and figure) can be adapted to show the specific substrate utilization rate versus the limiting nutrient with the same trends evidenced. Equation 4.2 can be simplified at extreme values of S. If S is much lower than K_s, the S term in the denominator can be ignored and the equation can be simplified to Equation 4.3, where K^1 is the simplified rate constant.

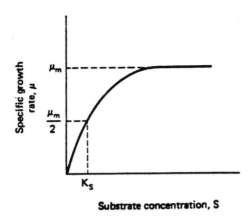

Figure 4.2. Relationship between specific biomass growth rate and growth Limiting nutrient (Benefield and Randall, 1980).

$$\mu = \frac{\mu_m K_s}{K_s} S = K^1 S \qquad (4.3)$$

In this equation biomass growth rate (and thus the specific substrate utilization rate) is a linear function of the concentration of the limiting nutrient (first order with respect to S); this linearity can be observed at low values of S in Figure 4.2. If S is much greater than K_s, then Equation 4.2 can be simplified by assuming the K_s to be negligible relative to S. In this case, the S terms cancel in the numerator and the denominator and the specific biomass growth rate (or substrate utilization rate) is observed to be zero order with respect to the limiting nutrient (a constant and a maximum), as observed in Equation 4.4.

$$\mu = \mu_m S = \mu_m \qquad (4.4)$$

This again can be observed in Figure 4.2 when the growth rate reaches a maximum value at large values of S. Thus, the specific biomass growth (substrate utilization) rate can be described over all concentrations by Equation 4.2, can be described by a first order reaction at low values of the limiting nutrient (Equation 4.3), or can be described by a zero order reaction at high values of the limiting nutrient (Equation 4.4). These concepts are important in the subsurface, as it will determine what order of reaction is necessary to describe the biodegradation.

In the above discussion, the growth kinetics have been assumed to be either zero or first order. As with all reactions, it is possible that a given metabolic reaction is of a higher order (second order, etc.). Discussions of higher order reactions (or reactions between zero and first order) have been discussed in the literature, but are beyond the scope of this document.

4.3 SUBSURFACE MICROBIOLOGY

For biodegradation to occur in the subsurface, microbiota must be present. Historically, it was felt that surface soils were rich in microbiota but that the aquifer zones were virtually devoid of microbiota. Since surface soils are rich in nutrients and food which are continually replenished, it was logical that these materials would be rich in microbial life. However, it was felt that the aquifer zones were sufficiently devoid of these elements such that they could not support microbial activity. Studies demonstrating that microbial numbers continuously decreased with depth seemingly supported the premise that aquifers were virtually devoid of microbial activities. It has only been in the last decade that the extent of subsurface microbiota has been realized, due largely to improvements in the procurement and handling of subsurface samples (Suflita, 1989a).

To obtain subsurface soil samples, it is necessary to reach the desired depth and collect the sample while minimizing the disturbances to the sample and the subsurface. This typically involves removal of the exterior of the collected sample with sterilized devices (e.g., paring devices) to eliminate microorganisms introduced to the exterior of the sample from higher zones or the sampling device itself. In saturated zones where the system is anaerobic, the sampling protocol must minimize the introduction of molecular oxygen to the sample in order to prevent the destruction of obligatory anaerobes. This may require use of aseptic glove boxes in a nitrogen environment for handling the samples. The focus of this discussion is not to elaborate on sampling devices or methods (see McNabb and Mallard, 1984; Leach et al., 1988; Ghiorse and Wilson, 1988; for discussions on this topic), but rather to provide a basis for understanding why the bioactivity of aquifer zones has not been realized until recently.

To document the bioactivity of aquifer zones, Ghiorse and Wilson (1988) summarized data from numerous researchers on the microbial populations in various geologic settings and at various depths for pristine and contaminated sites (Tables 4.4 and 4.5, respectively). The aquifers summarized are observed to range from shallow to deep, and from sandy alluvial to clayey sediments to consolidated materials. The microbial ecosystems are also observed to vary in terms of types of microorganisms, microbial systems (aerobic, sulfate reducing, etc.), and microbial counts. It is observed that high levels of microbiota are possible for pristine as well as for contaminated aquifer zones. Suflita (1989a) states that subsurface microorganisms tend to be small, mostly attached to the solid surfaces of the aquifer media, and ready to respond to increases in the limiting nutrients.

The presence of microbiota in subsurface samples is evidenced by the microbial activity observed in these samples. Suflita (1989a) summarized data from numerous researchers to demonstrate subsurface microbiota activity, as shown in Table 4.6. Note that the range of metabolic activity indicated in Table 4.6 corresponds to those suggested in Tables 4.1 through 4.3. Also note that it is possible that several of these individual metabolic processes may in fact occur sequentially (in series). For example, it is possible that when biodegradation of organic pollutants depletes the molecular oxygen, nitrate will act as the electron acceptor while undergoing denitrification to nitrogen gas.

4.4 SUBSURFACE METABOLIC PROCESSES

4.4.1 Introduction

Having discussed the fundamentals of microbiology and established that subsurface aquifers have viable microbial populations (see Table 4.4), it is appropriate to ask these questions: (1) what chemicals are biodegradable in the subsurface and under what conditions; (2) how fast is the biodegradation; and (3) how can we quantify (model) these reactions. These questions will be discussed in the following two sections.

The question of what chemicals are degradable in the subsurface and under what conditions is a difficult one to answer. A given chemical may be biodegradable under a number of subsurface conditions or may only be degraded under one set of conditions. A number of chemicals have been assumed to be recalcitrant (resistant to biodegradation) in the subsurface when subsequent research demonstrated that the chemicals were indeed biodegradable under conditions not evaluated in the previous research. Thus, determining the biodegradability of a compound in the subsurface requires detailed research. This section will document the results of research indicating the diversity of conditions resulting in the biodegradation of a variety of chemicals. This information is critical to determining the potential for biotransformation of a given compound under given conditions.

Table 4.4 Types, Numbers, and Activities of Organisms in Pristine Subsurface Samples

Location and Site Description	Type of Sample	Depth	Findings[a]	Reference
Saskatchewan, Canada	Groundwater from iron-bearing formations	Various depths	Diverse types of iron-depositing bacteria present in wells and cultures	Cullimore and McCann, 1977
Montana, U.S.	Groundwater from Madison limestone	10–263 and 1264–1752 m	Total counts $\approx 10^3$/mL Sulfate-reducing bacteria present	Dockins et al., 1980; Olson et at., 1981
Fulda Valley, Germany	Groundwater from sandy and gravelly sediments	3–6 m	Total counts 10^6–10^7/mL [^{14}C] glucose uptake correlated with C.O.D. and D.O.	Marxsen, 1981a
West Neidersachsen, Germany	Unsaturated and saturated zone sediments	10–90 m	Viable counts 10^3–10^6/g 11 physiological groups of bacteria detected	Hoos and Schweisfurth, 1982
Marmot, Basin, Alberta, Canada	Groundwater	1.5 m	Total counts 10^5–10^6/mL Viable counts 10^3–10^7/mL [^{14}C] ignocellulose degraded	Ladd et al., 1982
Southern Arizona U.S.	Groundwater	Various depths	Viable counts $\approx 10^3$/mL Aerobic bacteria isolated	Stetzenbach and Sinclair, 1983
Segeberger Forest, northern Germany	Groundwater and sandy aquifer sediments	5–31.5 m	Enrichment of diverse bacteria, fungi, protozoa Total counts 10^6–10^7/g Viable counts 10^3–10^4/g	Hirsch and Rades-Rohkohl, 1983; Rades-Rohkohl and Hirsch, 1985

Location	Setting	Depth	Observations	References
Fort Polk, Louisiana, U.S.	Unsaturated and saturated sandy and clayey sediments	1.2–6.7 m	Total counts 10^6–10^7/mL; Viable counts 10^2–10^5/g; Bacteria observed by TEM; Phospholipid 0.90–5.5 nmol/g; Muramic acid 2.0–11.3 nmol/g; Toluene, styrene degraded in microcosms	Ghiorse and Balkwill, 1983; Wilson et al., 1983b; White et al., 1983a,b; Balkwill and Ghiorse, 1985
Bucatunna aquifer near Pensacola, U.S.	Clayey sediment	410 m	Phospholipid 0.52 nmol/g; Muramic acid 0.62 nmol/g; Ratio higher in PHB and uronic acid/muramic acid than surface sediments	White et al., 1983a,b
Williamsburg and Sault Ste. Marie, Ontario, Canada	Groundwater	3–3.35 m	Total counts 10^6/mL; Viable counts 10^1–10^5/g; ^{14}C-glusose, -amino acids -benzoate mineralized	Larson and Ventullo, 1983; Ventullo and Larson, 1985
Butler and Dayton, Ohio, U.S.	Groundwater	10 and 12 m	Total counts 10^4–10^5/mL; Viable counts 10^2–10^4/mL; 14C-glucose, -amino acids mineralized	Ventullo and Larson, 1985

Table 4.4 (Continued)

Location and Site Description	Type of Sample	Depth	Findings[a]	Reference
Pickett and Lula, Oklahoma, U.S.	Unsaturated and saturated sandy, clayey sediments	1.2–8.0 m	Total counts 10^6–10^7/g. Viable counts 10^3–10^6/g. ATP and other biomass and activity estimates decreased with depth in unsaturated zone, but varied in saturated zone. Protozoa and fungi detected in some samples. Bacterial colony diversity decreased with depth. PHB-containing cells observed in saturated zone by TEM. Toluene degraded rapidly, cholorobenzene slowly in microcosms. Plasmids found in 2–8% of isolates	Wilson et al., 1983a; Bengtsson, 1984; Ghiorse and Balkwill, 1984, 1985; Balkwill and Ghiorse, 1985a,b; Webster et al., 1985; Sinclair and Ghiorse, 1987; Bone and Balkwill, 1986, 1988; Wilson, J.T. et al., 1986; Beloin et al., 1988; Ogunseitan et al., 1987; Thorn and Ventullo, 1988

Location	Geology	Depth	Observations	Reference
Kaiserslautern, Germany	Unsaturated sandstone rock over coal deposit	2–405 m	Viable counts decreased with depth. No viable counts at 24–343 m, 10^3/g at 405 coal layer	Weirich and Schweisfurth, 1983
Maryland, U.S.	Coastal plain sediments intervedded clay, silt, sand	14–182 m	Total counts 10^4–10^8/g. Viable counts 10^3–10^6/g. Methanogenic and sulfate-reducing bacteria found in <40% of samples. Heterotrophic bacteria may be a source of CO_2	Chapelle et al., 1987
Nemaha County, Kansas, U.S.	Clayey, sandy, and gravelly sediments in glacial buried-valley aquifers	26–86 m	Total counts 10^6–10^7/g. Viable counts <10^2–10^7. Protozoa found in very permeable layers	Sinclair et al., 1987

Source: Ghiorse and Wilson, 1988.

[a] Abbreviations: C.O.D., chemical oxygen demand; D.O., dissolve oxygen; TEM, transmission electron microscopy; PHB, poly-β-hydroxybutyrate.

Table 4.5 Types, Numbers, and Activities of Organisms in Subsurface Samples from Contaminated Sites

Location and Site Description	Type of Sample	Depth	Findings[a]	Reference
Sewage injection well Magothy aquifer, Bay Park, NY, U.S.	Groundwater	127–146 m	Viable counts 10^2–10/ml; Numbers of anaerobic and facultative bacteria declined with distance from injection well	Godsy and Ehrlich, 1978
Organically polluted and unpolluted sandy, gravelly aquifer, Fulda Valley, Germany	Groundwater	3 m	Total counts 10^6–10^7/ml; [^{14}C] glucose uptake highest in oxygenated part of contaminated zones; >90% bacteria attached to particles	Marxsen, 1981b
Hydrocarbon-polluted well, Shilo, Manitoba, Canada	Groundwater	27–32 m	Viable counts 10^1–10^5/ml; Seasonal variation in microbial numbers depending on D.O. and hydrocarbon concentration	Cullimore, 1983
Reclaimed coal strip mine, Canmore, Alberta, Canada	Groundwater	2–12 m	Total counts ~10^5/ml; Viable counts 10^2–10^4/ml; Seasonal variation in microbial population depended on flow	Wallis and Ladd, 1983
Creosote waste site, St. Louis Park, MN, U.S.	Groundwater from contaminated and uncontaminated zone	0–30 m	Total counts 10^6–10^7/ml; Viable counts 0–10^3/ml; Diverse anaerobic bacteria detected; Methanogens only in contaminated zone	Ehrlich et al., 1982, 1983

Site	Depth	Observations	References
Creosote waste site, Conroe, TX, U.S.	Unsaturated and saturated sediments contaminated and uncontaminated zones	2.0–9.0 m	Total counts 10^6–10^7/g Viable counts $<10^2$–10^5/g INT reducers 10^5–10^6/g ATP 0.01–0.17 ng/g Lower ratios of glycerol teichoic acid/phospholipid and PHB/phospholipid in contaminated zone Bacteria observed by TEM Rapid degradation of hydrocarbons at margin of waste plume 19% of isolates contained plasmids	Wilson, J.T. et al., 1985a, 1986; Smith et al., 1985; Webster et al., 1985; Ogunseitan et al., 1987
Sewage-contaminated aquifer, Cape Cod, MA, U.S.	Groundwater and sand and gravel sediments	5–40 m	Total counts declined from 10^7 to 10^5/ml and [^{14}C] glucose uptake rate declined with distance from infiltration beds $\approx$60% attached in contaminated zone; $\approx$95% attached in uncontaminated zone Estimated generation times 16–139 h	Harvey et al., 1984; Harvey and George, 1986, 1987
Sewage polluted aquifer,Templeton, New Zealand	Groundwater from glacial outwash gravel formation	12–24 m	10% of phreatic crustaceans contained sewage-derived bacteria in guts	Sinton, 1984

Table 4.5 (Continued)

Location and Site Description	Type of Sample	Depth	Findings[a]	Reference
Confined aquifers, Lincoln and Norfolk Counties, England	Limestone, glacial sand, and chalk cores from unsaturated and saturated zones	3–50 m	Viable counts of aerobic and anaerobic bacteria 10^3–10^5/ml Dentrifying bacteria found in enrichment cultures of most samples Presence of bacteria in chalk related to occurrence of fissures	Foster et al., 1985
Chalk aquifer affected by landfill contamination, England	Core samples from unsaturated and saturated zones in uncontaminated regions beneath landfills	2–35 m	Total counts 10^5–10^8/g Viable counts <10^2–10^7/g Bacteria unevenly distributed in depth profile Highest densities at interfaces	Towler et al., 1985
Groundwater treatment system southwest Germany	Hydrocarbon-polluted groundwater		Total counts 10^5–10^7/g Viable counts 10^4–10^5/ml Diverse bacteria detected at low population density Differences in metabolic activity of bacteria in polluted and nonpolluted zones	Dott et al., 1984; Frank and Dott, 1985

Septic tank tile field, Williamsburg and Sault Ste. Marie, Ontario, Canada	Surface and subsurface soil	1.2–3.0 m	Aerobic and anerobic ^{14}C mineralization rates lower in subsurface than in surface soil. Denitrification observed	Ward, 1985
Underground Iron and manganese removal system, Germany	Loose rock aquifer material from aerated and untreated zones	0.5–1.8 m	Viable counts 10^0–10^6/g. Diverse bacterial populations in both serated and untreated zones	Gottfreund et al., 1985a–c

Source: Ghiorse and Wilson, 1988.

[a] Abbreviations: D.O., dissolved oxygen; INT, 2-(*p*-iodophenyl)-3-(*p*-nitrophenyl)-5-phenyltertrazolium chloride; PHB, poly-β-hydroxybutyrate; TEM, transmission electron microscopy.

Table 4.6 Selected Subsufrace Processes Mediated by Microorganisms

Metabolic Process	Oxygen Requirement	Reference
I. Biodegradation of Organic Pollutants		
a. Petroleum hydrocarbons	Aerobic	Jamison, et al., 1975
		Lee and Ward, 1985
		McCarty et al., 1984
		Raymond et al., 1976
		Roberts et al., 1980
		Wilson et al., 1983
		Wilson et al., 1985
b. Alkylpyridines	Aerobic/anaerobic	Rogers et al., 1985
c. Creosote chemicals	Aerobic/anaerobic	Ehrlich et al., 1983
		Smolenski and Suflita, 1987
		Wilson et al., 1985
d. Coal gasification products	Aerobic	Humenick et al., 1982
e. Sewage effluent	Aerobic	Aulenbach et al., 1975
		Godsey and Ehrlich, 1978
		Harvey et al., 1984
f. Halogenated organic compounds	Aerobic/anaerobic	Gibson and Suflita, 1986
		McCarty et al., 1984
		Suflita and Gibson, 1985
		Suflita and Miller, 1985
		Ward, 1985
		Wilson et al., 1983
		Wood et al., 1985
g. Nitrilotriacetate (NTA)	Aerobic/anaerobic	Ventullo and Larson, 1985, Ward, 1985
h. Pesticides	Aerobic/anaerobic	Gibson and Suflita, 1986
		Suflita and Gibson, 1985
		Ventullo and Larson, 1985
		Ward, 1985
II. Nitrification	Aerobic	Barcelona and Naymic, 1984
		Idelovitch and Michail, 1980
		Preul, 1966

Table 4.6 (Continued)

Metabolic Process	Oxygen Requirement	Reference
III. Denitrification	Anaerobic	Ehrlich et al., 1983 Lind, 1975 Ward, 1985
IV. Sulfur oxidation	Aerobic	Olson et al., 1981
V. Sulfur reduction	Anaerobic	Beeman and Suflita, 1987 Bastin, 1926 Hvid-Hansen, 1951 Jacks, 1977 Olson et al., 1981 van Beek et al., 1962
VI. Iron oxidation	Aerobic	Olson et al., 1981 Hallburg and Martinell, 1976
VII. Iron reduction	Aerobic	Godsey and Ehrlich, 1978 Ehrlich et al., 1983
VIII. Manganese oxidation	Aerobic	Hallburg and Martinell, 1976
IX. Methanogenesis	Anaerobic	Beeman and Suflita, 1987 Belyaev and Ivanov, 1983 Davis, 1967 Gibson and Suflita, 1986 Godsey and Ehrlich, 1978 Suflita and Miller, 1985 van Beek et al., 1962

Source: Suflita, 1989a.

Space will not allow documentation of the detailed pathways (with all of the intermediate products) used by microorganisms for degradation of a given compound; the interested reader is directed to the references cited in this chapter for such details. The absence of documentation of the detailed pathways should not be interpreted as minimizing the importance of this information. To the contrary, this information is vital to a proper understanding of the factors affecting the mineralization of a compound and potential interferences

**Table 4.7 Aliphatic and Aromatic Compounds Utilized in Studies of
Aerobic and Anaerobic Degradation**

Chloroform	Ethylbenzene
Bromodichloromethane	Styrene
Dibromocholoromethane	Naphthalene
Bromoform	Chlorobenzene
1,1,1-Trichloroethane	1,2-Dichlorobenzene
Trichloroethylene	1,3-Dichlorobenzene
Tetrachloroethylene	1,4-Dichlorobenzene
	1,2,4-Trichlorobenzene

Source: "Microbiological Processes Affecting Chemical Transformations in Goundwater," McCarty, P. L., B. E. Rittman, and E. J. Bouwer in *Groundwater Pollution Microbiology,* G. Bitten and C. P. Gerba, Eds., Copyright 1984, John Wiley & Sons. Reprinted by permission of John Wiley & Sons, Inc.

that may occur. Also, metabolic pathways provide information as to metabolites that should be monitored in the event that incomplete mineralization is realized. This is especially important in cases where metabolites may be of equal or greater concern relative to the original compound.

As an introduction to subsurface metabolic processes, the work of McCarty et al. (1984) will be presented. Subsequent discussions will be organized around aerobic and anaerobic biodegradation processes and will be based largely on several excellent reviews which have appeared in the last few years (Lee et al., 1988; Ghiorse and Wilson, 1988; Suflita, 1989a,b; Sims et al., 1990; Suflita and Sewell, 1991; and Sims et al., 1991; to name several). The susceptibility of compounds to aerobic and anaerobic degradation is a function of a number of factors. In general, organics in reduced form tend to be susceptible to aerobic biodegradation. Conversely, compounds that are in a highly oxidized state tend to resist microbial degradation under aerobic conditions, but are susceptible to anaerobic degradation. Thus, the extent of oxidation/reduction of a parent compound will influence its susceptibility to aerobic vs anaerobic degradation (Rao and Davidson, 1980; Sims et al., 1991).

McCarty et al. (1984) stated that some chemicals appear to be only degradable under aerobic conditions, some appear to be only degradable under anaerobic conditions, some are degradable under both aerobic and anaerobic conditions, and some appear to be recalcitrant under both aerobic and anaerobic conditions. The researchers evaluated the biodegradability of a range of organic contaminants in laboratory batch and continuous flow studies to determine their potential for biodegradation under aerobic and anaerobic conditions. The organic chemicals evaluated were selected based on their occurrence in ground water, typically at low concentrations, and included halogenated aliphatic (one and two carbon), halogenated, and nonhalogenated aromatic compounds (see Table 4.7 for a list of the chemicals studied). Table 4.8 summarizes results for the biodegradation of five aromatic compounds (three halogenated) under aerobic and anaerobic conditions in batch reactors. Under aerobic conditions, the nonhalogenated compounds (naphthalene and styrene) approached complete biodegradation within one week with a minimal lag period at all concentrations evaluated. However, the halogenated compounds demonstrated a lag period in biodegradation, with 1,4-dichlorobenzene showing the fastest (and most complete) biodegradation. 1,3-Dichlorobenzene was not significantly degraded (except at the highest concentration) and was the most recalcitrant compound of the aromatics evaluated. Thus, the location of the second halogen significantly affected the biodegradability of the dichloroben-zene. It should be noted that none of the aromatic compounds studied showed significant biodegradation under anaerobic conditions (even though inoculated with an active methanogenic seed and incubated for eleven weeks). Thus, McCarty et al. (1984) observed that the aromatic compounds studied were biodegradable under aerobic conditions but not under anaerobic conditions. It should be noted that more recent research has revealed that a myriad of aromatic compounds are biodegradable under a variety of anaerobic conditions (see later discussions).

McCarty et al. (1984) noted that the halogenated aliphatic compounds appeared to be biodegradable only under anaerobic conditions, in contrast to the aromatic compounds which were only degradable under aerobic conditions. Table 4.9 summarizes the biodegradability of five halogenated aliphatics (three single carbon and two double carbon). The single carbon compounds (trihalomethanes) were observed to be more readily degraded than the double carbon compounds (chloroethylenes). Chloroform was degraded to a value of 0.2 μg/L within 16 weeks while the other two trihalomethanes were degraded to below detection limits within two weeks. The concentration of the brominated aliphatic compounds were observed to decrease even in sterile controls (suggesting chemical transformations); however, the rates of biodegradation were much more rapid in the seeded cultures. Chlorinated aliphatics were observed to only degrade under seeded conditions. From Table 4.9 it is observed that no biodegradation was observed during the 24 weeks of aerobic incubation for the halogenated one or two carbon aliphatics.

McCarty et al. (1984) utilized column studies to evaluate biodegradation in continuous flow systems. Figure 4.3 shows the aerobic degradation of chlorinated benzenes in column studies and supports the results obtained in the batch studies. From Figure 4.3, it is observed that 1,4-dichlorobenzene was most rapidly degraded, 1,3-dichlorobenzene was apparently recalcitrant to biodegradation, and chlorobenzene was intermediate, in agreement with the data in Table 4.8. It is interesting to note that the para and ortho dichlorobenzenes (1,4-DCB and 1,2-DCB, respectively) were the most readily degraded of the compounds investigated but the meta dichlorobenzene (1,3-DCB) was the least degradable. It is also interesting to note that the dichlo-

Table 4.8 Degradation of Aromatic Compounds Present at Low Concentrations under Aerobic and Anaerobic Conditions

Compound	Aerobic			Anaerobic	
	Concentration (µg/l)	Percent Degraded		Concentration (µg/l)	Percent Degraded 12 weeks
		1 week	11 weeks		
Naphthalene	190.0	>99	>99	93.0	0
	56.0	>99	>99	32.0	0
	24.0	>99	>99	10.0	0
	7.4	>99	>99		
Styrene	215.0	>99	>99	29.0	0
	51.0	>99	>99	7.8	0
	26.0	>99	>99	2.4	0
	5.0	>99	>99		
Chlorobenzene	280.0	23	>99	72.0	0
	81.0	0	99	31.0	0
	21.0	19	25	8.0	0
	8.0	22	90		
1,4-Dichlorobenzene	200.0	26	>99	74.0	0
	72.0	0	>99	29.0	6
	21.0	26	>99	7.4	18
	6.0	13	>99		

1,3-Dichlorobenzene	220.0	17	83	59.0	0
	79.0	0	36	23.0	0
	22.0	27	0	6.0	0
	7.0	19	0		

Source: "Microbiological Processes Affecting Chemical Transformations in Goundwater," McCarty, P. L., B. E. Rittman, and E. J. Bouwer in *Groundwater Pollution Microbiology*, G. Bitten and C. P. Gerba, Eds., Copyright 1984, John Wiley & Sons. Reprinted by permission of John Wiley & Sons.

Table 4.9 Degradation of Halogenated Aliphatic Compounds Present at Low Concentrations under Aerobic and Anaerobic Conditions

Compound	Aerobic		Anaerobic		
	Concentration (μg/l)	Percent Degraded 24 weeks	Concentration (μg/l)	Precent Degraded 4 weeks	16 weeks
Chloroform	80	0	118	24	71
	29	0	38	97	99
	13	0	14	86	99
Bromodichloromethane	85	0	161	99	99
	31	0	35	99	99
	12	0	16	99	99
Dibromochloromethane	84	0	205	99	99
	30	0	40	99	99
	12	0	19	99	99
Trichloroethylene	81	0	124	15	44
	31	0	38	0	21
	11	0	13	8	31
Tetrachloroethylene	74	0	118	21	53
	33	0	36	3	44
	9	0	12	8	42

Source: "Microbiological Processes Affecting Chemical Transformations in Groundwater," McCarty, P. L., B. E. Rittman, and E. J. Bouwer in *Groundwater Pollution Microbiology*, G. Bitten and C. P. Gerba, Eds., Copyright 1984, John Wiley & Sons. Reprinted by permission of John Wiley & Sons.

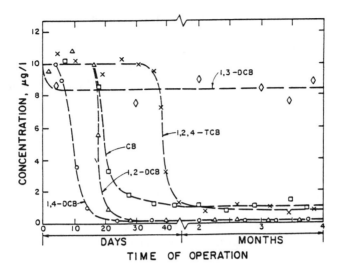

Figure 4.3.   Anaerobic degradation of chlorinated benzenes in laboratory
columns (From "Microbiological Processes Affecting Chemi-
cal Transformations in Goundwater," McCarty, P. L., B. E.
Rittman, and E. J. Bouwer in *Groundwater Pollution Microbiology*, G. Bitten and C. P. Gerba, Eds., Copyright 1984,
John Wiley & Sons. Reprinted by permission of John Wiley
& Sons).

robenzenes were more readily degraded than chlorobenzene. Table 4.10 summarizes the column results for the anaerobic degradation of halogenated
aliphatics and again supports the results of batch experiments (Table 4.9). It
is observed that longer acclimation periods were required for the chlorinated
aliphatics than for the brominated aliphatics, with the longest acclimation
periods required by the two carbon aliphatics. It is observed from Table 4.10
that all of the halogenated aliphatics had high removal efficiencies. The work
of McCarty et al. (1984) thus serves to illustrate that differing compounds
have differing susceptibilities to biodegradation under aerobic and anaerobic
conditions.

4.4.2 Aerobic Metabolism

A wide variety of aerobic metabolic processes are possible in the subsurface. Aerobic regions will be evidenced in unconfined aquifers where the
diffusion of oxygen from the vadose zone exceeds the rate of oxygen consumption in the subsurface. Confined aquifers will have dissolved oxygen if

Table 4.10 Anaerobic Degradation of Halogenated Aliphatics in Laboratory Columns

Compound	Acclimation Period in Weeks	Column Influent (μg/l)	Column Effluent (μg/l)	Removal %
Chloroform	4	33.3 ± 7.0	1.2 ± 0.6	96
Bromodichloromethane	0	30.3 ± 4.0	<0.1	99
Dibromochloromethane	0	34.1 ± 4.5	<0.1	99
Bromoform	0	33.8 ± 3.6	<0.1	99
1,1,1-Trichloroethane	8	25.4 ± 3.4	0.55 ± 0.3	98
Trichloroethylene	6	26.2 ± 4.0	1.6 ± 0.6	94
Tetrachloroethylene	16	18.3 ± 2.8	2.6 ± 1.3	86
Acetate	0	604 mg/L	42 mg/l	93

Source: "Microbiological Processes Affecting Chemical Transformations in Goundwater," McCarty, P. L., B. E. Rittman, and E. J. Bouwer in *Groundwater Pollution Microbiology,* G. Bitten and C. P. Gerba, Eds., Copyright 1984, John Wiley & Sons. Reprinted by permission of John Wiley & Sons.

the oxygen has not been depleted since recharge. Also, oxygen may be added to subsurface anaerobic regions in an effort to promote aerobic bioremediation of contaminants (Lee et al., 1988; Sims et al., 1990). Thus, aerobic metabolism plays an important role in assessing the subsurface transport and fate of contaminants, under natural conditions and enhanced remediation conditions. Table 4.6 summarizes a number of aerobic processes that take place in the subsurface; these processes include the degradation of organics (e.g., petroleum hydrocarbons, creosotes, sewage, halogenated organics, pesticides), nitrification, and sulfur, iron, and manganese oxidation.

In considering subsurface biotransformation of a compound (or compounds), it would be valuable to determine which contaminants have been documented as biodegrading in the subsurface. It would also be important to know under what conditions (electron acceptor, aquifer type, etc.) and how fast biotransformation occurred. Ghiorse and Wilson (1988) summarize the biodegradation of various organic compounds in subsurface materials from pristine sites and contaminated sites (Tables 4.11 and 4.12, respectively). These tables summarize both aerobic and anaerobic biotransformations. Ghiorse and Wilson (1988) indicated that many of the degradation rates shown in these tables are derived by the authors based on their best estimates of the data in the original references, often involving the reinterpretation of the data to generate first order rate constants. The authors state that the original researchers should not be held accountable for any error made in interpreting the data.

Table 4.11 Biodegradation of Various Organic Compounds in Subsurface Material from Pristine Sites

Compound	Dominant Electron Acceptor	Type of Sample and Location	Biodegradation Rate or Extent	Reference
Common Metabolites				
Arginine	Oxygen	Groundwater from 10–12 m in monitoring well in gravel aquifers, Hamilton and Dayton, OH, U.S.	Normalized to 10^7 cells/ml groundwater 27–130/week	Ventullo and Larson, 1985
Glucose			36–45/week	
Glutamate			120–300/week	
Glutamic acid	Oxygen	Groundwater from shallow soils under a subalpine forest in Marmot Basin, Alberta, Canada	Normalized to 10^7/ml groundwater, 61–200/week	Ladd et al., 1982
Glycolate			3.8/week	
Phenylalanine			115/week	
Volatile fatty acids acetic, propionic, butyric	Oxygen	Chalk form 10 m below water table, Ingram, Suffolk, England	0.3/week, no evidence of adaptation after 50 days	Kiene and Capone, 1986
Halogenated hydrocarbons				
Bromodichloromethane	Oxygen	Sandy clay just above the water table at Lula, OK, U.S.	Not detected, <0.03–0.02/week	Wilson et al., 1983a

Table 4.11 (Continued)

Compound	Dominant Electron Acceptor	Type of Sample and Location	Biodegradation Rate or Extent	Reference
Bromoform	Oxygen	Sand water table aquifer on Borden Air Force Base, Ontario, Canada	0.02/week, field experiment	Roberts et al., 1986
Carbon tetrachloride	Oxygen	Sand water table aquifer on Borden Air Force Base, Ontario, Canada	Not detected in 2 years, <0.002/week, field experiment	Barker et al., 1983a,b
Chlorobenzene	Oxygen	Sandy clay just above and within a water table aquifer at Fort Polk, LA and Pickett, OK, U. S.	<0.01/week, not greater than abiotic control	Wilson et al., 1983a,b
		Sandy clay within a shallow water table aquifer at Lula, OK, U.S.	Not detected, <0.01–0.05/week	
		Sand in a shallow confined aquifer at Lula, OK, U.S., 3–5 m deep	Not detected in 7 months, <0.009/week	
		Sand and gravel in a shallow confined aquifer at Lula, OK, U.S., 5.5–6.7 m	0.06–0.02/week	Wilson et al., 1986
1,2-Dibromoethane (ethylene dibromide)	Oxygen	Sand and sandy clay in and just above a shallow confined aquifer at Lula, OK, U.S.	Not detected, <0.03/week	Wilson et al., 1983a

Compound	Electron acceptor	Site	Rate/Observation	Reference
1,1-Dichloroethylene	Carbonate	Sand in a shallow confined aquifer at Lula, OK, U.S., 5.0 m deep	7/week	Aelion et al., 1987
		Sandy clay just above and within a water table aquifer at Fort Polk, LA, U.S.	Not detected, <0.012/week	Wilson et al., 1983a,b
		Sandy clay just above and within a water table aquifer at Pickett, OK, U.S.	Not detected <0.012/week	
		Sand water table aquifer on Borden Air Base, Ontario, Canada	0.03/week in field experiment, locally degradation was rapid and extensive	Roberts et al., 1986
cis-1,2-Dichloroethylene	Carbonate	Muck from Florida Everglades, U.S.	0.03–0.06/week, K_m=4 mg/L	Barrio-Lage et at., 1986
	Carbonate	Muck from Florida Everglades, U.S.	0.01–0.06/week, K_m=5mg/L	
trans-1,2-Dichloroethylene	Carbonate	Muck from Florida Everglades, U.S.	0.03–0.04/week, K_m = 3 mg/L	
Hexachlorothane	Oxygen	Sand water table aquifer on Borden Air Base, Ontario, Canada	0.1/week, field scale	Criddle et al., 1986
Tetrachloroethylene (perchloroethylene)	Oxygen	Sand water table aquifer on Borden Air Base, Ontario, Canada	Not detected in 2 years, <0.0009/week, field experiment	Roberts et al., 1986
		Sandy clay just above the water table at Fort Polk, LA, U.S.	0.009–0.002/week, not greater than abiotic control	Wilson et al., 1983a,b

Table 4.11 (Continued)

Compound	Dominant Electron Acceptor	Type of Sample and Location	Biodegradation Rate or Extent	Reference
Tetrachloroethylene (perchloroethylene) (continued)		Sand just above and within a water table aquifer at Pickett, OK, U.S.	0.01/week, not greater than abiotic control	
		Sandy clay just within a shallow water table at Lula, OK, U.S.	Not detected, <0.03/week	
1,2,4-Trichloro-benzene	Oxygen	Sand in shallow confined aquifer at Lula, OK, U.S.,5.o m deep	No adaption in microcosms after 7 months, <0.01/week	Aelion et al., 1987
1,1,1-Trichloroethane	Oxygen	Sandy clay just above and within a water table aquifer at Fort Polk, LA and Pickett, OK, U.S.	Not detected, <0.01/week	Wilson et al., 1983b
		Sandy clay and sand just above the water table and in a shallow confined aquifer at Lula, OK, U.S.	Not detected, <0.05/week	
Trichloroethylene	Oxygen	Sandy clay just above and within a water table aquifer at Fort Polk, LA and Pickett, OK, U. S.	Not detected, <0.01/week	Wilson et al., 1983a,b
Phenols				
m-Aminophenol	Oxygen	Sand in a shallow confined aquifer at Lula, OK, U.S., 5.0 m deep	0.3–0.5/week	Aelion et al., 1987

Compound	Oxygen	Location	Rate/Notes	Reference
2-Chlorophenol	Oxygen	Shallow water table aquifer near Pickett, OK, U.S.	50–100 µg/l disappeared in effluent of column microcosm in 2 weeks	Suflita and Miller, 1985
p-Chlorophenol	Oxygen	Sand in a shallow confined aquifer at Lula, OK, U.S., 5.0 m deep	3.3/week	Aelion et al., 1987
m-Cresol	Oxygen	Sandy clay, and gravel just above and within a water table at Lula, OK, U.S.	2.4–17/week	Dobbins and Pfaender, 1987
2,4-Dichlorophenol	Oxygen	Sand in a shallow confined aquifer at Lula, OK, U.S., 5.0 m deep	0.5/week	Aelion et al., 1987
p-Nitrophenol	Oxygen	Shallow water table aquifer near Pickett, OK, U.S.	50–100µg/L disappeared in effluent of column microcosm in 2 weeks	Suflita and Miller, 1985
Phenol	Oxygen	Sand in a shallow confined aquifer at Lula, OK, U.S.	<0.005/week; at 0.5 µg/liter, no adaptation in 90 days; <1.6 µg/liter, adaptation in 7–42 days	Aelion et al., 1987
	Oxygen	Shallow water table aquifer near Pickett, OK, U.S.	50–100 µg/liter disappeared in effluent of column microcosm in 2 weeks	Suflita and Miller, 1985
	Oxygen	Sand in a shallow confined aquifer at Lula, OK, U.S., 5.0 m deep	18/week	Aelion et al., 1987

Table 4.11 (Continued)

Compound	Dominant Electron Acceptor	Type of Sample and Location	Biodegradation Rate or Extent	Reference
2,4,6-Trichlorophenol	Oxygen	Shallow water table aquifer near Pickett, OK, U.S.	50–100 µg/liter disappeared in effluent of column microcosm in 2 weeks	Suflita and Miller, 1985
Petroleum-derived hydrocarbons				
Acenaphthene	Oxygen	Sand in shallow water table aquifer in Conroe, TX, U.S.	Not detected, <0.03/week	Wilson et al., 1985a
Benzene	Oxygen	Sand in shallow water table aquifer in Conroe, TX, U.S.	Not detected, <0.03/week	Wilson, J.T., et al., 1986
		Sand in water table aquifer in Borden Air Base, Ontario, Canada	0.1/week in laboratory, 0.2 mg/L-week in field	Barker and Patrick, 1986
		Sand just below the water table in a terrace of the South Canadian River in Norman, OK, U.S.	0.2/week	Mahadevaih and Miller, 1986
Dibenzofuran	Oxygen	Sand in shallow water table aquifer in Conroe, TX, U.S.	Not detected, <0.1/week	Wilson et al., 1985a
Ethylbenzene	Oxygen	Just below the water table in a terrace of the South Canadian River at Norman, OK, U.S.	0.2/week	Mahadevaiah and Miller, 1986

Compound	Electron acceptor	Site	Rate	Reference
Fluorene	Oxygen	Sand in shallow water table aquifer in Conroe, TX, U.S.	Not detected, <0.1/week	Wilson et al., 1985a
2-Methylnaphthalene	Oxygen	Sand in shallow water table aquifer in Conroe, TX, U.S.	Not detected, <0.1/week	
Naphthalene	Oxygen			
Styrene	Oxygen	Sand and sandy clay just above and within the water table aquifer at Pickett, OK, U.S.	0.02–0.04/week	
Styrene	Oxygen	Sandy clay just within a shallow water table aquifer at Fort Polk, LA, U.S.	0.1/week	Wilson et al., 1983a,b
Toluene	Oxygen	Sand just above the water table at Pickett, OK, U.S.	0.009 (abiotic)–0.52/week	
		Sand and sandy clay just above and within the water table aquifer at Lula, OK, U.S.	>2.5/week	
		Gravel in a shallow confined aquifer at Lula, OK, U.S.	Not detected, <0.01/week	Wilson, J.T. et al., 1986
		Sand in a shallow confined aquifer at Lula, OK, U.S., 5.5 m deep	3/week	
		Sand in water table aquifer on Borden Air Base, Ontario, Canada	0.1/week in laboratory, 0.3 mg/L-week in field	Barker and Patrick, 1985
		Sandy clay just above and within a shallow water table aquifer at Fort Polk, LA, U.S.	0.03/week, not greater than abiotic control	Wilson et al., 1983b

Table 4.11 (Continued)

Compound	Dominant Electron Acceptor	Type of Sample and Location	Biodegradation Rate or Extent	Reference
Toluene (continued)		Sand in shallow water table aquifer in Conroe, TX, U.S.	0.04/week, not greater than abiotic control	Wilson, J.T. et al., 1986
		Sand just below the water table in a terrace of the South Canadian River at Norman, OK, U.S.	0.2/week, adaptation occurred in 2–4 weeks	Mahadevaiah and Miller, 1986
o-Xylene	Oxygen	Sand in shallow water table aquifer in Conroe, TX, U.S.	Not detected, <0.03/week	Wilson, J.T. et al., 1986
		Sand just below the water table in a terrace of the South Canadian River at Norman, OK, U.S.	0.2/week	Mahadevaiah and Miller, 1986
		Sand in water table aquifer on Borden Air Base, Ontario, Canada	0.2/week in laboratory, 0.4 mg/L-week in field	Barker and Patrick, 1985
m-Xylene	Oxygen	Sand in shallow water table aquifer in Conroe, TX, U.S.	Not detected, <0.05/week	Wilson, J.T. er al., 1986
		Sand just below the water table in a terrace of the South Canadian River at Norman, OK, U.S.	0.2/week	Mahadevaiah and Miller, 1986
		Sand in water table on Borden Air Base, Ontario, Canada	0.3 mg/L-week in field	Barker and Patrick, 1985
p-Xylene	Oxygen	Sand just below the water table in a terrace of the South Canadian River at Norman, OK, U.S.	0.2/week	Mahadevaiah and Miller, 1986

	Electron acceptor	Location	Rate	Reference
	Oxygen	Sand in water table on Borden Air Base, Ontario, Canada	0.2/week in laboratory, 0.4 mg/liter-week in field	Barker and Patrick, 1985
Miscellaneous				
Aniline	Oxygen	Sand in a shallow confined aquifer at Lula, OK, U.S., 5.0 m deep	0.4–1.3/week	Aelion et al., 1987
Methanol	Oxygen	Sandy unsaturated material in PA, U.S.	40–46 mg/L-week	Novak et al., 1985
	Sulfate	Very shallow water table aquifer, black to gray marl soil in NY, U.S.	Lag for 30 days, then 20 mg/L-week	
	No oxygen, no nitrate little sulfate	Saturated sand and silty clay in VA, U.S.	6–16 mg/L-week	
Nitriloacetic acid	Carbonate	Anaerobic sands from the floodplain of the South Canadian River in Byers, OK, U.S.	4 mg/L-week	Dunlap et al., 1972
Tertiary buytl alcohol	Sulfate	Very shallow water table aquifer, black to gray marl soil in NY, U.S.	<0.02/week	Novak et al., 1985
	Oxygen	Sand unsaturated soil in PA, U.S.	4–12 mg/L-week	
	Oxygen depleted	Sandy unsaturated material in PA, U.S.	4 mg/L-week	

Source: Ghiorse and Wilson, 1988.

Table 4.12 Biodegradation of Various Organic Compounds in Subsurface Material from Contaminated Sites

Compound	Dominant Electron Acceptor	Type of Sample and Location	Biodegradation Rate or Extent	Reference
Common Metabolites				
Acetate	Oxygen	Unsaturated zone below septic tile field, Sault Ste. Marie, Ontario, Canada	14/week	Ward, 1985
Acetate	Nitrate	Shallow water table aquifer near domestic septic tile field and garden soil, Williamsburg, Ontario, Canada	8/week 84/week	
Benzoic acid (ring and carboxyl)	Oxygen	Unsaturated zone below septic tile field, Sault Ste. Marie, Ontario, Canada	6–30/week	Ward, 1985
Glucose	Nitrate	Shallow water table aquifer near domestic septic tile field and garden soil, Williamsburg, Ontario, Canada	4.5/week 94/week	
Glucose	Oxygen	Garden soil, Williamsburg, Ontario, Canada Unsaturated zone below septic tile field, Sault Ste. Marie, Ontario, Canada	25/week 8.6/week	Ward, 1985

Compound	Electron acceptor	Material and location	Rate	Reference
Glutamic acid	Oxygen	Shallow water table aquifer near domestic septic tile field and garden soil, Williamsburg, Ontario, Canada	6–7/week	Kiene and Capone, 1986
			47/week	
		Unsaturated zone below septic tile field, Sault Ste. Marie, Ontario, Canada	16/week	
Glutamic acid	Nitrate	Garden soil, Williamsburg, Ontario, Canada	69/week	
		Unsaturated zone below septic tile field, Sault Ste. Marie, Ontario, Canada	69/week	
Stearic acid	Oxygen	Shallow water table aquifer near domestic septic tile field and garden soil, Williamsburg, Ontario, Canada	0.34/week	
Stearic acid	Nitrate	Unsaturated zone below septic tile field, Sault Ste. Marie, Ontario, Canada	1.3/week	
Volatile fatty acids (acetic, propionic, butyric)	Oxygen	Chalk from 10 m below the water table, Ingram, Suffolk, England	1.0/week, no evidence of adaptation after 30 days	

Table 4.12 (Continued)

Compound	Dominant Electron Acceptor	Type of Sample and Location	Biodegradation Rate or Extent	Reference
Halogenated hydrocarbons				
Bromoform, chloroform, chlorodibromomethane	Carbonate	Confined aquifer injected with treated municipal wastewater, Palo Alto Baylands, CA, U.S.	0.2/week	Roberts et al., 1982
1,2-Dibromoethane (EDB)	Carbonate	Alluvium contaminated by landfill leachate, Norman, OK, U.S.	99% removed in 16 weeks	Wilson, B.H. et al., 1986
1,2-Dichlorobenzene	Oxygen	Field site on banks of Glatt River in Switzerland	1/week, exposed to 0.3 µg/L; 270/week, after acclimation to 30 µg/L	Kuhn et al., 1985
Dichlorobromomethane	Carbonate	Confined aquifer injected with treated municipal wastewater, Palo Alto Baylands, CA, U.S.	0.2/week	Roberts et al., 1982
1,1,1-Dichloroethane	Carbonate	Alluvium contaminated by landfill leachate, Norman, OK, U.S.	77–99% removed in 16–40 weeks	Wilson, B.H. et al., 1986
trans-1,2-Dichloroethane	Carbonate	Alluvium contaminated by landfill leachate, Norman, OK, U.S.	87% removed in 40 weeks	
2,4-Dichlorophenoxyacetic acid	Carbonate	Alluvium contaminated by landfill leachate, Norman, OK, U.S.	99% removed in 3 months	Gibson and Suflita, 1986

Compound	Electron acceptor	Site	Result	Reference
Tetrachloroethylene	Oxygen	Field site on banks of Glatt River in Switzerland	Not detected, <0.01/week	Kuhn et al., 1985
Tetrachloroethylene	Carbonate	Confined aquifer injected with treated municipal wastewater, Palo Alto Baylands, CA, U.S.	0.2/week	Roberts et al., 1982
1,1,1-Trichloroethande	Carbonate	Sand influenced by leachate from North Bay municipal landfill, Ontario, Canada	Field data showed disappearance	Barker et al., 1986
Trichloroethylene	Carbonate	Alluvium contaminated by landfill leachate, Norman, OK, U.S.	66–99% removed in 40 weeks; 90% removed in 8 weeks	Wilson, B.H. et al., 1986
Phenols				
2-Chlorophenol, 3-chlorophenol, 4-chlorophenol,	Sulfate	Alluvium contaminated by landfill leachate, Norman, OK, U.S.	100% removed in 3 months	Gibson and Suflita, 1986
2-Chlorophenol,	Carbonate	Alluvium contaminated by landfill leachate, Norman, OK, U.S.	Not detected, <0.02/week	Gibson and Suflita, 1986
o-Cresol	Sulfate or carbonate	Alluvium contaminated by landfill leachate, Norman, OK, U.S.	No acclimation within 90–100 d	Smolenski and Suflita, 1987
m-Cresol	Sulfate or carbonate	Alluvium contaminated by landfill leachate, Norman, OK, U.S.	43–90 d lag time for acclimation	
p-Cresol	Sulfate	River alluvium contaminated by landfill leachate, Norman, OK, U.S.	<10–46 d lag time; 180–3300 mg/L-week after acclimation, 3 mg sulfate required per mg p-Cresol consumed	

Table 4.12 (Continued)

Compound	Dominant Electron Acceptor	Type of Sample and Location	Biodegradation Rate or Extent	Reference
o-, m-, p-Cresol	Carbonate	Sand contaminated with creosote waste	Extensive degradation	Goerlitz et al., 1985
2,4-Dichlorophenol,	Sulfate	Alluvium contaminated by landfill leachate, Norman, OK, U.S.	Not detected, <0.04/week	Gibson and Suflita, 1986
2,5-Dichlorophenol,	Carbonate	Alluvium contaminated by landfill leachate, Norman, OK, U.S.	83% removed in 3 months	
3,4-Dichlorophenol,	Sulfate	Alluvium contaminated by landfill leachate, Norman, OK, U.S.	Not detected, <0.01/week	
Phenol	Sulfate or carbonate	Alluvium contaminated by landfill leachate, Norman, OK, U.S.	99–100% removed in 3 months	
Phenol	Carbonate	Water table aquifer contaminated with creosote, St. Louis Park, MN, U.S.	Extensive degradation to methane	Ehrlich et al., 1982
		Water table aquifer contaminated with creosote, Pensacola, FL, U.S.	>0.1 mg/L pentachlorophenol exhibited degradation	Godsy et al., 1983
2,4,5-Trichlorophenol	Sulfate	River alluvium contaminated by landfill leachate, Norman, OK, U.S.	Not detected, <0.04/week	Gibson and Suflita, 1986
Petroleum-derived hydrocarbons				
Acenaphthene	Oxygen	Shallow water table aquifer on Texas coast	>1.3/week	Wilson et al., 1985

Compound	Electron Acceptor	Site	Observation	Reference
Benzene	Oxygen	Sand 5 m below the water table in a shallow unconfined aquifer contaminated by a gasoline spill	2.3–2.7/week, degradation slowed at 4–5 µg/liter	Wilson, B. H. et al., 1987
Benzene	Carbonate	Sand 3 m below the water table in a shallow unconfined aquifer contaminated by a gasoline spill	0.9/week, degradation slowed at 12 µg/liter	
Benzene, ethylbenzene	Carbonate	Alluvium contaminated by landfill leachate, Norman, OK, U.S.	70–74% removed in 40 weeks, 99% in 120 weeks	
Dibenzofuran	Oxygen	Shallow water table aquifer on Texas coast	>1.8/week	Wilson et al., 1985a
Fluorene	Oxygen	Shallow water table aquifer on Texas coast	>0.9/week	
1-Methylnaphthene, 2-methylnaphthene naphthalene	Oxygen	Shallow water table aquifer on Texas coast	>1.6/week	
Naphthalene	Carbonate	Water table aquifer contaminated with creosote, St. Louis Park, MN, U.S.	No degradation detected	Ehrlich et al., 1982
Styrene	Carbonate	Alluvium contaminated by landfill leachate, Norman, OK, U.S.	90% removed in 8 weeks, 99% in 16 weeks	Wilson, B.H. et al., 1986
Toluene	Oxygen	Sand in shallow water table aquifer exposed to wood creosoting waste	0.2/week	Wilson, J.T. et al., 1986
		Sand 3–5 m below the water table in shallow unconfined aquifer contaminated by a gasoline spill	Degradation ceased at 3–5 µg-liter	
		Sand 3 m below the water table in plume of contamination	0.5/week, slowed at 56 µg/L	

Table 4.12 (Continued)

Compound	Dominant Electron Acceptor	Type of Sample and Location	Biodegradation Rate or Extent	Reference
1,2,4-Trimethyl-benzene	Carbonate	Alluvium contaminated by landfill leachate, Norman, OK, U.S.	99% removed in 40 weeks	Barker et al., 1986
		Sand influenced by leachate from North Bay municipal landfill, Ontario, Canada	No degradation	
o-Xylene	Oxygen or carbonate	Sand 3–5 m below the water table in shallow unconfirmed aquifer contaminated by a gasoline spill	2.6–4.8/week, degradation ceased at 1–3 µg/L	Wilson, B.H. et al., 1987
		Sand 3 m below the water table in plume of contamination	0.6/week	
o-Xylene	Nitrate	Field site on banks of Glatt River in Switzerland	160/week, after acclimation to 400 µg/L	Kuhn et al., 1985
o-Xylene	Carbonate	Sand influenced by leachate from North Bay municipal landfill, Ontario, Canada	0.01/week, estimated from field data	Barker et al., 1986
		Alluvium contaminated by landfill leachate, Norman, OK, U.S.	78% removed in 40 weeks, 99% removed in 120 weeks	Wilson, B.H. et al., 1986
m-Xylene	Oxygen	Field site on banks of Glatt River in Switzerland	>16/week, exposed to 0.2 µg/L	Kuhn et al., 1985

Compound	Electron acceptor	Location	Rate	Reference
p-Xylene	Nitrate	Sediments from banks of Glatt River in Switzerland	65/week, after acclimation to 400 µg/L	Wilson, B.H. et al., 1987
p-Xylene	Oxygen	Field site on banks of Glatt River in Switzerland	160/week, after acclimation to 400 µg/L	
m- and p-Xylene	Oxygen	Sand 3–5 m below the water table in shallow unconfined aquifer contaminated by a gasoline spill	3.0–4.8/week, degradation slowed at 2–3 µg/L	
	Carbonate	Sand 3 m below the water table in plume of contamination	0.4/week	
Miscellaneous				
Nirtilotriacetic acid	Oxygen or nitrate	Unsaturated zone below septic tile field, Sault Ste. Marie, Ontario, Canada	0.7–0.8/week	Ward, 1985
		Shallow water table aquifer near domestic septic tile field, Williamsburg, Ontario, Canada	1.3–2.3/week	
2,4-Dichlorophenoxy-acetic acid	Sulfate or carbonate	River alluvium contaminated by landfill leachate, Norman, OK, U.S.	<0.01/week, 99% removed in 3 months	Gibson and Suflita, 1986
Quinoline and isoquinoline	Carbonate	Sand contaminated with creosote wastes, Pensacola, FL, U.S.	Extensive degradation	Pereira et al., 1987

Source: Ghiorse and Wilson, 1988.

In Tables 4.11 and 4.12, the compounds that have been evaluated for aerobic subsurface biodegradation (as denoted by dominant electron acceptor of oxygen) are divided into the following groups: common metabolites (e.g., acetate, glucose), halogenated hydrocarbons, phenols, petroleum derived hydrocarbons, and miscellaneous (e.g., aniline, methanol). These tables thus provide valuable insights into the potential for biodegradation of these compounds under various conditions, based on currently available observations. The rate or extent of biodegradation is indicated in these tables with the rate constant expressed as a first order constant (t^{-1}). Larger values of the first order rate constant correspond to higher rates of biodegradation. From a relative standpoint, it is seen from Table 4.11 that simple organics (such as glucose) are more rapidly degraded under aerobic conditions (K = 36 to 45/week) than are the more complex organics (e.g., styrene, K = 0.1/week), as would be expected. It should be mentioned that comparative rates of biodegradation of compounds cannot be determined on an absolute basis from these tables due to potential variables between the studies (e.g., variations in biomass, nutrients, temperature, pH, and other environmental factors).

To get a relative feel for the amount of time necessary for these compounds to biodegrade, it would be useful to know the half-lives of these compounds. First order rate constants can be converted into half-lives (the time necessary for the compound to decrease in half due to a given reaction) as shown in Equation 4.5, where $t_{1/2}$ is the half-life (same time units as K) and K is the

$$t_{1/2} = 0.69 \, / \, K \tag{4.5}$$

first order rate constant. Referring back to glucose and styrene, the corresponding half-lives of glucose and styrene (based on the conditions under which the first order constants were determined) would be 0.025 and 10 weeks, respectively. Thus, the glucose concentration would decrease in half in approximately 4 hours while styrene would require 10 weeks.

It can thus be seen that a wide variety of contaminants are susceptible to aerobic biodegradation. The rates of degradation are, logically, a function of the complexity of the contaminant and the environmental conditions. In general, if a compound is susceptible to both aerobic and anaerobic degradation, the aerobic process will be more rapid. However, from a enhanced subsurface remediation perspective, this process may be limited by the ability to deliver oxygen to the subsurface (due to the low solubility of oxygen in water). For this reason, the use of electron acceptors which are more easily delivered (such as nitrate which is much more soluble than oxygen in water) may actually be more advantageous (Lee et al., 1988; Sims et al., 1990). This approach has received increased attention recently (e.g., Hutchins et al., 1989). Subsurface anoxic/anaerobic metabolism of contaminants will be discussed in more detail subsequently.

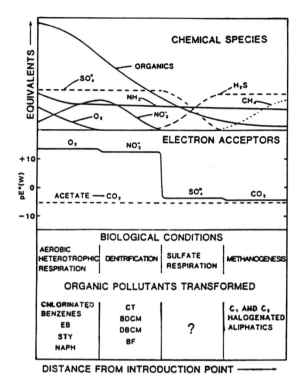

Figure 4.4. Possible biodegradation processes in the subsurface as a function of time (distance) and redox potential (Bouwer and McCarty, 1984) (EB, ethylbenzene; STY, styrene; NAPH, napthalene; CT, carbon tetrachloride; BDCM, bromodichloromethane; DBCM, dibromochloro-methane; BF, bromoform).

4.4.3 Anoxic/Anaerobic Metabolism

Under many natural conditions, especially in response to contamination, the molecular oxygen available in an aquifer is depleted. Under these conditions, an alternative electron acceptor is required. Figure 4.4 demonstrates how the electron acceptor can change in response to depletion of preferred electron acceptors and the types of pollutants susceptible to biodegradation under these conditions. Thus, for the release of a waste containing a variety of compounds degradable under a variety of redox potentials (electron acceptors), the scenario depicted in Figure 4.4 would be advantageous to the degradation of these wastes.

In Tables 4.7, 4.11, and 4.12 research on the susceptibility and rates of biodegradation of a wide variety of organics under anoxic/anaerobic conditions are summarized (dominant electron acceptors other than free molecular oxygen). Several excellent reviews discuss recent findings on subsurface anoxic/anaerobic metabolic processes (Ghiorse and Wilson, 1988; Suflita, 1989a; Suflita and Sewell, 1991; and Sims et al., 1991). Summaries of these findings are presented in this section, the interested reader is referred to these documents for additional detail.

Suflita (1989a) indicated that under anaerobic conditions microorganisms may enter into very tightly linked metabolic consortia comprised of a variety of microorganisms. Different members of the consortium are responsible for specific reactions which cumulatively may result in the complete mineralization of the compound. Without the individual members of the consortium, degradation of the compound may be inhibited, giving the appearance that the compound is recalcitrant. Suflita and Sewell (1991) indicate that the evolution of microbial consortia has several advantages including: microenvironments are created that allow survival of organisms in what would otherwise be a hostile environment, metabolically linking reactions may allow otherwise thermodynamically unfavorable reactions to occur, the removal of toxic or inhibitory compounds may be accomplished by members of the consortium, and the diverse metabolic capabilities of the members of the consortium may allow more rapid processing of inputs than would be possible by any individual microorganism. Suflita and Sewell (1991) discussed anaerobic biodegradation of several classes of organic compounds including the following: heterocycles; anilines, benzenesulfonamides and benzamides; aromatics; and halogenated aliphatics and aromatics. Anaerobic degradation of each of these classes of organics will be discussed in subsequent paragraphs.

Two thirds of the approximately 4 million known organic compounds have a chemical structure made up of carbon, and either oxygen, nitrogen or sulfur atoms arranged in a ring, known as heterocycles (Suflita and Sewell, 1991). These compounds are widely used inputs for the synthesis of pharmaceuticals, pesticides, explosives, dyes, and food additives; they are also found in fossil fuel deposits and serve as important components of metabolic processes (e.g., vitamins, nucleic acids, proteins, and carbohydrates). Subsurface heterocyclic contaminants are typically derivatives of pyridines, furans, and thiophenes (dioxane is an example of a saturated heterocycle). Suflita and Sewell (1991) presented data indicating that most of the heterocylces have been shown to be biodegradable under methanogenic or sulfate reducing conditions. It is observed from Table 4.13 that substitution of the heterocycle with a carboxyl group increased the biodegradability of the O, N, and S heterocylces and that the oxygen and nitrogen heterocycles were more susceptible than sulfur heterocycles to anaerobic degradation.

Suflita and Sewell (1991) reported that substituted anilines, benzene-sulfonamides and benzamides appear commonly as subsurface contaminants,

Table 4.13 Biological Degradation of Heterocyclic Compounds in Methanogenic (Anaerobic) Conditions

Compound	Chemical Structure	R	Biotrans-formation[a]
Pyridine		H	+
4-Picoline		CH_3	+
3-Picoline	R (ring with N)	CH_3	−
2-Picoline		CH_3	−
Nicotinic acid		COOH	+
Furan		H	+
2-Methylfuran	R (ring with O)	CH_3	−
2-Furoic acid		COOH	+
Thiophene		H	−
2-Methylthiophene		CH_3	−
3-Methylthiophene	R (ring with S)	CH_3	−
2-Thiophene carboxylic acid		COOH	+

Source: Suflita and Sewell, 1991.

[a] Loss of test compound and production of methane relative to abiotic controls during 8 month anaerobic incubation in aquifer derived methanogenic microcosms.

with the methylated or higher alkylated derivatives of these compounds being of particular importance. The amino substituted benzenes have been shown to be biodegradable when the aromatic nucleus is substituted with a carboxyl group; however, aniline and methylated anilines (toluidines) were observed to be recalcitrant under methanogenic conditions (Table 4.14). Benzamides were biodegraded under sulfate-reducing and methanogenic conditions with methyl group substituents in certain locations not hindering this biodegradation; however, the presence of multiple methyl groups or complicated alkylation patterns was observed to hinder the biodegradation (Table 4.15). The aryl sulfonates were reported to be resistant to biodegradation under sulfate reducing and methanogenic conditions, with the exception of carboxyl substituted compounds, as observed in Table 4.16.

While early research indicated that aromatic compounds were not degradable under anaerobic conditions, more recent research has suggested several anaerobic systems which promote the anaerobic degradation of aromatic compounds. Suflita and Sewell (1991) summarized a number of these studies (Table 4.17). Aromatic compounds listed as undergoing anaerobic degradation include alkylbenzenes and oxygen substituted aromatics. Anaerobic conditions promoting the biodegradation of aromatic compounds include

Table 4.14 Biological Degradation of Amino-Substituted Benzenes under Anaerobic Conditions

R_1—⬡—NH_3

Compound	R group	Biotrasnformation[a] M[b]	SR[c]
Aniline	H	–	+
o-Toluidine	CH_3	–	–
m-Toluidine	CH_3	–	+
p-Toluidine	CH_3	–	–
o-Aminobenzoate	COOH	+	+
m-Aminobenzoate	COOH	+	+
p-Aminobenzoate	COOH	+	+

Source: Suflita and Sewell, 1991.

[a] Loss of target compound relative to abiotic controls during 10 month anaerobic incubation.

[b] Methanogenic aquifer microcosms.

[c] Sulfate-reducing aquifer microcosms.

Table 4.15 Biological Degradation of Alkylated Benzamides under Anaerobic Conditions

R_1—⬡—$\overset{\overset{O}{\|}}{C}$—$N\overset{R_2}{\underset{R_3}{}}$

Compound	Substituents R_1	R_2	R_3	Biotransformation[a] M[b]	SR[c]
Benzamide	H	H	H	±	+
N-Methyl-benzamide	H	H	CH_3	±	+
N,N-Dimethyl-benzamide	H	CH_3	CH_3	–	–
p-Toluamide	CH_3	H	H	+	+
N,N-Diethyl-m-toluamide	CH_3	C_2H_5	C_2H_5	–	–

Source: Suflita and Sewell, 1991.

[a] Loss of target compound relative to abiotic controls during 11 month anaerobic incubation.

[b] Methanogenic aquifer microcosms.

[c] Sulfate-reducing aquifer microcosms.

Table 4.16 Biological Degradation of Benzenesulfonic Acids and Benzenesulfonamides under Anaerobic Conditions

Compound	Substituents R_1	R_2	Biotransformation[a] M^b	SR^c
Benzenesulfonic acid	H	OH	–	–
Orthanilic acid	NH_2 *(o)*	OH	–	–
Metanilic acid	NH_2 *(m)*	OH	–	–
Sulfanilic acid	NH_2 *(p)*	OH	–	–
p-Toluenesulfonic acid	CH_3	OH	–	–
p-Phenosulfonic acid	OH	OH	–	–
p-Benzosulfonic acid	COOH	OH	+	+
Benzene-sulfonamide	H	NH_2	±	–
p-Toluene-sulfonamide	CH_3	NH_2	–	–
N-Methylbenzene-sulfonamide	H	NHCH3	–	–
N,N-Diethyl- *p*-toluene-sulfonamide	CH_3	$N(C_2H_5)_2$	–	–
N-Ethyl-p/m-toluene-sulfonamide	CH_3	$N(C_2H_5)_2$	–	–

Source: Suflita and Sewell, 1991.

[a] Loss of target compound relative to abiotic controls during anaerobic incubation (13 months for benzenesulfonic acids and 10 months for benzenesulfonamides).

[b] Methanogenic aquifer microcosms.

[c] Sulfate-reducing aquifer microcosms.

methanogenic, sulfate reducing, nitrate reducing, and iron reducing conditions. Thus, it can be seen that it is difficult to totally eliminate the possibility that a compound will biodegrade based on limited research.

Halogenated compounds (e.g., chlorinated solvents) have proven to be widespread ground water contaminants. While resistant to biodegradation under aerobic conditions, reductive dehalogenation has been evidenced as a mechanism for degradation under anaerobic conditions (Sims et al., 1991). In the process of reductive dehalogenation, the halogenated compound acts as an electron acceptor, coupled with the oxidation of an electron donor, with the biomass gaining energy in the process. The compounds resulting from the reduction are generally more susceptible to subsequent transformations (Suflita and Sewell, 1991). Table 4.18 lists several halogenated aliphatic and aromatic compounds which have proven to be susceptible to anaerobic degradation. For a more detailed discussion of the reductive dehalogenation process, the reader is directed to Sims et al. (1991).

Table 4.17 Reported Biological Degradation of Aromatic Compounds under Anaerobic Conditions by Subsurface Mircoorganisms or with Pure Cultures

Compound	Inoculum Source[a]	Incubation Conditions[b]	Reference
Toluene	AqM	M	Wilson et al., 1986
Toluene, xylenes	AqM	NR	Kuhn et al., 1988
Toluene	PC	NR, IR	Lovely and Lonegran, 1990
Toluene, xylenes	AqM	NR	Hutchins et al., 1990
Cresols	AqM	M, SR	Smolenski and Suflita, 1987
Phenol, benzoate, hydroxybenzoate	AqM	M, SR, NR	Kuhn et al., 1989
Phenol, cresol, benzoate, hydroxybenzoate	PC	IR, NR	Lovely and Lonegran, 1990
Phenoxy-acetate	AqM	M	Gibson and Suflita, 1986
Methoxy-benzoate	PC	An	DeWeerd et al., 1988
Hydroxy-biphenyl	AqM	SR	Suflita et al., 1990

Source: Suflita and Sewell, 1991.

[a] Source of microorganisms in laboratory transformation experiments: AqM, aquifer material, PC, model pure culture.

[b] Incubation conditions: M, methanogenic; SR, sulfate-reducting; NR, nitrate-reducing; IR, iron-reducing; An, anaerobic (fermentation).

4.4.4 Field Scale Metabolic Processes

This section has addressed the question of what chemicals may biodegrade in the subsurface and under what conditions. Results from a multitude of researchers have been summarized to document the biodegradation of compounds in the presence of aquifer materials. The data presented has indicated that some compounds are more susceptible to aerobic degradation while others are more susceptible to anaerobic degradation. The nature and complexity of the compound (including the location and type of substituents on the

**Table 4.18 Reported Biological Degradation of Haloorganic
Compounds under Anaerobic Conditions by
Subsurface Mircoorganisms or with Pure Cultures**

Compound	Inoculum Source[a]	Incubation Conditions[b]	Reference
Cl_{1-2}-Phenols, Cl_{1-3}-Benzoates, Cl_{2-3}-Phenoxyacetates	AqM	M	Suflita et al., 1988
Cl_{2-4}-Anilines	AqM	M, SR	Kuhn et al., 1990
Bromacil	AqM	M	Adrian and Suflita, 1990
Tetrachloroethene, Trichloroethene (PCE, TCE)	AqM, PC	M, SR	Suflita et al., 1988
PCE, TCE	PC	M	Fatherpure et al., 1987
TCE cis-1,2-Dichloroethene, trans-1,2-Dichloroethene, 1,1-Dichloroethene, 1,2-Dibromoethane,	AqM	M	Wilson et al., 1986
Tetrachloromethane, 1,1,1-Trichloroethane (CT, TCA)	AqM	An	Parsons et al., 1985
Trichloroethane (CF), TCA, CT	PC	An	Gälli and McCarty, 1989
1,2-Dichloroethene, CT	PC	M, SR	Egli et al., 1987
Bromoform, CF, CT	PC	M	Mikesell and Boyd, 1990
Vinyl Chloride	AqM	An	Barrio-Lage et al., 1990

Source: Suflita and Sewell, 1991.

[a] Source of microorganisms in laboratory transformation experiments: AqM, aquifer material, PC, model pure culture.

[b] Incubation conditions: M, methanogenic; SR, sulfate-reducing; IR, iron-reducing; An, anaerobic (undefined).

molecule) has been observed to be important in determining the susceptibility of the compound to degradation. Environmental factors (inadequate nutrients, unacceptable pH, etc.) may also limit the biodegradation.

It is encouraging to see the multitude of compounds that have been documented as undergoing biodegradation in the presence of aquifer materials.

However, it should be emphasized that much of the research reported has been conducted in controlled laboratory settings. Field systems are extremely complicated, and much remains to be learned about *in situ* subsurface biodegradation of compounds. Care must be taken in extrapolating the results summarized in this section to conditions other than those in which the data was collected. Variations in environmental conditions can alter the types and rates of biodegradation processes observed. It should be realized that abiotic processes may affect the biodegradation of the compound (adsorption/desorption, hydrolysis, etc.).

Care must also be exercised in analyzing data from field observations. In certain instances, chemical loss may be due to abiotic transformations rather than biotic transformations. These processes may even produce similar intermediates, further complicating the determination of the controlling process. Madsen (1991) discussed the difficulties in assessing *in situ* biodegradation processes from laboratory and field studies and proposed a strategy for determining the nature and extent of *in situ* biodegradation processes (as shown in Table 4.19).The first step in the strategy involves preliminary site characterization. The second step identifies the key chemical and biological criteria for the site that can aid in determining the occurrence and extent of *in situ* biodegradation. The third step refines the site characterization based on Step 2. In the fourth step, the subsurface system is analyzed to distinguish biotic from abiotic processes. Finally, Step 5 seeks confirmatory microbiological evidence to support the occurrence of *in situ* biodegradation. The significance of each step and examples illustrating the importance of and usefullness/reliability of each step (including some references) are included in Table 4.19.

The strategy of Madsen (1991) will enhance our ability to definitively state that *in situ* biodegradation is responsible for contaminant transformation. This is important as we evaluate the ability of the subsurface to naturally cleanse itself or as we consider altering the system to speed up the *in situ* bioremediation. Attractive aspects of this approach are the possibility of rendering the contaminant innocuous (especially if mineralization is realized) without requiring removal of the contaminant and risking increased exposure. However, a thorough understanding of the *in situ* biodegradation process is necessary before natural or enhanced biodegradation can be anticipated/ designed with confidence.

In summary, microorganisms associated with subsurface materials have demonstrated the ability to biodegrade a wide variety of compounds under numerous environmental conditions. Our understanding of metabolic processes for organic contaminants has seen significant advances in the past decade. However, the complexity of field systems make interpretation and prediction of *in situ* biodegradation a formidable challenge. In addition, the presence of mixed wastes (organics and inorganics) further complicates this system. These challenges must be addressed in future research.

Table 4.19 A Strategy for Assessing In Situ Biodegradation Processes

Step	Objective/Approaches	Explanation/Examples	Usefulness/Reliability
1. Conduct preliminary site investigation	Characterize field system to assess the following: Spatial, physical, and hydrologic constraints for successfully understanding the site	A homogeneous geologic setting may be amenable to a sampling scheme that provides interpretable data demonstrating clear concentration gradients and hydrologic flow paths; however, a hydrologically and geologically complex setting may prevent interpretable measurements from being obtained	Site characterization methods are well proven; geostatistics may be used
	Type and degree of contamination that has occurred	The identity of chemicals present and their concentrations should pose a a limited set of biodegradation possibilities, based on established physiological and biochemical knowledge	Biodegradation principles and techniques of analytical chemistry are well proven
	Predominant physiological regimes that are already established by the indigenous microorganisms	If oxygen is already depleted, then attention to anaerobic processes may be warranted	Physiological principles for surmising on-site microbial processes are well established

Table 4.19 (Continued)

Step	Objective/Approaches	Explanation/Examples	Usefulness/Reliability
1. Conduct preliminary site investigatoion (continued)		If the site is marine, then sulfate reduction may be important, and sulfides should be abundant If the site is at high altitude or latitude, then cold temperatures may severely impair microbial activity	
2. Develop *in situ* biodegradation assessment strategy	Based on existing biochemical knowledge, identify key chemical and biological criteria that matchsite characteristics with biodegradation processes	PCBs can be reductively dechlorinated (Quensen et al., 1988) Many petroleum-derived hydrocarbons are readily metabolized aerobically (Ward et al., 1980; Atlas, 1981; and Cerniglia, 1984) Toxic chemicals in high concentrations may preclude biodegradation entirely or restrict it to zones where the contaminant(s) are diluted	Genetic, biochemical, and physiological principles, as they relate to biodegradation, are well established
3. Conduct detailed site characterization	Refine the site characterization sampling plan to provide optimally interpretable chemical and microbiological measurements	Site characterization and/or statistical designs must provide the framework for sampling; analytical and sampling procedures should allow trends indicative of on-site processes to be discerned	Highly feasible but not well proven

4. Seek proof of *in situ* biodegradation	Distinguish biotic from abiotic processes by seeking chemical and biological criteria that are uniquely indicative of microbial metabolic processes: Production of metabolic intermediate compounds, especially as they contrast with contaminants originally released at the site		
		Trans-dichloro-ethylene oxide (Semprini et al., 1990)	This metabolic intermediate may be an unequivocal indicator of trichloroethylene metabolism
		N_2O during denitrification (Smith et al., 1991)	This is a useful indicator under anaerobic conditions; however, denitrification may or may not be tied to contaminant loss Furthermore, trace amounts of N_2O may be produced during nitrification (Robertson and Tiedje, 1987); thus, caution should be exercised
		CH_4 during methanogenesis (Beeman and Suflita, 1990)	Reliable if other methane sources are absent; however, methanogenesis may or may not be tied to containment loss

Table 4.19 (Continued)

Step	Objective/Approaches	Explanation/Examples	Usefulness/Reliability
4. Seek proof of *in situ* biodegradation (continued)	Selective disappearance of microbiologically labile isomers especially as they contrast with conserved nonbiodegradable tracers fortuitously present on site	C_{17}/pristane and C_{18}/phytane ratios (Pritchard and Costa, 1991; Ward et al., 1980; and Atlas, 1981)	Not always reliable because pristane and phytane are biodegradable (Westlake et al., 1974; Cooney et al., 1985)
		Ratio of biodegradable petroleum components to hopanes; using source crude oil as an initial standard	Proposed for Exxon *Valdez* oil spill (Prince et al., 1990; Faller et al., 1991), but limitations must be acknowledged
	Selective metabolism of contaminant stereoisomers	The enantiomeric ratio of a chiral organic pollutant, α-hexachlorocyclohexane, changes during microbial metabolism (Chapelle et al., 1988)	Promising; likely to be highly effective if contaminant-specific biochemistry is thoroughly documented
	Stable isotope fractionation patterns	Some metabolic processes such as methanogenesis are selective between light and heavy isotopes (Grossman et al., 1989; Trudell et al., 1986)	Not yet reliable for biodegradation; useful for geochemical reactions

Loss of coreactants that participate in the biodegradation process	O_2 depletion NO_3^- depletion NO_3^- Br$^-$ ratio (Lee et al., 1985)	Confirmatory only Confirmatory only This decreasing ratio has been used to indicate *in situ* denitrification, which may or may not be tied to contaminant loss
Metabolic adaptation	Metabolic adaptation may be indicated by using biodegradation flask assays to compare samples from contaminated and uncontaminated zones	Reliable if used in conjunction with enhanced numbers of protozoa or other metabolically stimulated members of the microbial community (Madsen et al., 1991)
Manipulate the field site to elicit unique biotic responses	Add nutrients or other key metabolic stimulants to a portion of the site or as a pulse in time so a relative response can be detected (Lamar and Dietrich, 1990; Westlake et al., 1978; Raymond et al., 1976; Spain et al., 1984; and Semprini et al., 1990)	Reliable

Table 4.19 (Continued)

Step	Objective/Approaches	Explanation/Examples	Usefulness/Reliability
4. Seek proof of *in situ* biodegradation (continued)		Release of isotopically labeled (stable or radioactive) contaminant at the site, followed by the monitoring of metabolite production (Cheng and Lehmann,1985; Führ, 1985; and Ridgway et al., 1990)	Reliable if labeled metabolites are unique to metabolic pathways
5. Gather microbiological evidence that confirms in situ biodegradation	Although many microbiological assays are individually of limited relevance to *in situ* processes, convergent lines of indirect evidence lend strength to arguments addressing *in situ* biodegradation	Perform laboratory tests demonstrating that the biodegradation occurs in biotic, but not abiotic, samples from the site (Alexander, 1981; Leahy and Colwell, 1990; and Metcalf, 1977) Isolate and enumerate microorganisms from the site which catalyze biodegradation (Sayler and Layton, 1990) Use DNA or RNA gene probes to assess distribution of potential catabolic expression (Olson, 1991; Wong and Crosby, 1978)	Limited relevance to *in situ* biodegradation, but supportive auxiliary physiological and microbiological data can be very useful

Source: Madsen, 1991.

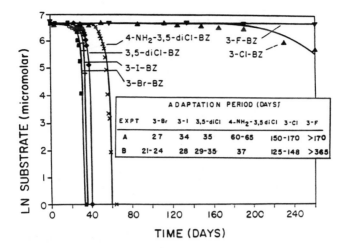

Figure 4.5. **Acclimation period for anaerobic degradation of halobenzoates with Initial concentration of 800 μm (Suflita, 1989b).**

4.5 RATES OF SUBSURFACE METABOLIC PROCESSES

Rates of subsurface metabolic processes can be characterized by the lag period before the bioreaction begins and the rate of the bioreaction. The extent of the lag period is a function of several factors, as discussed below. The reaction rate can be described by a myriad of reaction rate models; the discussion below will consider zero order and first order reactions, and will also discuss the concepts of a biofilm model. Certainly other reaction models exist, but they are beyond the scope of this discussion.

Two majors factors controlling the length of the lag period are the chemical structure of the compound and the concentration of the compound. The acclimation period for biodegradation of a series of halobenzoates under methanogenic conditions are illustrated in Figures 4.5 and 4.6. The compounds were degraded via reductive dehalogenation in a specific order based on the chemical structure of the compound (Figure 4.5). For example, 3-bromobenzoate degraded faster than 3-iodobenzoate which degraded faster than 3,5-dichlorobenzoate, etc. Thus, the acclimation period for this series of compounds was observed to be affected by the chemical structure (the insert to Figure 4.5 indicates that the results were reproducible). Figure 4.6 demonstrates the impact of concentration variations on the degradation of 3,5-dichlorobenzoate and 3-chlorobenzoate. It is observed that the acclimation period is relatively constant for the dichlorobenzoate but increases at higher concentrations for chlorobenzoate. In contrast, lower concentrations of 4-

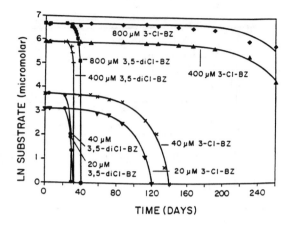

Figure 4.6. **Acclimation period for chlorobenzoates as a function of concentration (Suflita, 1989b).**

amino-3,5-dichlorobenzoate were observed to result in increased acclimation periods — in excess of one year (Suflita, 1989b; Linkfield et al., 1989). Thus, variations in chemical structure and concentration (too high or two low) can result in increasing acclimation periods.

A reaction is zero order with respect to a chemical (substrate) when the rate of the reaction is independent of the concentration of the chemical. Thus, the chemical is not limiting the rate of the reaction and the reaction rate does not decrease as the concentration of the chemical further decreases. Equation 4.4 is an example of a zero order rate expression. A reaction is first order with respect to a chemical (substrate) when the rate of the chemical reaction proceeds in direct proportion to the chemical concentration (to the first power). In this case, as the concentration of the chemical decreases, the rate of the reaction decreases (or increases, in the case of production of the chemical), resulting in the exponential decay of the chemical. Equation 4.3 is an example of a first order rate expression. Frequently, biodegradation is described by a first order rate expression when the compound of interest is limiting. However, it is possible that a zero order rate expression will be more appropriate if the compound of interest is present in excess and some other compound is limiting. As discussed previously, Equation 4.2 combines the zero and first order rate expressions into a single expression. Incorporation of zero and first order biodegradation into one-dimensional solute transport models will be discussed below.

In the previous chapter, the governing equation for one-dimensional solute transport including advection, dispersion, and sorption was shown (Equation 3.10). This equation can be expanded to include zero order and first order biodecay coefficients. When using zero and/or first order rate coefficients, it is assumed that the conditions under which the rate coefficients were determined correspond to the conditions for which the model is to be used; violation of this assumption will render the model predictions invalid. Equation 4.6 shows the governing equation expanded to include zero order and first order reactions (van Genuchten and Alves, 1982), where μ is the first order decay coefficient (1/t) and γ is the zero order production coefficient (M/L^3–t).

$$r_f \frac{\partial C}{\partial t} = D_x \frac{\partial^2 C}{\partial x^2} - V_x \frac{\partial C}{\partial x} - \mu C + \gamma \qquad (4.6)$$

van Genuchten and Alves (1982) chose to express the zero order reaction as a production reaction; input of a negative value for γ will be equivalent to making this a decay term. Analytical solutions for this governing equation are more complicated than for Equation 3.10 and are summarized for various boundary conditions by van Genuchten and Alves (1982). Solutions to these equations are also included in the program CXTFIT (Parker and van Genuchten, 1984). Of course, input of a zero value for either coefficient will eliminate that reaction in the expression.

Rate constants for a myriad of chemicals have been presented above; Tables 4.11 and 4.12 included information on rates of biodegradation for a wide variety of compounds. Data from these tables can be incorporated into the model described above for assessing transport and fate of compounds. Also, Rao and Davidson (1980) summarized information on the rates of biodegradation for a number of pesticides; this information can also be used in assessing the fate of the pesticides.

The use of zero and first order reactions to describe biotransformations is a simplified approach. McCarty et al. (1984) and Bouwer and McCarty (1984) discussed the development of a biofilm model to describe the biodegradation of compounds in the subsurface. The biofilm model is presented here to provide additional insights into biotransformations in the subsurface (where microbes grow attached to the media).

The biofilm is conceptualized to be a homogeneous matrix of bacteria and extracellular polymers which bind the bacteria together and to the surface. Figure 4.7 is a schematic of this conceptualization. As ground water flows past the biofilm, flux of the chemical occurs toward the biofilm in response to a concentration gradient, which is maintained by the biodegradation of the compound within the biofilm. Thus, the chemical undergoes diffusion across a hydrodynamic boundary layer (exterior to the biofilm), diffusion within the biofilm, and biodegradation. The biofilm model describes the diffusion processes (boundary layer and internal diffusion) by Fick's Law and the biodegradation by a Monod type relationship (as shown in Equation 4.2).

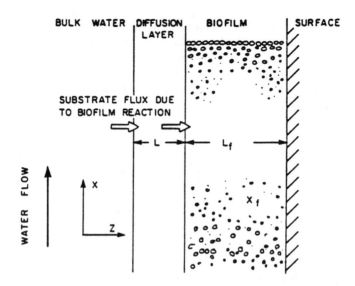

Figure 4.7. Conceptualization of substrate flux into subsurface biofilm
of uniform cell density (x_f), biofilm thickness (L_f) and
boundary layer (L) (From "Microbiological Processes Affecting Chemical Transformations in Goundwater,"
McCarty, P. L., B. E. Rittman, and E. J. Bouwer in *Groundwater Pollution Microbiology*, G. Bitten and C. P. Gerba,
Eds., Copyright 1984, John Wiley & Sons. Reprinted by
permission of John Wiley & Sons).

Under steady state conditions, the rate of diffusion of chemicals into the
biofilm will equal the rate of biodegradation of the chemical within the biofilm.
Examples of results from such a biofilm model are given below.

In microbial biofilms, it is necessary for the substrate(s) and the electron
acceptor to diffuse into the biofilm and for the end products of the metabolism to diffuse out of the biofilm. Figure 4.8 depicts the diffusion of a substrate and oxygen into a biofilm, and it depicts examples where the biodegradation is substrate limited, oxygen limited, and limited by both substrate
and oxygen (Harris and Hansford, 1976). In Figure 4.8 (a), the substrate
concentration is at its lowest (300 mg/L) and the biodegradation is observed
to be substrate limited; the substrate concentration reaches zero while there
is still an oxygen concentration of over 2 mg/L at a depth of approximately
60 μm into the biofilm. In Figure 4.8 (c), the substrate concentration is at its
highest (500 mg/L) and the biodegradation is observed to be oxygen limited;
the oxygen concentration reaches zero while there is still a substrate concentration of 40 mg/L at a depth of approximately 60 μm into the biofilm. In
Figure 4.8 (b), the substrate concentration is intermediate (400 mg/L) and the
biodegradation is observed to be substrate and oxygen limited; they both

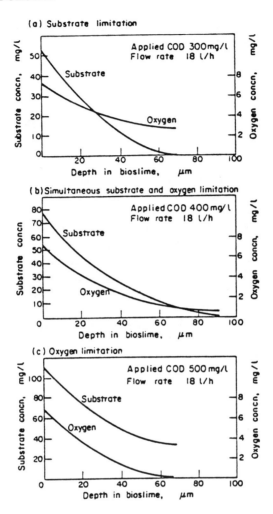

Figure 4.8.   Oxygen and substrate profiles in biofilm (bioslime) with various limiting conditions (Reprinted with permission from *Water Research*, Vol. 10, N. P. Harris and H. G. S. Hansford, "A Study of Substrate Removal in a Microbial Film Reactor", Copyright 1976, Pergamon Press).

approach a zero concentration at approximately 80 µm into the biofilm. While the work of Harris and Hansford (1976) was oriented towards fixed film biological wastewater treatment processes, the concepts of biofilm kinetics and substrate/electron acceptor limitations in biofilms also occurs in subsurface microbial systems.

It should be noted that biofilm models of microbial processes are in the developmental stages. While biofilm models more accurately describe subsurface biotransformation, they also require input data which is not readily available. For this reason, the use of first order expressions is more common in modeling subsurface biotransformations.

4.6 SUBSURFACE TRANSPORT OF MICROORGANISMS

The previous discussions have focused on the biodegradation of contaminants by subsurface microbiota. In these discussions, it has been assumed that the microbiota were attached to the media surfaces (immobile). It is also possible for the microorganisms to migrate through the subsurface, which has implications relative to human health if the microorganisms are pathogenic. In this case, the microorganisms themselves become the contaminants of concern and biotic processes controlling their migration become of interest. Subsurface migration of microorganisms is of concern when evaluating impacts on the subsurface of land treatment of wastewater, recharge of ground water with tertiary treated wastewater, septic tank drainage fields, etc. Factors affecting the survival and transport of bacteria and viruses in the subsurface will be introduced.

Table 4.20 summarizes several factors affecting the survival of enteric bacteria in soil systems, including moisture content, moisture holding capacity, temperature, pH, sunlight, organic matter, and antagonism from soil microflora (Gerba and Bitton, 1984; Gerba, 1985). Bacterial survival is dependent upon sufficient moisture content to promote bioactivity. Thus, moist soils will be more likely to evidence bacterial activity than will dry soils. The moisture holding capacity of the soil will affect the moisture content between recharge events. For example, sandy soils have less ability to retain moisture than the finer soils. Thus, sandy soils will dry out more quickly between recharge events and will provide a poor environment for bacterial activity. Cold temperatures tend to favor the survival of microorganisms, with certain bacteria capable of surviving freezing conditions. Bacteria generally have longer survival times at elevated values of pH as opposed to low values of pH. Thus, acidic peats are less favorable for survival of bacteria than are limestone soils (high pH). Sunlight has been observed to be lethal to certain bacteria; however, this will not affect bacteria once they have reached aquifer zones. Organic matter has been observed to promote bacteria survival in soils. Organic matter has high surface area and thus adds to the moisture holding capacity of the soil. Components of the organic matter may also promote bacterial survival. Competition of bacteria with other subsurface microbiota can reduce their survival time (Gerba and Britton, 1984; Gerba, 1985).

Table 4.20 Factors Affecting Survival of Bacteria in Subsurface Media

Factor	Remarks
Moisture content	Greater survival time in moist soils and during times of high rainfall
Moisture-holding capacity	Less survival time in sandy soils with greater water-holding capacity
Temperature	Longer survival at low temperatures; longer survival in winter than in summer
pH	Shorter survival time in acid soils (pH 3–5) than in alkaline soils
Sunlight	Shorter survival time at soil surface
Organic matter	Increased survival and possible regrowth when sufficient amounts of organic matter are present
Antagonism from soil microflora	Increased survival time in sterile soil

Source: "Microbial Contamination in the Subsurface", C. P. Gerba, in *Ground Water Quality,* C. H. Ward, W. Giger, and P. L. McCarty, Eds., Copyright 1985, John Wiley & Sons. Reprinted by permission of John Wiley & Sons.

Table 4.21 Factors Affecting Virus Survival in Subsurface Media

As virus adsorption to soil increases, virus survival is prolonged
Virus survival increases with increasing levels of exchangeable aluminum
Virus survival decreases with increasing pH and resin-extractable phosphorus
As temperatures increases, survival decreases
Aerobic soil microorganisms adversely affect virus survival while anaerobic microorganisms have no effect
In general, virus survival is less at lower moisture levels
Fulvic and humic acids may mask virus infectivity

Source: Gerba, 1985.

Survival of viruses will also be affected by the factors listed in Table 4.20. In addition, the adsorption of viruses to soil particles appears to offer protection against those forces which are responsible for viral inactivation in the subsurface. Thus, those factors that promote adsorption of viruses will also increase their survival (see Table 4.21).

The survival of bacteria and viruses in the subsurface raises concerns as to their potential for migration and contamination of ground water resources. The main mechanism limiting migration of viruses in the subsurface is adsorption to the media surface. The larger size of bacteria as well as the structure of the subsurface medium will affect the degree of filtration as a migration limiting mechanism. The rate of ground water movement through the media will also affect the extent of bacteria filtration. Factors that reduce the repulsive forces between the bacteria and the soil will act to increase the adsorption of bacteria onto the media surfaces. Increasing the concentration and valency of cations in the ground water medium will decrease the diffuse double layer and promote the accumulation of the bacteria at the solid surfaces. Of course, each type of virus and bacteria is affected to differing degrees by filtration and/or adsorption. It should be mentioned that filtration and adsorption of bacteria and viruses are reversible; the microorganisms are not permanently attached to the media and under the proper conditions additional migration is possible (Gerba and Bitton, 1984; Gerba, 1985).

Alexander et al. (1991) conducted a recent study on the movement of bacteria through soil and aquifer sand. This study evaluated the migration of 19 strains of bacteria. The microorganisms and their interactions with the subsurface media were characterized by net surface electrostatic charge, zeta potential, cell size, encapsulation, flagellation, hydrophobicity, and sorption partition coefficient. Of these factors, the transport of the bacteria was observed to be most significantly affected by the sorption and cell length factors. The ionic strength and flow velocity of the ground water in column studies was also observed to affect the migration of bacteria in column studies. These observations are consistent with the mechanisms of filtration and adsorption discussed above and their impact on bacteria migration.

Tim and Mostaghimi (1991) describe a model (VIROTRANS) for predicting virus migration in the subsurface. The model incorporates virus adsorption and inactivation (die-off) as reaction terms and is capable of handling transient flow in both saturated and unsaturated flow conditions. Solution of the governing partial differential equations was accomplished by a Galerkin finite element method. The model was compared to an analytical solution and to measured data on water flow and virus transport. These comparisons served to verify and validate the model's accuracy. Sensitivity analyses showed the virus distribution to be more sensitive to changes in the sorption parameter relative to the inactivation parameter.

REFERENCES

Alexander, M. *Introduction to Soil Microbiology,* 2nd ed. (New York: John Wiley & Sons, Inc., 1977).

Alexander, M., R. J. Wagenet, P. C. Baveye, J. T. Gannon, U. Migelgrin, and Y. Tan "Movement of Bacteria Through Soil and Aquifer Sand," EPA/ 600/52-91/010, U.S. Environmental Protection Agency, Ada, Oklahoma (1991).

Benefield, L. D. and C.W. Randall *Biological Process Design for Wastewater Treatment* (Englewood Cliffs, NJ: Prentice-Hall, 1980).

Bouwer, E. J. and P. L. McCarty "Modeling of Trace Organics Biotransformation in the Subsurface," *Ground Water,* 22(4) July-August):433–440 (1984).

Devinny, J. S. "Microbiology of Subsurface Wastes," in *Subsurface Migration of Hazardous Wastes,* J. S. Devinny, L. G. Everett, J. C. S. Lu, and R. L. Stollar, Eds. (New York: Van Nostrand Reinhold, 1990).

DeWeerd, K. et al. "The Relationship Between Reductive Dehalogenation and Other Aryl Substituent Removal Reactions Catalyzed by Anaerobes," *FEMS Microb. Ecol.* 38:331–339 (1986).

Edmonds, P. Microbiology: *An Environmental Perspective* (New York: MacMillan Publishing Co., Inc., 1978).

Furukawa, K., K. Tonomura, and A. Kamibayashi "Effect of Chlorine Substitution on the Biodegradability of Polychlorinated Biphenyls," *Appl. Environ. Microbiol.* 35(2):223–227 (1978).

Furukawa, K. N. Temizuka, and A. Kamibayashi "Effect of Chlorine Substitution on the Bacterial Metabolism of Various Polychlorinated Biphenyls," *Appl. Environ. Microbiol.* 38(2): 301–310 (1979).

Gerba, C. P. "Microbial Contamination of the Subsurface," in *Ground Water Quality,* C. H. Ward, W. Giger, and P. L. McCarty, Eds. (New York: John Wiley & Sons, Inc., 1985).

Gerba, C. P. and G. Bitton "Microbial Pollutants: Their Survival and Transport Pattern to Groundwater," in *Groundwater Pollution Microbiology,* G. Bitton and C.P. Gerba, Eds. (New York: John Wiley & Sons, Inc., 1984).

Ghiorse, W. C. and J. T. Wilson "Microbial Ecology of the Terrestrial Subsurface," in *Advances in Applied Microbiology,* Vol. 33, A. I. Laskin, Ed. (San Diego, CA: Academic Press, 1988).

Grady, C. P. L. and H. C. Lim *Biological Wastewater Treatment: Theory and Applications,* (New York: Marcel Dekker, 1980).

Harris, N. P. and H. G. S. Hansford "A Study of Substrate Removal in a Microbial Film Reactor," *Water Res.* 10:935–943 (1976).

Hutchins, S. R., W. C. Downs, D. H. Kampbell, J. T. Wilson, D. A. Kovacs. R. H. Douglas, and D. J. Hendrix "Pilot Project on Biorstoration of Fuel-Contaminated Aquifer Using Nitrate: Part II — Laboratory Microcosm Studies and Field Performance", in *Petroleum Hydrocarbons and Organic Chemicals in Ground Water: Prevention, Detection and Restoration*, (Dublin, OH: National Water Well Association, 1989).

Leach, L. E., F. P. Beck, J. T. Wilson, and D. H. Kampbell "Aseptic Subsurface Sampling Techniques for Hollow-Stem Auger Drilling," in *Proceedings of the National Outdoor Action Conference on Aquifer Restoration, Ground Water Monitoring and Geophysical Methods*, Columbus, OH: National Water Well Association, 1988).

Lee, M. D. et al. "Biorestoration of Aquifers Contaminated With Organic Compounds," *CRC Crit. Rev. Environ. Control* 18(1):29–89 (1988).

Linkfield, T. G., J. M. Suflita, and J. M. Tiedje "Characterization of the Acclimation Period Prior to the Anaerobic Biodegradation of Haloaromatic Compounds," *Appli. Environ. Microbiol.* (1989).

Madsen, E. L. "Determining In Situ Biodegradation: Facts and Challenges," *Environ. Sci. Technol.*, 25(10), 1991, 1663–1673.

McCarty, P. L., B. E. Rittman, and E. J. Bouwer "Microbiological Processes Affecting Chemical Transformations in Groundwater," in *Groundwater Pollution Microbiology*, G. Bitten and C. P. Gerba, Eds. (New York: John Wiley & Sons, Inc., 1984).

McNabb, J. F. and G. E. Mollard "Microbiological Sampling in the Assessment of Groundwater Pollution," in *Groundwater Pollution Microbiology*, G. Bitten and C. P. Gerba, Eds. (New York: John Wiley & Sons, Inc., 1984).

Metcalf and Eddy, Inc. *Wastewater Engineering: Treatment, Disposal and Reuse*, 3rd Ed., G. Tchobanoglous and F.L. Burton, Eds. (New York: McGraw-Hill Book Co., 1991).

Parker, J. C. and M. Th. Van Genuchten "Determining Transport Parameters from Laboratory and Field Tracer Experiments," Bulletin 84-3, Virginia Agricultural Experiment Station, Blacksburg, Virginia (1984).

Rao, P. S. C. and J. M. Davidson "Estimation of Pesticide Retention and Transformation Parameters Required in Nonpoint Source Pollution Models," in *Environmental Impact of Nonpoint Source Pollution*, M. R. Overcash and J. M. Davidson, Eds. (Ann Arbor, MI: Ann Arbor Science, 1980) pp. 23–67.

Reynolds, J. D. *Unit Operations and Processes in Environmental Engineering* (Monterey, CA: Brooks-Cole Engineering Division, 1982).

Sims, J. L., R. C. Sims, and J. E. Matthews "Approach to Bioremediation of Contaminated Soil," *Haz. Waste Haz. Mat.* 7(2):117–149 (1990).

Sims, J. L., J. M. Suflita, and H. H. Russell "Reductive Dehalogenation of Organic Contaminants in Soils and Ground Water," EPA/540/4-90/054, U.S. Environmental Protection Agency, Ada, Oklahoma (1991).

Suflita, J. M. "Microbial Ecology and Pollutant Biodegradation in Subsurface Ecosystems," in *Transport and Fate of Contaminants in the Subsurface,* EPA/625/4-89/019, U.S. Environmental Protection Agency, Ada, Oklahoma (1989a), pp. 67–84.

Suflita, J. M. "Microbiological Principles Influencing the Biorestoration of Aquifers," in *Transport and Fate of Contaminants in the Subsurface,* EPA/625/4-89/019, U.S. Environmental Protection Agency, Ada, Oklahoma (1989b), pp. 85–99.

Suflita, J. M. and G. W. Sewell "Anaerobic Biotransformation of Contaminants in the Subsurface," EPA/600/M-90/024, U.S. Enviromental Protection Agency, Ada, Oklahoma (1991).

Tim, U. S. and S. Mustaghimi "Model for Predicting Virus Movement Through Soils," *Ground Water,* 29(2, March-April):251–259 (1991).

Van Genuchten, M. Th. and W. J. Alves "Analytical Solutions of the One-Dimensional Convective-Dispersive Solute Transport Equation," Technical Bulletin Number 1661, U.S. Department of Agriculture (1982).

ADDITIONAL REFERENCES (CITED IN TABLES)

Aelion, C. M., C. M. Swindoll, and F. K. Pfaender *Appl. Environ. Microbiol.* 53:2212–2217 (1987).

Adrian, N. R. and J. M. Suflita "Reductive Dehalogenation of a Nitrogen Heterocyclic Herbicide in Anoxic Aquifer Slurries," *Appl. Environ. Microbiol.* 56:292–294 (1990).

Alexander, M. *Science* 211:132–138 (1981).

Atlas, R. M. *Microbiol. Rev.* 45:180–209 (1981).

Aulenbach, D. B., N. L. Clesceri, and T. J. Tofflemire "Water Renovation by Discharge into Deep Natural Sand Filters," Proceedings of AICHE Conference, Chicago, IL, May 4–8, 1975.

Balkwill, D. E. and W. C. Ghiorse *Planetary Ecology,* D. E. Caldwell, J. A. Brierley, and C. L. Brierley, Eds. (New York: Van Nostrand Reinhold, 1985a), pp. 399–408.

Balkwill, D. L. and W. C. Ghiorse *Appl. Environ. Microbiol.* 50:580–588 (1985b).

Barcelona, M. J. and T. G. Naymik "Dynamics of a Fertilizer Plume in Ground Water," *Environ. Sci. Technol.* 18:257–261 (1984).

Barker, J. F. and G. C. Patrick *Proc. Petroleum Hydrocarbons and Organic Chemicals in Ground Water — Prevention, Detection, and Restoration Conference and Exposition,* Houston, 1985, pp. 160–177.

Barker, J. F., J. S. Tessmann, P. E. Plotz, and M. Reinhard *J. Contam. Hydrol.* 1:171–189 (1986).

Barrio-Lage, G. A., F. Z. Parsons, R. M. Narbaitz, and P. A. Larenzo "Enhanced Anaerobic Biodegradation of Vinyl Chloride in Ground Water," *Environ. Toxicol. Chem.* 9:403–415 (1990).

Barrio-Lage, G. A., F. Z. Parsons, R. S. Nazzar, and P. A. Larenzo *Environ. Sci. Technol.* 20:96–99 (1986).

Beeman, R. E. and J. M. Suflita *J. Ind. Microbiol.* 5:45–58 (1990).

Beeman, R. E. and J. M. Suflita "Microbial Ecology of a Shallow Unconfined Ground Water Aquifer Polluted by Municipal Landfill Leachate," *Microbiol. Ecol.* 14:39–54 (1987).

Beloin, R. M., J. L. Sinclair, and W. C. Ghiorse *Microbiol. Ecol.* 16:1988.

Belyaev, S. S. and M. V. Ivanov "Bacterial Methanogenesis in Underground Waters," *Ecol. Bull.* 35:273–280 (1983).

Bengtsson, G. *Ground Water Quality,* C. H. Ward, W. Giger, and P. L. McCarty, Eds. (New York: John Wiley & Sons, 1985), pp. 330–341.

Bone, T. L. and D. L. Balkwill *Appl. Environ. Microbiol.* 51:462–468 (1986).

Bone, T. L. and D. L. Balkwill *Microbiol. Ecol.* 16:(1988).

Cerniglia, C. E., in *Petroleum Microbiology,* R. M. Atlas, Ed. (New York: Macmillan, 1984), pp. 99–128.

Chapelle, F. H. et al. *Geology* 16:117–121 (1988).

Chapelle, F. H., J. L. Zelibor, D. J. Grimes, and L. L. Knobel *Water Resour. Res.* 23:1625–1632 (1987).

Cheng, H. H. and R. G. Lehmann *Weed Sci.* 33 (Suppl. 2):7–10 (1985).

Cooney, J. J., S. A. Silver, and E. A. Beck *Microbiol. Ecol.* 11:127–137 (1985).

Criddle, C. S., P. L. McCarty, M. C. Elliott, and J. F. Barker *J. Contam. Hydrol.* 1:133–142 (1986).

Cullimore, D. R. *Ground Water* 21:558–563 (1983).

Cullimore, D. R. and A. E. McCann in *Aquatic Microbiology,* F. A. Skinner and J. M. Shewan, Eds. (New York: Academic Press, 1977), pp. 263–274.

Davis, J. B. *Petroleum Microbiology* (Amsterdam: Elsevier Publishing Co., 1967), p. 604.

DeWeerd, K. A., A. Saxena, D. P. Nagle, Jr., and J. M. Suflita "Metabolism of the O Methoxy Substituent of 3-methoxybenzoic Acid and other Unlabeled Methoxybenzoic Acids by Anaerobic Bacteria," *Appl. Environ. Microbiol.* 54:1237–1242 (1988).

Dobbins, D. C. and F. K. Pfaender *Microbiol. Ecol.* 15:(1987).

Dockins, W. S., G. J. Olson, G. A. McFeters, and S. C. Turbak *Geomicrobiol. J.* 2:83–97 (1980).

Dott, W., C. Frank, and P. Werner *Zentral Bakteriol. Hyg. I A:bt. Orig. B* 180:62–75 (1984).

Dunlap, W. J., R. L. Cosby, J. F. McNabb, B. E. Bledsoe, and M. R. Scalf *Ground Water* 10:107–117 (1972).

Egli, C., R. Scholtz, A. M. Cook, and T. Leisinger "Anaerobic Dechlorination of Tetrachloromethane and 1,2-dichloroethane to Degradable Products by Pure Cultures of Desulfobacterium sp. and Methanobacterium sp.," *FEMS Microbiol. Lett.* 43:257–261 (1987).

Ehrlich, G. G., D. F. Goerlitz, E. M. Godsey, and M. F. Hult *Ground Water* 20:703–710 (1982).

Ehrlich, G. G., E. M. Godsy, D. F. Goerlitz, and M. F. Hult *Dev. Ind. Microbiol.* 24:235–245 (1983).

Faller, J., et al. *Environ. Sci. Technol.* 25:676–678 (1991).

Fathepure, B. Z., J. P. Nengu, and S. A. Boyd "Anaerobic Bacteria that Dechlorinate Perchloroethene," *Appl. Environ. Microbiol.* 53:2671–2674 (1987).

Foster, S. S. D., D. P. Kelly, and R. James in *Planetary Ecology,* D. E. Caldwell, J. A. Brierley, and C. L. Brierley, Eds. (New York: Van Nostrand Reinhold, 1985), pp. 356–382.

Frank, C. and W. Dott *Zentral Bakteriol. Hyg.I. Abt. Orig. B* 180:459–470 (1985).

Führ, F. *Weed Sci.* 33(Suppl. 2):11–17 (1985).

Gälli, R. and P. L. McCarty "Biotransformation of 1,1,1-trichloroethane, Trichloromethane, and Tetrachloromethane by a Clostridium sp.," *Appl. Environ. Microbiol.* 55:837–844 (1989).

Ghiorse, W. C. and D. L. Balkwill *Dev. Ind. Microbiol.* 24:213–224 (1983).

Ghiorse, W. C. and D. L. Balkwill in *Progress in Chemical Disinfection II: Problems at the Frontier,* J. E. Janauer and W. C. Ghiorse, Eds. (New York: Binghamton SUNY, 1984), pp. 91–106.

Ghiorse, W. C. and D. L. Balkwill in *Ground Water Quality,* C. H. Ward, W. Giger, and P. L. McCarty, Eds. (New York: Wiley & Sons, Inc., 1985), pp. 387–401.

Gibson, S. A. and J. M. Suflita "Extrapolation of Biodegradation Results to Groundwater Aquifers: Reductive Dehalogenation of Aromatic Compounds," *Appl. Environ. Microbiol.* 52:681–688 (1986).

Godsy, E. M., D. E. Troutman, and G. G. Ehrlich *Proc. Natl. Meet. Am. Chem. Soc. Div. Environ. Chem., 186th* 23:288–289 (1983).

Godsey, E. M. and G. G. Ehrlich "Reconnaissance for Microbial Activity in the Magothy Aquifer, Bay Park, New York, Four Years After Artificial Recharge," *J. Res. U.S. Geol. Sur.* 6:829–836 (1978).

Goerlitz, D. F., D. E. Troutman, E. M. Godsy, and B. J. Franks *Environ. Sci. Technol.* 19:955–961 (1985).

Gottfreund, E., I. Gerber, and R. Schweisfurth *Landwirtsch. Forsch.* 38:72–79 (1985a).

Gottfreund, E., J. Gottfreund, I. Gerberg, G. Schmitt, and R. Schweisfurth *Water Supply* 3:109–115 (1985b).

Gottfreund, E., J. Gottfreund, and R. Schweisfurth *Forum Städte-Hyg.* 36:178–183 (1985c).

Grossman, E. L., et al. *Geology* 17:495–499 (1989).

Hallburg, R. O. and R. Martinell "Vyredox-In Situ Purification of Ground Water," *Ground Water* 14:88–93 (1976).

Harvey, R. W. and L. H. George *Appl. Environ. Microbiol.* 53:2992–2996 (1987).

Harvey, R. W., D. L. Smith, and L. George "Effect of Organic Contamination Upon Microbial Distribution and Heterotrophic Uptake in a Cape Cod, Massachusetts Aquifer," *Appl. Environ. Microbiol.* 48:1197–1202 (1984).

Hirsch, P. and E. Rades-Rohkohl *Dev. Ind. Microbiol.* 24:183–200 (1983).

Hoos, E. and R. Schweisfurth *Vom Wasser* 58:103–112 (1982).

Humenick, M. J., L. N. Bitton, and C. F. Mattox "Natural Restoration of Ground Water in UCG," *In Situ* 6:107–125 (1982).

Hutchins, S. R., G. W. Sewell, D. A. Kovacs, and G. A. Smith "Biodegradation of Aromatic Hydrocarbons by Aquifer Microorganisms Under Denitrifying Conditions," *Environ. Sci. Technol.* 1990.

Idelovitch, E. and M. Michail "Treatment Effects and Pollution Dangers of Secondary Effluent Percolation to Ground Water," *Prog. Wat. Tech.* 12:949–966 (1980).

Jamison, V. W., R. L. Raymond, and J. O. Hudson "Biodegradation of High-octane Gasoline in Ground Water," *Dev. Ind. Microbiol.* 16:305–312 (1975).

Kiene, R. P. and D. G. Capone *Appl. Environ. Microbiol.* 51:1247–1251 (1986).

Kuhn, E. P., G. T. Townsend, and J. M. Suflita "Effect of Sulfate and Organic Carbon Supplements on Reductive Dehalogenation of Chloroanilines in Anaerobic Aquifer Slurries," *Appl. Environ. Microbiol.* 56:2630–2637 (1990).

Kuhn, E. P., J. M. Suflita, M. D. Rivera, and L. Y. Young "Influence of Alternate Electron Acceptors on the Metabolic Fate of Hydroxybenzoate Isomers in Anoxic Aquifer Slurries," *Appl. Environ. Microbiol.* 55:590–598 (1989).

Kuhn, E. P., J. Zeyer, P. Eicher, and R. P. Schwarzenbach "Anaerobic Degradation of Alkylated Benzenes in Denitrifying Laboratory Aquifer Columns," *Appl. Environ. Microbiol.* 54:490–496 (1988).

Kuhn, E. P., P. J. Colberg, J. L. Schnoor, O. Nanner, A. J. B. Zehnder, and R. P. Schwarzenbach *Environ. Sci. Technol.* 19:961–968 (1985).

Ladd, T. I., R. M. Ventullo, P. M. Wallis, and J. W. Costerton, *Appl. Environ. Microbiol.* 44:321–329 (1982).

Lamar, R. T. and D. M. Dietrich *Appl. Environ. Microbiol.* 56:3093–3100 (1990).

Larson, R. J. and R. M. Ventullo *Proc. Natl. Symp. Aquifer Restor. Ground Water Monitor* (Worthington, OH: 3rd, National Water Works Association, 1983), pp. 402–408.

Leahy, J. G. and R. R. Colwell *Microbiol. Rev.* 54:305–315 (1990).

Lee, K., et al. *Microb. Ecol.* 11:337–351 (1985).

Lee, M. D. and C. H. Ward "Biological Methods for the Restoration of Contaminated Aquifers," *Environ. Toxicol. Chem.* 4:721–726 (1985).

Lind, A. M. "Nitrate Reduction in the Subsoil," *Proc. Int. Assoc. Water Pollut. Res. Copenhagen* 1(August):14 (1975).

Lovely, D. R. and D. J. Lonergan "Anaerobic Oxidation of Toluene, Phenol, and p-cresol by the Dissimilatory Iron-reducing Organism, GS-15," *Appl. Environ. Microbiol.* 56:1858–1864 (1990).

Madsen, E. L., J. L. Sinclair, and W. C. Ghiorse *Science* 52:830–833 (1991).

Mahadevaiah, B. and G. D. Miller *Proc. Natl. Symp. Expo. Aquifer Restor. Ground Water Monitor* (Columbus, OH: 6th National Wate Works Association, 1986), pp. 384–412.

Marxsen, J. *Int. J. Speleol.* 11:173–201 (1981a).

Marxsen, J. *Verh. Int. Verein. Limnol.* 21:1371–1375 (1981b).

McCarty, P. L., B. E. Rittmann, and E. J. Bouwer "Microbiological Processes Affecting Chemical Transformations in Ground Water," in *Ground Water Pollution Microbiology*, G. Bitton and C. P. Gerba, Eds. (New York: John Wiley & Sons, 1984).

Metcalf, R. L. *Ann. Rev. Entomol.* 22:241–261 (1977).

Mikesell, M. D. and S. A. Boyd "Dechlorination of Chloroform by Methanosarcina Strains," *Appl. Environ. Microbiol.* 56:1198–1201 (1990).

Novak, J. T., C. D. Goldsmith, R. S. Benoit, and J. H. O'Brien *Nat. Sci. Technol.* 17:71–85 (1985).

Ogunseitan, O. A., E. T. Tedford, D. Pacia, K. M. Sirotkin, and G. A. Sayler *J. Ind. Microbiol.* 1:311–317 (1987).

Olson, B. H. *Environ. Sci. Technol.* 25:604–611 (1991).

Olson, G. J., W. S. Dockins, G. A. McFeters, and W. P. Iverson *Geomicrob. J.* 2:327–340 (1981).

Olson, G. J., G. A. McFeters, and K. L. Temple "Occurrence and Activity of Iron and Sulfur-oxidizing Microorganisms in Alkaline Coal Strip Mine Spoils," *Microb. Ecol.* 7:40–50 (1981).

Parsons, F., G. Barrio-Lage, and R. Rice "Biotransformation of Chlorinated Organic Solvents in Static Microcosms," *Environ. Toxicol. Chem.* 4:739–742 (1985).

Pereira, W. and C. E. Rostad *Environ. Toxicol. Chem.* 6:163–176 (1987).

Preul, H. C. "Underground Movement of Nitrogen," *Adv. Water Pollut. Res. Proc. Third Int. Conf.* 1966, pp. 309–328.

Prince, R. C., J. R. Clark, and J. E. Lindstrom in *Bioremediation Montioring Program Report* (Annandale, NJ: Exxon Research and Engineering Co., 1990).

Pritchard, P. H. and C. F. Costa *Environ. Sci. Technol.* 25:372–379 (1991).

Quensen, J. R., III, J. M. Tiedje, and S. A. Boyd *Science* 242:752–754 (1988).

Rades-Rohkohl, E. and P. Hirsch *Abstr. Int. Symp. Environ. Biogeochem.*, 6th, Santa Fe, October 1983, p. 56.

Raymond, R. L., J. O. Hudson, and V. Jamison *Appl. Environ. Microbiol.* 31:522–535 (1976).

Raymond, R. L., V. W. Jamison, and J. O. Hudson "Beneficial Stimulation of Bacterial Activity in Ground Water Containing Petroleum Products," *Aice Symp. Ser.* 73:390–404 (1976).

Ridgway, H. F., et al. *Appl. Environ. Microbiol.* 56:3565–3575 (1990).

Roberts, P. V., J. Schreiner, and G. Hopkins *Water Res.* 16:1025–1035 (1982).

Roberts, P. V., et al. "Organic Contaminant Behavior During Ground Water Recharge," *J. Water Pollut. Control Fed.* 52:161–172 (1980).

Roberts, P. V., M. N. Goltz, and D. M. Mackay *Water Resour. Res.* 22:2047–2058 (1986).

Robertson, G. P. and J. M. Tiedje *Soil Biol. Biochem.* 19:187–193 (1987).

Rogers, J. E., et al. "Microbial Transformation of Alkylpyridines in Ground Water," *Wat. Air Soil Poll.* 24:443–454 (1985).

Sayler, G. S. and A. C. Layton *Annu. Rev. Microbiol.* 44:625–648 (1990).

Semprini, L., et al. *Ground Water* 28:715–727 (1990).

Sinclair, J. L. and W. C. Ghiorse *Appl. Environ. Microbiol.* 53:1157–1163 (1987).

Sinton, L. W. *Hydrobiologia* 119:161–169 (1984).

Smith, G. A., J. S. Nickels, J. D. Davis, R. H. Findlay, P. S. Vashio, J. T. Wilson, and D. C. White *2nd Int. Cong. Ground Water Quality*, Tulsa, March 1984.

Smith, G. A., J. S. Nickels, B. D. Kerger, J. D. Davis, S. P. Collins, J. T. Wilson, J. F. McNabb, and D. C. White *Can. J. Microbiol.* 32:104–111 (1986).

Smith, R. L., R. W. Harvey, and D. R. LeBlanc *J. Contam. Hydrol.* 7:285–300 (1991).

Smolenski, W. J. and J. M. Suflita "The Microbial Metabolism of Cresols in Anoxic Aquifers," *Appl. Environ. Microbiol.* 53:710–716 (1987).

Spain, J. C., et al. *Appl. Environ. Microbiol.* 48:944–950 (1984).

Stetzenbach, L. D. and N. A. Sinclair *6th Abstr. Int. Symp. Environ. Biogeochem.* Santa Fe, October 1983, p. 57.

Suflita, J. M., L. Liang, and L. Shi "The Anaerobic Metabolism of 2-Hydroxybiphenyl by Sulfate-reducing Bacterial Enrichments," *Current Microbiol.* 1990.

Suflita, J. M., S. A. Gibson, and R. E. Beeman "Anaerobic Biotransformation of Pollutant Chemicals in Aquifers," *J. Ind. Microbiol.* 3:179–194 (1988).

Suflita, J. M. and S. A. Gibson "Biodegradation of Haloaromatic Substrates in a Shallow Anoxic Ground Water Aquifer," *Proc. Second Intl. Conf. Ground Water Quality Res.*, N. N. Durham and A. E. Redelfs, Eds. (Stillwater, OK: National Center for Ground Water Research, 1985), pp. 30–32.

Suflita, J. M. and G. D. Miller "The Microbial Metabolism of Chlorophenolic Compounds in Ground Water Aquifers," *Env. Toxicol. Chem.* 4:751–758 (1985).

Thorn, P. M. and R. M. Ventullo *Microb. Ecol.* 16:1988.

Towler, P. A., N. C. Blakey, T. E. Irving, L. Clark, P. J. Maris, K. M. Baxter, and R. M. MacDonald *18th Proc. Congr. Int. Assoc. Hydrogeol.*, Cambridge (154), 1985, pp. 84–97.

Trudell, M. R., R. W. Gillham, and J. A. Cherry *J. Hydrol.* 83:251–268 (1986).

van Beek, C. G. E. M. and D. van der Kooij "Sulfate-Reducing Bacteria in Ground Water from Clogging and Nonclogging Shallow Wells in the Netherlands River Region," *Ground Water* 20:298–302 (1962).

Ventullo, R. M. and R. J. Larson "Metabolic Diversity and Activity of Heterotrophic Bacteria in Ground Water," *Env. Toxicol. Chem.* 4:321–329 (1985).

Wallis, P. M. and T. I. Ladd *Geomicrob. J.* 3:49–78 (1983).

Ward, D. M., et al. *AMBIO* 9:277–283 (1980).

Ward, T. E. "Characterizing the Aerobic and Anaerobic Microbial Activities in Surface and Subsurface Soils," *Environ. Tox. Chem.* 4:727–737 (1985).

Webster, J. J., G. J. Hampton, J. T. Wilson, W. C. Ghiorse, and F. R. Leach *Ground Water* 23:17–25 (1985).

Weirich, G. and R. Schweisfurth *Geomicrob. J.* 4:1–20 (1983).

Westlake, D. W. S., et al. *Can. J. Microbio.* 20:915–928 (1974).

Westlake, D. W. S., A. M. Jobson, and F. D. Cook *Can. J. Microbio.* 24:254–260 (1978).

White, D. C., G. A. Smith, M. J. Gehron, J. H. Parker, R. H. Findlay, R. F. Martz, and H. L. Fredrickson *Dev. Ind. Microbiol.* 24:201–211 (1983).

White, D. C., J. T. Wilson, J. F. McNabb, W. C. Ghiorse, and D. L. Balkwill *Abstr. Intl. Symp. Environ. Biogeochem.* 6th, Santa Fe, October 1983b, pp. 56–57.

Wilson, B. H., G. B. Smith, and J. F. Rees "Biotransformation of Selected Alkylbenzenes and Halogenated Aliphatic Hydrocarbons in Methanogenic Aquifer Material: A Microcosm Study," *Environ. Sci. Technol.* 20:997–1002 (1986).

Wilson, B. H., B. E. Bledsoe, D. H. Kampbell, J. T. Wilson, and J. H. Armstrong in *Petroleum Hydrocarbons and Organic Chemicals in Ground Water: Prevention, Detection, and Restoration*, (Houston, TX: National Water Well Association and American Petroleum Institute, 1986).

Wilson, J. T., J. F. McNabb, D. L. Balkwill, and W. C. Ghiorse *Ground Water* 21:134–142 (1983a).

Wilson, J. T., J. F. McNabb, B. H. Wilson, and M. J. Noonan *Dev. Ind. Microbiol.* 24:225–233 (1983b).

Wilson, J. T., J. F. McNabb, J. W. Cochran, T. H. Wang, M. B. Tomson, and P. B. Bedient *Environ. Tox. Chem.* 4:721–726 (1985a).

Wilson, J. T., M. J. Noonan, and J. F. McNabb in *Ground Water Quality*, C. H. Ward, W. Giger, and P. L. McCarty, Eds. (New York: John Wiley & Sons, Inc., 1985b) pp. 483–492.

Wilson, J. T., G. D. Miller, W. C. Ghiorse, and F. B. Leach *J. Contam. Hydrol.* 1:163–170 (1986).

Wong, A. S. and D. G. Crosby in *Pentachlorophenol*, K. Ranga Rao, Ed. (New York: Plenium, 1978), pp. 19–25.

Wood, P. R., R. F. Lang, and I. L. Payan "Anaerobic Transformation, Transport and Removal of Volatile Chlorinated Organics in Ground Water," in *Ground Water Quality*, C. H. Ward, W. Giger, and P. L. McCarty, Eds. (New York: John Wiley & Sons, Inc., 1985), pp. 493–511.

5

PHYSICAL MODELS OF
CONTAMINANT TRANSPORT

5.1 INTRODUCTION

Numerous reports of ground water pollution have caused considerable attention to be given to assessing and/or quantifying subsurface transport and fate processes. A large number of research-oriented field studies of the transport and fate of selected contaminants have been conducted or are ongoing. Laboratory-oriented microcosm studies can be used in lieu of, or as a supplement to, field studies. In this context, a microcosm can be defined as a controlled laboratory system which attempts to simulate, on a small scale, a portion of a real world subsurface environment. Pritchard and Bourquin (1983) have indicated that a microcosm study involves the establishment of a physical model or simulation of part of the ecosystem in the laboratory within definable physical and chemical boundaries under a controlled set of experimental conditions. Microcosm studies have been conducted on terrestrial, surface water, and ground water systems; the emphasis of this chapter will be on the use of such studies for examining organic contaminants in ground water and subsurface environment systems.

Several microcosm designs have been used in studies of the subsurface movement of synthetic organic contaminants. In addition, these studies have been conducted for several purposes. Accordingly, this chapter will review various microcosm designs and delineate the advantages and limitations of microcosm studies directed toward developing a better understanding of subsurface transport and fate processes. Specific sections are included on the

Table 5.1 Natural Processes that Affect Subsurface Contaminant Transport

Physical processes
  Advection (porous media velocity)
  Hydrodynamic dispersion
  Molecular diffusion
  Density stratification
  Immiscible phase flow
  Fractured media flow
Chemical processes
  Oxidation-reduction reactions
  Radionuclide decay
  Ion-exchange
  Complexation
  Co-solvation
  Immiscible phase partitioning
  Sorption
Biological processes
  Microbial population dynamics
  Substrate utilization
  Biotransformation
  Adaptation
  Co-metabolism

Source: U.S. Environmental Protection Agency, 1987.

objectives of microcosm studies: four classifications of microcosm designs along with several examples of each, the advantages and limitations of microcosm studies, pertinent design considerations, and key issues and concerns. The final section of this chapter discusses recent novel applications of microcosm technology for studying hydrodynamic processes and the use of other physical models for studying multiphase flow processes.

5.2 OBJECTIVES OF MICROCOSM STUDIES

Fundamentally, microcosm studies have been conducted to gain a better understanding of subsurface transport and fate processes. Table 5.1 lists the processes in three categories; namely, physical, chemical, and biological processes (U.S. Environmental Protection Agency, 1987). Typical studies

Table 5.2 Classification of Microcosms for Subsurface Transport and Fate Studies

Classification of Microcosm	Form of Subsurface Material in Microcosm	Status of Extracted Subsurface Material	Type of Microcosm	System Dynamics
Slurry (SL)	Slurry	Homogenized	Test Tube or serum bottle	Static
Homogenized subsurface (HS)	Solid	Homogenized	Column	Flow-through
Incremental subsurface (IS)	Solid	Incremental sections	Column	Flow-through
Subsurface core (SC)	Solid	Intact	Column	Flow-through

focus on determining the relative importance of these processes for particular contaminants under a given set of subsurface media and environmental conditions. Process information can be used for several purposes including:

1. Development of rate information for inclusion in ground water flow and solute transport models
2. Planning of source-oriented ground water quality monitoring programs
3. Delineation of appropriate ground water protection strategies for pollution source categories in specific hydrogeological settings
4. Site selection for waste disposal operations
5. Determination of potential plume management and ground water pollution remediation measures for locations with contaminated soil and/or ground water.

5.3 CLASSIFICATION OF MICROCOSMS

Microcosms for usage in studies of the subsurface transport and fate of contaminants can be divided into several categories as shown in Table 5.2. The simplest category is the slurry microcosm (S1) which is comprised of subsurface material from either the unsaturated or saturated zones which has been subsequently mixed with water (Ashford, 1987). The SL microcosms are typically established in test tubes or containers without flow-through capabilities. Sealed SL microcosms are often referred to as "batch" studies.

If no attempt is made to preserve the orientation of the subsurface stratum which was sampled for the study, the microcosm unit may be defined as a homogenized subsurface microcosm (HS) in which the subsurface material has been reorganized by either mixing or sieving or both. The incremental subsurface microcosm (IS) is characterized by the placement of subsurface material in the microcosm in increments in the same order as found in the sampled subsurface environment. Care is generally taken to not homogenize the incremented sections. The subsurface core microcosm (SC) is established to preserve the spatial configuration of subsurface material by the utilization of intact cores of sampled subsurface material. The latter three classes of microcosms are typically used in flow-through experiments to simulate sub-surface flows in either the unsaturated or saturated zones. These types of experiments are often referred to as "column" studies.

5.4 EXAMPLES OF UTILIZED MICROCOSMS

The use of batch and column microcosm studies for assessing subsurface transport and fate processes is increasing and finding wide applicability. This section highlights examples of each of the above-listed four classes of microcosms used in studies of the subsurface environment.

5.4.1 Slurry Microcosms

Slurry microcosms have typically been used to determine if indigenous microbes in the subsurface environment are capable of degrading organic contaminants. These types of studies use either test tubes or serum bottles.

The static SL microcosm used by Wilson et al. (1981) is depicted in Figure 5.1. Intact aseptic samples of the deeper subsurface were acquired from Lula and Pickett, Oklahoma, and Fort Polk, Louisiana, at two depths for each collection site. A slurry was formed by mixing the sample with sterile water, then the slurry was poured into a 35 mL screw cap test tube. Ground water containing selected concentrations of organic chemicals was then added to each tube after which they were sealed with a Teflon®-lined screw cap. The dose solution was dispersed throughout the slurry with a vortex mixer.

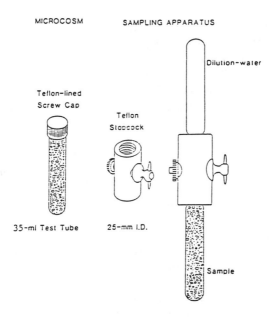

MICROCOSM   SAMPLING APPARATUS

Figure 5.1.   Slurry microcosm used to determine degradation of pollutants in the subsurface environment (Wilson et al., 1981).

To provide an abiotic control, a portion of the slurry was autoclaved prior to the transfer into the test tubes. The microcosms were sampled after an incubation period by diluting the pore water. The cap was removed and replaced by a Teflon® stopcock which had connected at the other end a 35 mL test tube filled with water. The contents of the tubes were allowed to mix together with a vortex mixer. The tubes were then separated, sealed, and centrifuged. Microcosms were analyzed after 0, 1, 3, and 9 weeks of incubation. At the end of each period, at least two nonautoclaved microcosms and one autoclaved microcosm were analyzed. The results obtained indicated that all the chlorinated alkanes and alkenes studied, with the exception of bromodichloromethane, were stable in material from the deeper subsurface.

In a follow-on study using the same test tube design approach as depicted in Figure 5.1, Wilson et al. (1983) extracted samples at 1.2, 3.0, and 5.0 m depth from Lula, Oklahoma, to determine the degradation of organic pollutants. The contents of the SL microcosms were allowed to incubate for 0, 1, 2, 4, 8 and 16 weeks at 20°C. In this study also, at least one autoclaved and two nonautoclaved microcosms were sampled at each time period. Wilson, et al. (1983) found that the indigenous bacteria degraded toluene rapidly, and that chlorobenzene was degraded in the vadose (unsaturated) zone while dichloromethane was degraded in the saturated zone.

Test-tube SL microcosms as depicted in Figure 5.1 were also used in a study to evaluate the biodegradation potential of a subsurface alluvial material exposed to selected petroleum hydrocarbons (Mahadevaiah, 1985; and Mahadevaiah and Miller, 1986). The material was aseptically obtained from an alluvial area near the South Canadian River in central Oklahoma. The selected petroleum hydrocarbons included benzene, toluene, ethyl benzene, and para-, meta-, and orth-xylene in concentrations from about 50 to 600 ppb. Triplicate nonautoclaved and autoclaved SL microcosms were tested at 0, 1, 2, 4, and 8 weeks. The test results from one experimental run are shown in Table 5.3 (Mahadevaiah and Miller, 1986). All of the tested hydrocarbons were rapidly degraded in the nonautoclaved alluvial material to below detection limits. However, the hydrocarbons in the autoclaved material also disappeared, but at a slower rate, most likely due to volatilization and diffusion through the Teflon® lining of the screw caps. The higher rates of decreases in the concentrations of the petroleum hydrocarbons in the nonautoclaved material relative to the autoclaved controls documents the presence of microorganisms and their ability to biodegrade the test contaminants.

Wilson (1985) studied the fates of trichloroethylene, 1, 1-dichloroethylene, *trans*-1,2-dichloroethylene, and *cis*-1,2-dichloroethylene in anaerobic aquifer material in 125 mL serum bottle microcosms. Each microcosm contained approximately 100 g of aquifer material (dry weight) plus 15% by weight additional aquifer water. These components were added in slurry form; the slurry was aseptically poured into the 125 mL serum bottles to a premarked level. The lip of the bottle was wiped clean with a sterile chem-wipe, dosed with the appropriate solution, and immediately capped with a Teflon®-coated silicon septum and a crimp cap seal as shown in Figure 5.2 (Wilson, 1985). The approximate initial concentrations were trichloroethylene, 150 μg/L; and *trans*-1,2-dichloroethylene, 125 μg/L. Triplicates of autoclaved and nonautoclaved microcosms were sampled at 0, 3, 7, 16, and 40 weeks. The microcosms were stored upside-down (to reduce the loss of volatiles) in the dark at 17°C.

The results of the study by Wilson (1985) indicated anaerobic degradation for the four compounds, although the disappearance was not rapid. Initial lag times from 7 to 16 weeks were required before significant degradation began to occur. Even after 40 weeks of incubation, the concentration of trichloroethylene in some nonautoclaved microcosms remained similar to that of the autoclaved control microcosms, while other nonautoclaved microcosms had concentrations below the limit of detection. The compound 1,1-dichloroethylene was not present above the limit of detection after 40 weeks of incubation in most of the samples analyzed, either nonautoclaved or autoclaved, thus suggesting a possible abiotic fate.

Significant degradation of *trans*-1,2-dichloroethylene did not occur until

Table 5.3 Biodegradation Rate and Loss of Volatiles in Test Tube
SL Microcosms

Experiment Parameter	Compounds						
	Benzene	Toulene	Ethyl-Benzene	p-Xylene	m-Xylene	o-Xylene
Incubation time, weeks	8	8	8	8	8	8
Expected initial pore water concentration, ppb	73–91	41–52	120–150	147–184	245–307	228–285
Final concentration after incubation in nonautoclaved material, ppb	0[a]	0	0	0	0	0
Average % decrease/week						
Total loss	21.20	20.15	16.72	20.49	19.71	19.70
Loss of volatiles	4.38	1.52	0.89	2.81	2.09	1.45
Biodegradation	16.82	18.63	15.83	17.68	17.62	18.25

Source: Mahadevaiah and Miller, 1986.
[a] Results indicate that the final concentrations were below the dectection limit for all the parameters.

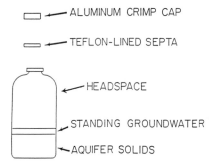

Figure 5.2.   Configuration of serum bottle SL microcosm (Wilson, 1985).

after 7 weeks of incubation. At 40 weeks of incubation, the compound was detected at 50% of its original concentration in some nonautoclaved microcosms or to the level of detection in others. The concentration in the autoclaved microcosms at 40 weeks was approximately 70% of the original value, once again suggesting abiotic losses. The disappearance of *cis*-1,2-dichloroethylene was rapid. The concentration of the compound was reduced to less than 5% of its original concentration by the 16th week of incubation. Autoclaved microcosms at this sampling interval remained at 60% of the original concentrations.

The microbial process of reductive dehalogenation was thought to explain the disappearance of these compounds from the subsurface material of this study (Wilson, 1985). It should be noted that the slow movement of ground water increases the residence times of the chemicals in the subsurface. These extended residence times increases the chance for microbial adaption and allows these slow removal rates to be environmentally significant.

Finally, in terms of slurry microcosms, Wilson et al. (1986) also studied the disappearance of alkylbenzenes and halogenated aliphatic hydrocarbons in serum bottle SL microcosms constructed with aquifer material maintained under methanogenic conditions. The intact aquifer solid sample was slurried by the addition of a predetermined amount of aquifer water. The slurry was then transferred into a 160 mL serum bottle. The microcosms were then dosed with the chemicals of interest and sealed by a cap coated with Teflon®. Prior to dosing, sterile controls were autoclaved. Both the sample and control SL microcosms were stored upside down in the dark at 17°C. Two to four replicates of each treatment were analyzed at each sampling interval. Incubation periods were 0, 6, 12, 20, 40, and 120 weeks. All of the alkylbenzenes were biodegraded in the study. Of the halogenated aliphatic hydrocarbons, only *cis*-1,2-dichloroethylene and 1,2-dibromoethane did not require long lag times before degradation began.

5.4.2 Homogenized Subsurface Microcosms

Homogenized subsurface (HS) microcosms are those in which extracted subsurface material is mixed or sieved prior to placement in columns. These microcosms provide opportunities for the study of subsurface transport and fate processes using flow through systems.

The displacement of 1,2-dibromo-3-chloropropane (DBCP) in saturated soil laboratory columns was studied by Biggar et al. (1984). Samples of Panoche clay loam were obtained from the West Side Field Station of the University of California in the San Joaquin Valley. The soil samples were air dried and sieved, and the sieved soil was packed in four plastic columns 7.5 cm i.d. by either 30 or 60 cm in length. Column numbers 1, 2, and 3 had lengths of 30 cm. These HS microcosms recovered 40% of the original DBCP introduced, with the reduction due to either volatilization or microbial degradation. Column 4, with a length of 60 cm showed an 85% recovery of DBCP. The linear plot of percent recovery vs the ratio of the mass of DBCP applied to the residence time of the chemical in the columns demonstrates that the dissipation rate is proportional to the mass applied and the rate of displacement.

Stuanes and Enfield (1984) utilized HS microcosms to determine if an analytical, one-dimensional convective dispersive solute transport equation could be used to approximate phosphorous (P) transport. Four Norway soils were tested for P removal. The soils were air dried and passed through a sieve. Duplicate samples were packed in small columns 4.2 cm i.d. and 1.0, 2.6, 3.0, and 3.9 cm in length. The experimental arrangement is depicted in Figure 5.3 (Stuanes and Enfield, 1984). The columns were supported on sintered glass plates and maintained at a constant 7.5 kPa negative pressure. A solution of 0.01 M $CaCL_2$ containing various amounts of KH_2PO_4 was applied to the columns. A solution without P was first added to attain steady state hydraulic conditions after which a solution with P was introduced to the column continuously until the effluent reached a constant level. Measurement of the P concentration in the effluent was performed by an automatic sampler. The column studies demonstrated an agreement between the proposed model and the experimental data. The proposed model adequately describes the movement of P in the column, but it does not address the processes governing P behavior in the soil.

Kuhn et al. (1985) researched microbial transformations of substituted benzenes during infiltration of river water to ground water. The experiments were conducted with two different HS microcosm column designs. The first one was a preliminary design of 4 cm i.d. by 20 cm in length composed of borosilicate glass. It contained three sampling ports positioned at 1.5, 4.7, and 9.1 cm from the inlet. The ports were made of stainless steel tubes projecting

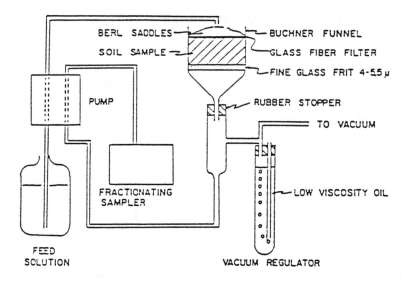

Figure 5.3.   HS microcosm design to test for phosphorus removal (Stuanes and Enfield, 1984).

2 cm into the center of the column. Nylon screens were placed at the ends of the tubes to prevent clogging. The inlet of the column was constructed with a glass fitted filter and membrane filter. The column was then wet packed with wet sieved aquifer material from the interface of a river/ground water infiltration site. This column design was primarily used to evaluate the feasibility of HS microcosms in investigating the microbial transformations taking place, and for this reason they were inoculated with xylene-degrading bacteria.

The second column design, a 5.0 cm i.d. by 100 cm length, was constructed of transparent plexiglas which does not adsorb hydrophobic organic compounds (Kuhn et al., 1985). Sampling ports were made of plexiglas stoppers fitted with stainless steel, capillary tubes with headpieces of sintered stainless steel positioned in the center of the column. The first port was located 2.7 cm from the inlet, with 19 additional ports at 5 cm intervals. These second columns were packed similarly to the first columns, but they were not inoculated and the sediment was air dried, dry sieved, and then wet packed.

Figure 5.4 depicts the overall experimental setup (Kuhn et al. 1985). The $CaCO_3/CO_2$ waters prepared in carboy A rinsed the columns before each experiment. A high performance liquid chromatography (HPLC) pump was used for the low flow rates in the experiment. The velocity of the water solution passing through column 1 (first design) was 4.0 cm/hr, and in column 2 (second design) it was 7.6 cm/hr. A syringe pump was used to deliver the appropriate dosage of the chemicals and ammonium into a stainless steel T-joint connection to the $CaCO_3/CO_2$ water at the base of the columns. The

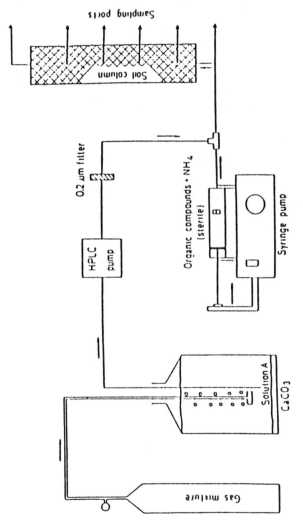

Figure 5.4. Saturated upflow HS microcosm of Kuhn et al. (1985); Solution A: reservoir containing mineral salts; Solution B: syringe containing concentrated organic compounds and ammonia. (Reprinted with permission from Kuhn, E. P., P. J. Colberg, J. L. Schnoor, O. Nanner, A. J. B. Zehnder, and R. P. Schwarzenbach, "Microbial Transformations of Substituted Benzenes During Infiltration of River Water to Groundwater: Laboratory Column Studies," *Environ. Sci. Technol.*, Vol. 19, No. 10, pp. 961–968, Copyright 1985, American Chemical Society).

columns were then shifted to 9 months of anoxic, denitrifying conditions. Aromatic hydrocarbons such as xylene were degraded under these conditions. These types of compounds were previously believed to be inert in the absence of molecular oxygen.

Laboratory HS microcosm experiments to study the sorption behavior of nonpolar organic compounds such as halogenated alkenes and benzenes in a river water-ground water infiltration system were performed by Schwarzenbach and Westall (1981). A 1.2 cm i.d. by 29 cm long glass column was wet packed with fine sand and a HPLC pump was used to attain low flow rates. The microcosms were incubated at a temperature of $20 \pm 0.5°C$. The experimental set-up is shown in Figure 5.5 (Schwarzenbach and Westall, 1981). Water ($CaCO_3/CO_2$) in bottle 1 rinsed the column prior to each experiment. The $CaCO_3/CO_2$ aqueous solution was spiked with the organic chemicals, and volatiles loss was minimized by the small headspace. Effluent samples were taken at the outlet of the column by immersing a stainless steel tubing into a preweighed gas washing flask containing volatile free water. The samples were then spiked by standards and analyzed immediately by a closed loop gaseous stripping/adsorption/elution procedure. The column influent was monitored by collecting samples by siphon at a higher flow rate. Partition coefficient values determined from column experiments were found to be similar to those determined from batch experiments performed by the authors. Column experiments conducted at velocities approximating 10^{-2} cm/sec showed the effect of slow sorption kinetics. Therefore, these HS microcosm studies indicated that sorption kinetics may have an effect on contaminant transport over the range of flow velocities encountered in aquifers.

5.4.3. Incremental Subsurface Microcosms

Incremental subsurface (IS) microcosms are those in which segments of the extracted subsurface material are packed into columns so as to maintain each section in the same relative position as in the original subsurface profile. The *in situ* lithology of the native subsurface material is approximated, but the material is still considered to be a disturbed sample.

Wilson et al. (1981) utilized IS microcosms to investigate the transport and fate of 13 organic pollutants in an Oklahoma sandy soil with low organic content. Columns of borosilicate glass 5 cm i.d. by 150 cm in length were packed with the soil to a depth of 140 cm. Packing was done in 10 increments so as to maintain the same relative position each increment occupied in the original soil profile. The saturated hydraulic conductivity of the packed columns ranged from 120 to 190 cm/d. The columns received 14 cm of water each day, which resulted in unsaturated flow through almost the entire length of

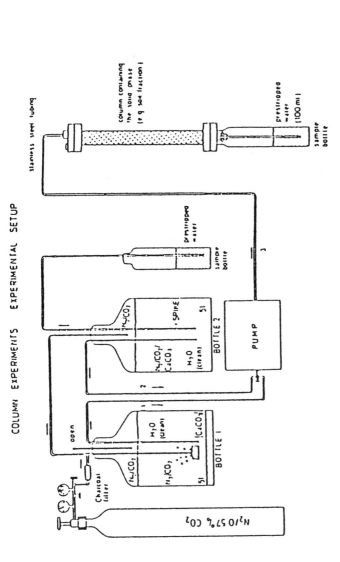

Figure 5.5.   HS microcosm system utilized by Schwarzenbach and Westall (1981) to study the behavior of the nonpolar organic compounds. The diagram includes: (1) clean-water input for column; (2) spiked-water input for column; (3) column influent (either 1 or 2); (4) sample of spiked water representative of column influent; and (5) sample of column effluent (Reprinted with permission from Schwarzenbach, R. P. and J. Westall, "Transport of Nonpolar Organic Compounds from Surface Water to Groundwater," *Environ. Sci. Technol.*, Vol. 15, No. 11, pp. 1360–1367, Copyright 1981, American Chemical Society).

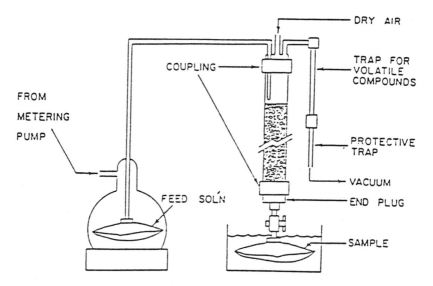

Figure 5.6.   Assembly used to study transport and fate of organic pollutants in an IS microcosm (Wilson et al., 1981).

the columns. The soil was not dried or sieved in order to protect the indigenous biota. To minimize potential sorption and/or leaching, all materials contacting the feed solution, the soil material, and the column effluent were made of either borosilicate glass or Teflon®.

The experimental apparatus used by Wilson et al. (1981) is shown in Figure 5.6. The feed solution was stored inside a Teflon® gas sampling bag to minimize the loss of volatile compounds. The bag was fitted inside a sealed glass container filled with water. Water was pumped into the container around the bag with a peristaltic metering pump, thus forcing the feed solution from the bag onto the IS microcosm. Air was delivered to the headspace above the soil at a pressure of 2 cm of water. Air was removed from the headspace at a rate of 12 mL/min which produced an air exchange once every 8 min. The bottom of the column had a port cap packed with pyrex glass wool. Effluent from the column was collected in a Teflon® gas sampling bag immersed in water in order to prevent air from the headspace from passing through the soil and inflating the bag. All experiments were conducted at a temperature of 20 ± 1°C.

The columns were dosed with spring water (Wilson et al., 1981). One column in each experiment received spring water alone and served as a control. Three replicate columns received a mixture of spring water with the organic compounds. In the first experiment, the water was dosed with 1 mg/L of the organic chemicals. In the second experiment, the concentration was 0.2 mg/L. The feed solution was applied for 45 d and all compounds reached steady state concentrations in the column effluent after 25 d. The quantities of compounds that volatilized from the soil surface were measured as well as the concentrations in the column effluent.

Most of the organic pollutants examined by Wilson et al. (1981) were readily transported through soil, which helps explain the occurrence of these compounds in ground water. The results of this study indicate that ground water underlying soils with low organic matter content are vulnerable to pollution by chloroform, 1-2-dibromo-3-chloropropane, dichlorobromometh-ane, 1,2-dichloroethane, tetrachloroethene, 1,1,2-trichloroethane, trichloroethene, toluene, nitrobenzene, and bis(2-chloroethyl) ether; ground water is vulnerable to a lesser extent to chlorobenzene, 1,2-dichlorobenzene, and 1,2,4-trichlorobenzene.

Piwoni et al. (1986) studied the behavior of organic pollutants during rapid infiltration of wastewater into soils using IS microcosms. The experimental design allowed for a direct measurement of the amount of volatilization and determination of the amount degraded. A fine sandy soil from the Lincoln series was collected in 10 cm increments from Ada, Oklahoma. The soil was packed into seven columns in the same relative position occupied in the original profile. The soil was not dried or sieved in order to protect the soil biota. The columns were planted with Reed Canary Grass, and the temperature was maintained at $20 \pm 1°C$. The experimental setup is shown in Figure 5.7 (Piwoni et al., 1986). All materials used were either borosilicate glass or Teflon®. The glass columns were 15 cm i.d. by 150 cm long, with sampling ports at 7.5, 15, 30, and 60 cm. The ports were fitted with suction samplers composed of a cylinder of sintered glass 1 cm i.d. and 2 cm long. The samplers were positioned at the mid line of the soil columns.

To allow determination of the rate of volatilization of the compounds, the top of each column was enclosed in a greenhouse 45 cm high. Room air was flushed through the greenhouse, resulting in replacement of the air every 8 min. The greenhouses were illuminated with white fluorescent lamps., and the microcosms were illuminated 12 hr every day. The soil columns were wrapped in foil to prevent algae growth on the walls of the columns (Piwoni et al. 1986).

Three of the columns received primary municipal wastewater from Ada, Oklahoma. Three other columns received wastewater dosed with organic chemicals. The concentrated dosing solution of the organic chemicals was stored without headspace in a Teflon® bag fitted inside a sealed glass container filled with water. Water was pumped into the container around the bag with a peristaltic metering pump, thus forcing the dosing solution from the bag into a chamber where it was mixed with the wastewater. This amended wastewater was delivered to the soil column using a siphon. A seventh column served as a control by receiving spring water dosed with the organic compounds. Each of the columns received 4.4 ± 0.17 cm of water per day.

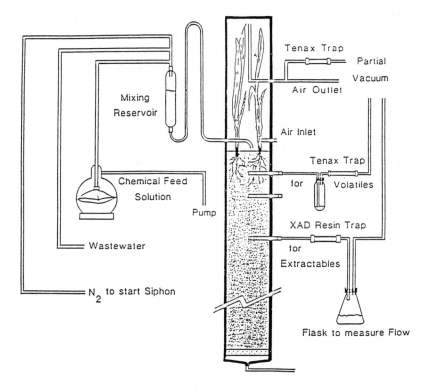

**Figure 5.7. IS microcosm utilized by Piwoni et al. (1986) to study
behavior of organic pollutants during rapid infiltration of
wastewater into soil.**

An equal volume application of wastewater to each column was made every
4 hours (Piwoni et al. 1986).

The study showed chloroform, 1,1-dichloroethane, 1,1,1-trichloroethane,
trichloroethene, and tetrachloroethene to be extensively volatilized. There
was observable degradation of 1,1-dichloroethane but no degradation of other
volatile compounds. Small proportions of chlorobenzene, 1,2-dichloroben-
zene, and toluene volatilized. The remaining portion was degraded extensively. The concentration of 1,2,4-trichlorobenzene in the effluent was ±0.7%
of that applied indicating extensive degradation. Degradation was also extensive for nitrobenzene, phenol, 2-chlorophenol, 2,3-dichlorophenol, and 2,3,6-
trichlorophenol (Piwoni et al. 1986).

Nichols (1987) studied the transport and fate of phenol released from the
land application of U.S. Navy aircraft paint stripping sludges by using IS
microcosms. The columns were made of beaded glass 5 cm. i.d. and 91 cm.
long as shown in Figure 5.8 (Nichols, 1987). Glass and caps with 7 mm i.d.
ports were used to gather the leachate at the base of the column. Stainless

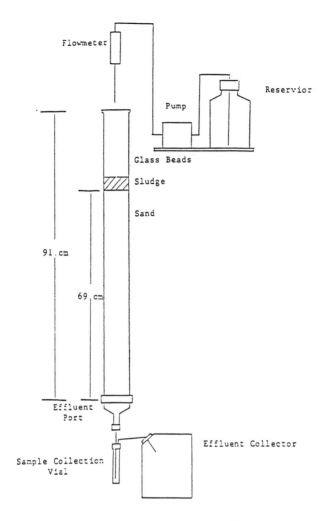

Figure 5.8.   Schematic of IS microcosm used in ^{14}C-labeled phenol study (Nichols, 1987).

steel couplings with high temperature viton rubber gaskets were used to connect the glass columns and end caps together. Teflon® valves and tubing were used to plumb the columns for sample collection. A Monostat cassette pump was used to pump tritiated water influent from a nalgene carboy reservoir to the column. A flowmeter in the line between the pump and influent discharge was used to measure the flowrate. Approximately 5 cm of boiling beads were placed on top of the sludge to better distribute the influent.

The subsurface media used in the columns was Lincoln sand taken from a quarry in Ada, Oklahoma (Nichols, 1987). The samples were taken in successive 10 cm increments from an exposed vertical face. A portion of the sample from each 10 cm increment was then passed through a number 200 sieve (2 mm openings). The sieved Lincoln sand increments were then placed in the columns at the same relative position they occupied in the original profile. The increments were packed in the column in 2 to 3 cm lifts. The columns were then saturated from the bottom up using capillary action to draw in the water. Water was drawn through Teflon® tubing from a nalgene carboy up into the columns. The water level in the carboy was slowly raised as the wetting front in the columns advanced, but was always kept below the wetting front until the entire column was saturated. After the column was saturated, the water level was raised above the top of the soil so as to produce a column of free standing water above the soil. Wetting the columns slowly in this manner ensures complete saturation of the soil pore spaces.

Phenol was chosen as the study pollutant within the paint stripping sludge (Nichols, 1987). Radio-labeled (^{14}C-labeled) phenol was used along with tritiated (^{3}H) water to facilitate sample analyses via radioisotopic techniques. A sample of sludge was obtained from a U.S. Navy facility in Pensacola, Florida, and spiked with 0.103 mg ^{14}C phenol per kg of sludge. A total of 50 gm of spiked sludge was mixed with 250 gm of sand and the sludge-sand mixture was then applied to each of four IS microcosms. Prior to the application, two of the four microcosms were sterilized with a 2% formaldehyde solution.

The results from the IS microcosm study by Nichols (1987) indicated that phenol in the sludge is indeed mobile and can be leached from the sludge. The leaching was represented by a two-stage process. The first stage involved displacement for ^{14}C phenol spiked pore water with the tritiated influent and accounted for approximately 30 to 50% of the ^{14}C phenol removal. After most of the spiked pore water had been removed, ^{14}C phenol that had been adsorbed to the sludge particulates began to desorb and was leached from the sludge. Comparisons of data from the sterile and nonsterile IS microcosms indicated that there was no significant removal of phenol from the leachate due to biological degradation. This was probably a result of minimal indigenous bacteria in both the paint stripping sludge which is generated by chemical precipitation, and in the Lincoln sand in the IS microcosms. Nichols (1987) noted that when considering land application of paint stripping sludge an effort should be made to avoid sandy soils to prevent the rapid transport of phenol through the vadose zone. In addition, it was recommended that the sludge be mixed with a source of bacteria, such as the waste sludge from an activated sludge reactor or an attached growth operation, to promote biodegradation of the organics in the paint stripping sludge. Land application on fine grained soils of paint stripping sludge mixed with a source of bacteria will reduce both the rate of transport and the mass of phenol available for movement to the ground water.

Nanjundeswar (1988) used IS microcosms in a study focused on *in situ* bioreclamation of subsurface gasoline spills. Aquifer material samples and ground water samples from saturated subsurface material were obtained from the alluvium and terrace deposits of the South Canadian River, located in McClain County in central Oklahoma. The aquifer material samples were collected using aseptic techniques. Two IS microcosms were used — one for studying biotic processes and the other for abiotic processes. The columns were designed so that all surfaces in contact with water, soil and gasoline were either glass or Teflon®. The columns consisted of three lengths of glass tubes (internal diameter of 0.4 cm), connected with Teflon®-lined watertight couplings to form a column of an overall length of 30 cm. The columns had sampling ports at the top, middle, and bottom as shown in Figure 5.9 (Nanjundeswar, 1988). Glass fiber was placed inside the columns near each port. Glass beads were placed over the glass fiber at the bottom influent port of each column. This was done to ensure uniform distribution of feed water over the entire area of each column. The aquifer media was packed into each column in 2 to 3 cm lifts. The columns were then vibrated to ensure proper compaction. After the columns were completely filled with soil, Teflon® tubes were fixed to the influent and effluent ports. Both columns were completely wrapped with aluminum foil to prevent light from affecting the microflora in the aquifer media. Both IS microcosms were saturated by slowly wetting with ground water from the bottom up using a Monostat cassette pump.

A known volume of gasoline was allowed to infiltrate the soil from the top of each column using inverted volumetric flasks (Nanjundeswar, 1988). Ground water was then allowed to infiltrate into each column to flush out free gasoline through the bottom. Ground water, enriched with oxygen and nutrients, was then pumped from the bottom of each column at a rate of about 10 to 15 pore volumes per day. This rate is fairly typical for *in situ* bioreclamation projects. Influent ground water to the abiotic column received 1 mg/L of mercuric sulfate as a sterilizing agent. The increased oxygen concentration in the feed water was achieved by diffusing pure oxygen through the ground water in influent reservoirs.

Table 5.4 contains a summary of the processes apparently affecting the benzene, toluene, and xylene concentrations in each column (Nanjundeswar, 1988). The dominant removal process for each constituent appeared to be physical displacement. Biological degradation seemed to account for some removal for toluene and xylene, with the major zone of occurrence being at the bottom portion of the biotic column (the nutrients and oxygen were introduced at the bottom). Adsorption and subsequent desorption appeared to account for observed initial increases in the concentrations of toluene, and *p*, *m*, and *o*-xylenes in the top portions of the columns. More specifically, it was found that hydrocarbons at high concentrations in the aquifer media were

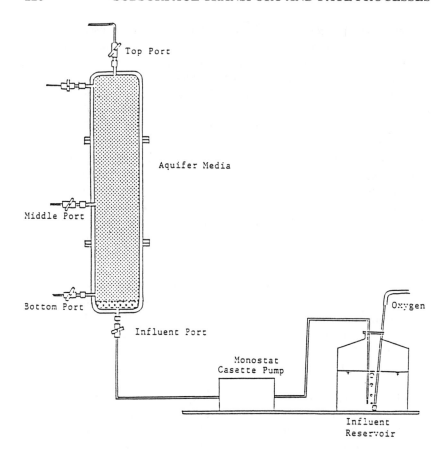

Figure 5.9.   **Experimental design of IS microcosms for *in situ* bio-degradation study (Nanjundeswar, 1988).**

rapidly mobilized (physical displacement) by the infiltrating ground water. The physical displacement process depends on several properties of the hydrocarbon including the octanol/water partition coefficient. The octanol/water partition coefficients for benzene, toluene, and *p*-xylene are 1.56 to 2.15, 2.69, and 3.15, respectively. The lower octanol/water partition coefficient of benzene may partially explain the more rapid physical displacement of benzene when compared to toluene or xylene. At low concentrations, the rate of physical displacement decreased. Under such conditions, biodegradation can become a major mechanism for removing hydrocarbons.

Nanjundeswar (1988) noted that IS microcosms are useful in developing design information for *in situ* biorestoration projects. Factors which can be studied or information which can be developed include:

Table 5.4 Summary of Observed Removal Processes in IS Microccosms

	Top Portion of Column	Middle Portion of Column	Bottom Portion of Column
Benzene			
Abiotic	PD[a]	PD	PD
Biotic	PD	PD	PD
Toluene			
Abiotic	A, PD	PD	PD
Biotic	A, B (?), PD	B (?), PD	B, PD
p-Xylene			
Abiotic	A, PD	PD	PD
Biotic	B, A, PD	B (?), PD	B, PD
m-Xylene			
Abiotic	A, PD	PD	PD
Biotic	B, A, PD	B (?), PD	B, PD
o-Xylene			
Abiotic	A, PD	PD	PD
Biotic	B, A, PD	B (?), PD	B, PD

Source: Nanjundeswar, 1988.

a PD, decrease in concentration due to physical displacement (deporption and dissolution; B, decrease in concentration due to biological degradation; and A, initial increase in concentration due to movement of previously adsorbed or retained contaminant into that portion of the column.

1. The influence of ground water flow rate
2. Transport of oxygen and nutrients to the zone of contamination
3. Oxygen and nutrients requirements
4. Physical and biological processes for removal of different types of organics
5. The time needed for complete restoration of the aquifer
6. The costs for *in situ* bioreclamation

Huff (1988) used IS microcosms as a screening tool for determining the mobility of hazardous wastes in the vadose zone. The soil was collected at a site near Ada, Oklahoma, and it was a mixed (thermic type ustifluent) of the Lincoln series. Successive subsamples were taken by removing soil in 10 cm increments from the surface down to 150 cm. The IS microcosms consisted of borosilicate glass 5 cm i.d. by 150 cm in length as shown in Figure 5.10 (Huff, 1988). Soil from each of the 10 cm increments was packed into the columns in the relative position it occupied in the original environment.

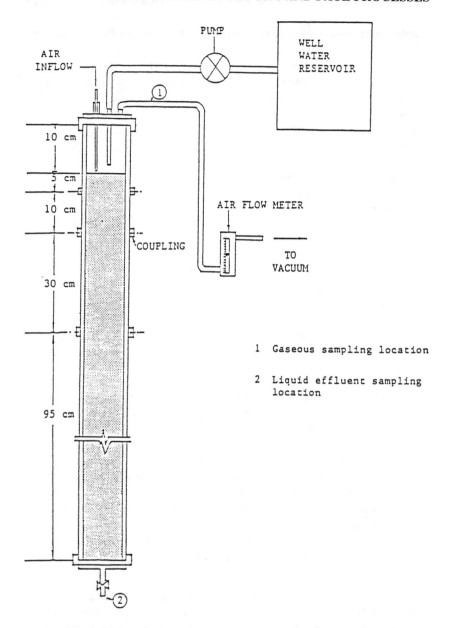

Figure 5.10. Vadose zone IS microcosm for determining waste constituent mobility (Huff, 1988).

The soil was packed into the columns in 2 to 3 cm increments. Each increment was packed by raising and dropping (approximately 30 times) a small solvent cleaned spatula (3 cm × 15 cm) attached to a stainless steel rod. The spatula was slowly rotated during the raising and dropping action to allow for better packing. The soil from the 140 cm depth level up to the 10 cm level was packed before wetting. The top 10 cm increment was packed at the time of dose application.

Each column was wetted from the bottom up using a standing head reservoir connected to the bottom cap port (Huff, 1988). The water table was raised slowly (20 to 30 cm/d), allowing water to first enter the soil by capillary action. The water table was kept below the wetting front until the entire column was wet. After the water table rose above the surface of the soil, saturated downward flow was initiated and maintained with a constant head of at least 5 cm above the soil surface.

A total of 14 IS microcosms were assembled by Huff (1988), and these were divided into 2 sets of 7. One column from each set remained undosed and served as an analytical control. Six columns of one set received a pulse dose of five target organic compounds, the other six columns received a single dose of a complex petroleum refinery API separator sludge waste. The five target organic compounds were also added to the sludge waste columns. The target organic compounds used by Huff (1988) included the volatiles benzene and o-xylene; and the base neutrals 2-methylnaphthalene, phenanthrene, and pyrene. The microcosms were designed so that gaseous samples could be collected from the headspace at the top of the columns and liquid effluent samples could be collected from the bottom of the columns. The base neutrals were monitored in the liquid effluent only. A flow approximately equal to 10% of the saturated hydraulic conductivity was maintained in each column from a constant input of feed water metered by cassette pumps.

The results of the volatile and base neutral analyses from the headspace and effluent of the IS microcosms indicated that vadose zone soil columns can be successfully used to determine the movement of organic compounds. Only a small degree of variability was observed in the hydraulic characteristics and transport data collected from replicate microcosms. Benzene and o-xylene were released from the microcosms in both liquid and gaseous phases. The transport of benzene and o-xylene appeared to be differentially affected by the presence of the organic materials in the API separator sludge. Pyrene, phenanthrene, and 2-methylnaphthalene were not detected in liquid effluent samples. Huff (1988) concluded that although significant progress was made in developing the use of microcosms as a screening tool, more experiments need to be conducted using different soils and organics under various conditions in order to adequately validate this procedure and to correlate microcosm results to field data.

5.4.4. Subsurface Core Microcosms

Subsurface material cores excised intact and transferred to a microcosm are considered the best mimic of the environment since the soil structure is retained and the abiotic and biotic complexity of the ecosystem is well simulated. These types of microcosms are called subsurface core (SC) microcosms and are often referred to as undisturbed samples.

The biodegradation of selected volatile organic pollutants was studied by Wilson et al. (1985) in SC microcosms. Aseptic subsurface material from below the water table at Pickett, Oklahoma was transferred into glass columns 7.0 cm i.d. by 40 cm in length; the columns had sampling ports at 20 cm and at the effluent end. Figure 5.11 depicts the design used in the study (Wilson et al., 1985). A feed solution for the columns was prepared from municipal wastewater that had been renovated by passage through 1.5 m of unsaturated soil. The feed solution was stored in a Teflon® bag placed in a plexiglas box filled with water. The bag was void of headspace to prevent loss of volatiles. Pressure from a reservoir forced the feed solution onto three replicate SC microcosms. A peristaltic pump maintained a downward flow of 2 cm/d.

Sampling for volatile components was performed by Wilson et al. (1985) by inserting a sterile gas tight syringe into the sampling port. Sampling for nonvolatile components was done by opening the port and collecting the liquid effluent. The SC microcosms were incubated at 17°C. The aquifer material had little organic material and, as a result, retardation of the selected organic compounds was found to be very low. In addition, there was minimal evidence of biodegradation.

The contamination of aquifers with trichloroethylene and some of its related compounds have caused researchers such as Wilson et al. (1987) to study the chemical fate and biodegradability in SC microcosm studies. The results of these studies are relevant to the development of treatment technologies. In oxygenated ground water, these compounds are resistant to biodegradation, but it has been shown that they can be co-metabolized by bacteria that are able to oxidize either methane or propane.The authors developed an SC microcosm with undisturbed soil cores acquired from four bore holes at the Moffett Naval Station in California. The core material was then packed into glass columns of 4 cm i.d. by 24 cm length. Duplicate columns were packed for each borehole. The columns were first perfused with water from the Moffett site sparged with oxygen. An aqueous solution of trichloroethylene and 1,1,1-trichloroethane and methane or propane was then added to the columns. The columns were sealed and incubated for ten days. Finally, the columns were perfused with 150 mL of deoxygenated Ada, Oklahoma tap water to collect 150 mL of effluent for the analysis of oxygen, methane, propane, trichloroethylene, and 1,1,1-trichloroethane.

Two sets of duplicate SC microcosms containing propane and two dupli-

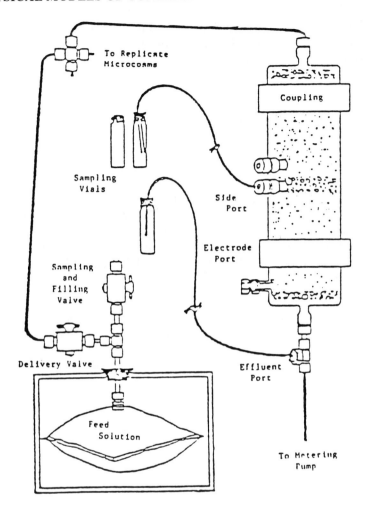

Figure 5.11.   SC microcosms used to determine the transport and fate of volatile organic pollutants ("Biodegradation of Contaminants in the Subsurface," Wilson, J. T., M. J. Noonan, and J. F. McNabb in *Ground Water Quality*, C. H. Ward, W. Giger, and P. L. McCarty, Eds., Copyright 1985, John Wiley and Sons. Reprinted by permission of John Wiley & Sons).

cate sets containing methane were perfused during the study (Wilson et al., 1987b). To provide a control for nonbiological removal, one SC microcosm for each set of duplicates was poisoned with sodium azide. Trichloroethylene in the SC microcosms containing propane disappeared gradually while the microcosm poisoned with azide showed a 40% recovery. The 1,1,1-trichloroethane in the propane-acclimated SC microcosms disappeared rapidly, whereas the poisoned microcosm had a 100% recovery. The extent of biodegradation in the methane-treated columns was lower than the propane columns.

5.5 ADVANTAGES AND LIMITATIONS OF MICROCOSM STUDIES

As illustrated by the examples of different types of microcosms, usage of these systems has primarily focused on developing a better understanding of subsurface transport and fate processes. The generic advantages of microcosm studies have been enumerated by several authors as follows (Gillet et al., 1977; Ausmus et al., 1980; Gillet and Witt, 1980; Pritchard, 1982; and Pritchard and Bourquin, 1983):

1.   Microcosms are convenient to use; their small size allows for a control of the environmental variables; and they permit replication and the use of controls.
2.   The experimental variables, both physical and chemical, can be varied in different ways and the effects monitored and analyzed.
3.   The structure of the system can be made to represent the structure of the environment (i.e., SC microcosms).
4.   Microcosm studies represent realistic interactions among physical, chemical, and biological processes.
5.   Microcosms provide indices of temporal and spatial distributions that are of interest when studying the transport, fate, and effect of chemicals on the environment.
6.   Microcosms provide an intermediate step in verifying the effects of certain actions on the subsurface environment.
7.   Microcosm systems are practical, time effective, and cost effective.
8.   Microcosms provide quantitative background information for regulatory decisions.

Microcosm studies also have limitations that should be considered. These limitations include

1.   Not all processes of a given environment or ecosystem may be included in a single microcosm study — this technique is then restricted by its size.
2.   Microcosms are not self sustaining.

3.   Containerization of environmental components may result in structural and functional changes that do not correspond to the environment from which the samples were taken.
4.   Biological portions of microcosms must be carefully introduced into the microcosms and controlled, since otherwise they might be destroyed.
5.   Laboratory microcosms have unnaturally high surface area to volume ratios due to large vessel wall surfaces.
6.   Microcosm studies may lack estimated confidence intervals around their measurements of the different parameters.
7.   Statistical analysis needed for microcosm results is not well documented.
8.   A lack of demonstrated applicability of microcosm studies is due to the lack of verification from laboratory conditions to the field situation.

One of the most significant issues related to microcosm studies is whether or not microcosm-based transport and fate processes match those processes at field sites. Wilson et al. (1987) conducted a field evaluation of a SC microcosm used for simulating the behavior of volatile organic compounds in subsurface materials. The microcosm was evaluated by comparing the behavior of tetrachloroethylene, bromoform, carbon tetrachloride, 1,2-dichlorobenzene, and hexachloroethane in: (1) the microcosm, (2) the microcosm constructed with autoclaved material, and (3) an experimental plume developed in a joint field study conducted by Stanford University and the University of Waterloo at a site on the Canadian Forces Base, Borden, Ontario, Canada. The microcosm study adequately simulated the nonbiological removal of these compounds from solution in the experimental plume. However, it failed to detect biotransformation of 1,2-dichlorobenzene despite extensive degradation of this compound in the field. One of the samples of aquifer material used to construct the microcosms had a much greater capacity to biotransform bromoform and carbon tetrachloride than did the other samples of aquifer material, or the aquifer itself. The microcosm did accurately simulate the behavior of tetrachloroethylene and hexachloroethane.

5.6 DESIGNING A MICROCOSM STUDY

The first step in designing a microcosm study requires outlining the goals of the study and identification of the natural ecosystem being modeled (Callendar and Canter, 1980). Thereafter, attention is given to the materials

used in construction of the physical system, stability and sensitivity of analytical equipment, quality controls for biotic and abiotic components, and
finally, the costs and feasibility of the above parameters (Gillet and Witt,
1980).

When designing the physical portion of the microcosm (container, fittings,
tubing, valves, etc.), the main considerations in the selection of proper materials
are that the substances be biologically and chemically inert, that the encasement
structure maintains the subsurface profile structure, and that channeling of
water down the sides of the microcosm be minimized. Generally, glass and
Teflon® will satisfy these requirements, and consequently, they are used
extensively in microcosm studies. However, all components should be
thoroughly cleaned to ensure that no organic compounds will be present to
cause interferences or false readings (Huff, 1988).

The number of experimental units included in the study should permit
replicates for the purpose of verifying the validity and accuracy of test data.
Analytical control units for each parameter are essential to proper experimental
design. Other design factors that must be considered in a microcosm study
include features such as temperature, which can be controlled by water baths
or temperature controlled rooms. Lighting can be provided, when necessary,
by a bank of cool white fluorescent lights, thus avoiding undue heat problems.
Various types of monitoring equipment are often incorporated into a microcosm
system to provide constant measurements of dissolved oxygen, pH, carbon
dioxide, and temperature (Pritchard and Bourquin, 1983).

The duration of the microcosm study will depend on the particular project
goal. For this reason, pretest studies are often performed to establish preliminary
information based on various flow rates and contaminant concentrations. The
incubation period can, therefore, vary from hours to months depending on the
study. In addition, contaminants can be introduced to a microcosm in a single
dose or by continuous dosing. Clearly then, there is not a single microcosm
or study design, but rather, a variety of systems that can be designed depending
on the parameters of study, the goals of the program, and the hypotheses to
be tested.

5.7 ISSUES AND CONCERNS

This review of microcosms used in studies of the subsurface transport and
fate of organic contaminants demonstrates that there is current emphasis toward
the development of a better understanding of relevant physical, chemical, and
biological processes in the subsurface. Information developed via microcosm
studies can be used in ground water modeling and monitoring, the develop-

ment of protection strategies, site selections for waste disposal, and contaminant plume management and ground water remediation decisions. Microcosm designs have ranged from static test tubes to flow-through columns containing intact cores of subsurface materials.

Two fundamental needs relative to microcosm studies of the subsurface environments can be identified. One is related to the development of standard protocols for microcosm studies so as to provide quality control relative to generated data. The second, and perhaps most important need, is for additional studies to demonstrate the relationships between microcosm study results and the experienced transport and fate processes actually occurring in the field.

5.8 INNOVATIONS IN MICROCOSM TECHNOLOGY

Many different types of physical models exist, of which soil microcosms are but one class. Canter et al. (1987) give overview discussions of the various types of physical models and their applications. The earliest physical models were the analog models used primarily to assess the response of saturated formations to various hydraulic stresses. The early applications of scale models (microcosms) also focused on quantitative (flow) rather than qualitative (contaminant transport) processes. In fact, the experiments of Henri Darcy were conducted using simple sand columns.

As discussed in Chapter 2, many of the early applications of soil microcosms (soil columns) focused on the hydrodynamic transport processes of advection and dispersion. Much effort was exerted toward quantifying the alleged porous media characteristic called dispersivity through use of tracer column studies. The consensus opinion to date is that the dispersive characteristics of a porous medium are a function of the scale of the investigation. Hence, the results from a laboratory soil column study have little applicability to a field scale problem.

As outlined in Chapter 4 and in the earlier sections of this chapter, soil microcosm studies of the biological effects on transport and fate of contaminants have seen increased popularity. However, novel applications of microcosm technology are also being used to assess other transport processes. Gronow et al. (1988) discuss accelerated testing through use of centrifuged soil columns. If successful, accelerated testing methods could significantly reduce the inherent time consuming nature of most soil column studies.

The sophistication of soil column studies has also increased rapidly. West (1990) discusses the results of laboratory soil column studies aimed at assessing transport of macromolecules and colloids through saturated porous media.

The hypothesis that part of solid subsurface matrix is not immobile is relatively new, as are the applications of microcosm technology designed to study the phenomenon. The report outlines meticulous laboratory preparation procedures and detailed analytical methods that are now standard practice for most soil column studies.

5.9 OTHER PHYSICAL MODELS

Probably the area receiving the most attention in terms of physical modeling is multiphase flow processes. The behavior of multiple immiscible fluids in the saturated and unsaturated zones is of interest because of the large number of releases of nonaqueous phase liquids now being discovered. The intense activity associated with the nation's underground storage tanks has identified the need for better descriptions of the movement of both light and dense nonaqueous phase liquids in the subsurface. Physical models are currently being used to help develop these descriptions.

Outside of work done in the petroleum industry, the most renowned study of nonaqueous fluids in porous media is that of Schwille (1988). This study used a variety of physical models that the author refers to as trough (two-dimensional sand tank) experiments, column experiments, glass frit (glass filter) experiments, and lysimeter (unsaturated soil) experiments. The unique aspects of these experiments included the use of dyes to visually distinguish the various phases and the use of macroscopic photography to capture phase behavior at the pore size level. Time varying photographs show the dyed solvents moving through the media under various hydrogeologic scenarios. The microscopic photographs depict the position of both the water and the solvent within the pore network of packed glass beads. The ability to visualize enhances the verbal descriptions of multiphase processes.

Other studies have been conducted using physical models to assess multiphase flow in porous media. Schiegg and McBride (1987) describe a two-dimensional model for studying oil propagation in simulated vadose and water- saturated zones. The model is equipped with instruments to develop liquid pressure and fluid saturation data. Milligan and Durnford (1989) used laboratory soil columns to validate a theoretical equation for determining actual petroleum thickness in soils based on observed thicknesses in monitoring wells. The computed actual thicknesses agreed reasonably well with the values observed in the soil columns.

A study by Wilson et al. (1990) on residual liquid organics in soils used a variety of physical models and also developed some novel innovations to microcosm technology. The four different experimental approaches are described below (Wilson et al., 1990):

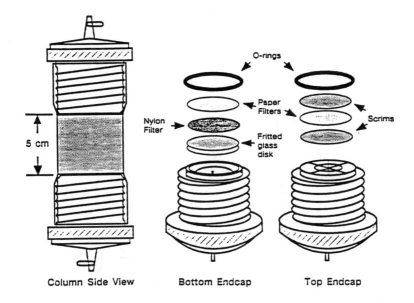

O-rings

Nylon
Filter

Paper
Filters

Scrims

Fritted
glass
disk

5 cm

Column Side View   Bottom Endcap   Top Endcap

Figure 5.12.   Short column apparatus, with blow-up views of the endcaps
and filters (Wilson et al., 1990).

1.   **Quantitative displacement experiments using short columns**
were performed to relate the magnitude of residual organic liquid
saturation to fluid and soil properties and to the number of fluid
phases present (i.e., both saturated and vadose zone conditions —
see Figure 5.12)

2.   **Quantitative displacement experiments using long columns**
were performed under two-phase saturated zone conditions,
yielding water and organic liquid relative permeabilities; reductions
of residual organic saturation were correlated to the pressure
gradient applied in hydraulic sweeps, and the potential for
hydraulic mobilization of residual "blobs" was investigated —
see Figure 5.13

3.   **Pore and blob casts** were produced for saturated zone conditions
by a technique in which the organic liquid was solidified in place
within a soil column at the conclusion of a displacement
experiment, allowing the distribution of organic liquid to be
observed. The polymerized organic phase was rigid and chemically
resistant. Following polymerization, the water phase was removed
and replaced by an epoxy resin. The solid core, composed of soil,
solidified styrene (the organic phase), and epoxy resin (the water
phase), was cut into sections to show the organic liquid phase in
relation to the soil and the water phase. The sections were pho-

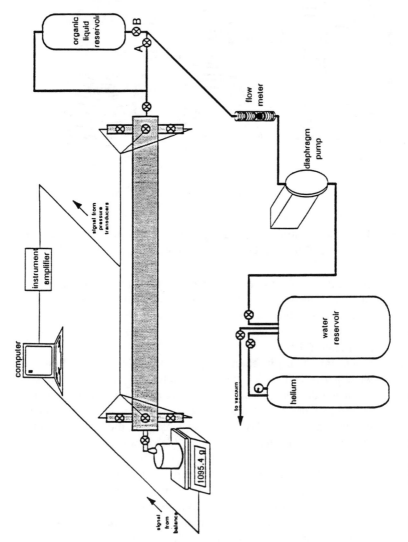

Figure 5.13.   Long column experimental setup (Wilson et al., 1990).

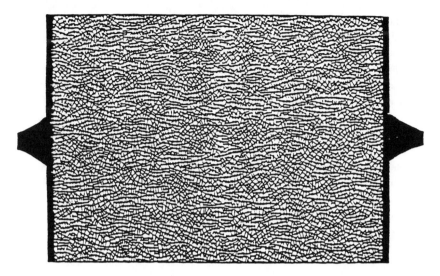

Figure 5.14.   Pore-network pattern for the homogeneous model (Wilson et al., 1990).

tographed under an optical microscope. Although polymerization only gave a "snapshot" of the displacement process, it offered the advantage of seeing organic liquid in its "natural habitat" (i.e., within a soil) as compared to that observed in etched glass micromodels. Sometimes, instead of replacing the water with epoxy resin, the solid matrix of the soil column was dissolved with hydrofluoric acid, leaving only the hardened organic liquid. The solidified organic phase was then observed under a scanning electron microscope (SEM) and photographed. For vadose zone conditions, styrene and epoxy liquids were sequentially applied, drained and hardened in an attempt to simulate proper fluid distributions above the water table. The resulting pore casts were photographed under an optical microscope.

4.   **Etched glass micromodels** were used to observe dynamic multiphase displacement processes. Micromodels (Figure 5.14) provide two-dimensional networks of three-dimensional pores. They offer the ability to actually see fluids displace one another in both a bulk sense and also within individual pores. Although displacements are known to be dependent upon a variety of factors, this report describes micromodel experiments that focused on only three: (1) the fluid flow rate, (2) the presence of heterogeneities, and (3) the number of fluid phases present. The experiments were photographed and videotaped.

The authors concluded that the flow visualization techniques above were useful tools for examining multiphase displacement processes. They also stress that flow visualization approaches could be applied to other pollutant transport problems including: diffusion, dispersion, and macrodispersion; bacteria and colloid adhesion and migration; biotransformation of organic pollutants; adsorption of dissolved organic pollutants; and the possible alteration of soil wetting properties (Wilson et al., 1990).

In summary, the use of physical models for studying subsurface transport and fate processes is becoming increasingly sophisticated. Ground water contamination problems have traditionally been misunderstood because of the "hidden" nature of the problem. Recent innovations in physical modeling now allow visualization of fluid(s) behavior in the soil matrix, thus increasing our immediate understanding of the underlying processes governing the behavior.

REFERENCES

Ashford, A. "Microcosms for Subsurface Pollutant Transport and Fate Studies," Research Report, University of Oklahoma, Norman, OK (1987).

Ausmus, B. S., P. van Voris, and D. R. Jackson "Terrestrial Microcosms: What Questions Do They Address?," in *Microcosms in Ecological Research*, J. P. Giesy, Ed., Technical Information Center, U.S. Department of Energy, Washington, D.C. (1980), pp. 937–953.

Biggar, J. W., D. R. Nielsen, and W. R. Tillotson "Movement of DBCP in Laboratory Soil Columns and Field Soils to Groundwater," *Environ. Geol.* 5(3):127–131 (1984).

Callender, A. B. and L. W. Canter "Microcosm Studies in Environmental Research," Research Report, University of Oklahoma, Norman, OK (1980).

Canter, L. C., R. C. Knox, and D. Fairchild "Flow and Solute Transport Modeling," in *Ground Water Quality Protection*, (Chelsea, MI: Lewis Publishers, Inc., 1987), pp. 209–276.

Giles, J. D., J. C. Collins, and J. W. Gillet "The Soil Core Microcosm — A Potential Screening Tool," EPA 600/3-79-089, U.S. Environmental Protection Agency, Corvallis, OR (1979).

Gillet, J. W. and J. M. Witt "Chemical Evaluation: Projected Application of Terrestrial Microcosm Technology," in *Microcosms in Ecological Research*, J. P. Giesy, Ed., Technical Information Center, U.S. Department of Energy, Washington, D.C. (1980), pp. 1008–1033.

Gillet, J. W., J. M. Witt, and C. J. Wyatt "Terrestrial Microcosms," in *The Proceedings of the Workshop on Terrestrial Microcosms*, J. M. Witt, J.

W. Gillet, and C. J. Wyatt, Eds.,(Washington, D.C.: National Science Foundation, 1977), pp. 1–18.

Gronow, J. R., R. I. Edwards, and N. A. Schofield "Drum Centrifuge Study of the Transport of Leachates from Landfill Sites," Cambridge University, England (1988).

"Handbook: Ground Water," Center for Environmental Research Information, Cincinnati OH: U.S. EPA Report-625/6-87/016 (March, 1987).

Huff, D. R. "Transport of Benzene, o-Xylene, 2-Methylnaphthalene, Phenanthrene, and Pyrene in Vadose Zone Microcosms," PhD Dissertation, University of Oklahoma, Norman, OK (1988).

Kuhn, E. P., P. J. Colberg, J. L. Schnoor, O. Nanner, A. J. B. Zehnder, and R. P. Schwarzenbach. "Microbial Transformations of Substituted Benzenes During Infiltration of River Water to Groundwater: Laboratory Column Studies," *Environ. Sci. Technol.* 19(10):961–968 (1985).

Mahadevaiah, B. "Transport and Fate of Petroleum Hydrocarbons from Leaking Underground Storage Tanks," MS Thesis, University of Oklahoma, Norman, OK (1985).

Mahadevaiah, B. and G. D. Miller "Application of Microcosm Technology to Study the Biodegradation Potential of a Subsurface Alluvial Material Exposed to Selected Petroleum Hydrocarbons," in *Proceedings of Sixth National Symposium and Exposition on Aquifer Restoration and Ground Water Monitoring,* (Columbus, OH: National Water Well Association, 1986).

Milligan, J. D. and D. Durnford "Petroleum Thickness in Groundwater — A Laboratory Study," ESL-TR-89-53, Engineering and Services Laboratory, Tyndall Air Force Base, FL (1989).

Nanjundeswar, B. V. "*In Situ* Bioreclamation for Subsurface Gasoline Cleanup," MS Thesis, University of Oklahoma, Norman, OK (1988).

Nichols, R. L. "Transport and Fate of Phenol in Paint Stripping Sludges Applied to Soil Columns," MS Thesis, University of Oklahoma, Norman, OK (1987).

Piwoni, M. D. et al. "Behavior of Organic Pollutants During Rapid Infiltration of Wastewater into Soil: I. Process Definition and Characterization Using a Microcosm," *J. Haz. Waste* 3:43–56 (1986).

Pritchard, P. H. "Model Ecosystems," in *Environmental Risk Analysis for Chemicals,* R. A. Conway, Ed. (New York: Van Nostrand Reinhold Company, 1982), pp. 257–353.

Pritchard, P. H. and A. W. Bourquin "The Use of Microcosms for Evaluation of Interactions Between Pollutants and Microorganisms," EPA-600/D-83-050, U.S. Environmental Protection Agency, Gulf Breeze, FL (June, 1983).

Schiegg, H. O. and J. F. McBride "Laboratory Setup to Study Two-Dimensional Multiphase Flow in Porous Media," in *Petroleum Hydrocarbons and Organic Chemicals in Ground Water Conference,* (Columbus, OH: National Water Well Association, 1987).

Schwarzenbach, R. P. and J. Westall "Transport of Nonpolar Organic Compounds from Surface Water to Groundwater," *Environ. Sci. Technol.* 15(11 November):1360–1367 (1981).

Schwille, F. *Dense Chlorinated Solvents in Porous and Fractured Media* (Chelsea, MI: Lewis Publishers, Inc., 1988).

Stuanes, A. O. and C. G. Enfield "Prediction of Phosphate Movement Through Some Selected Soils," *J. Environ. Qual.* 13(2):317–320 (1984).

U.S. Environmental Protection Agency "Handbook: Ground Water," Center for Environmental Research Information, U.S. EPA Report-625/6-87/016, Cincinnati, OH (March, 1987).

West, C. C. "Transport of Macromolecules and Humate Colloids Through a Sand and a Clay Amended Sand Laboratory Column," EPA/600/52-90/020, U.S. Environmental Protection Agency, Ada, OK (1990).

Wilson, B. H. "Behavior of Trichloroethylene, 1,1-Dichloroethylene, *cis*-1,2-Ichloroethylene, and *trans*-1,2-Dichloroethylene in Anaerobic Subsurface Environments," MS Thesis, University of Oklahoma, Norman, OK (1985).

Wilson, B. H., G. B. Smith, and J. F. Rees "Biotransformations of Selected Alkylbenzenes and Halogenated Aliphatic Hydrocarbons in Methanogenic Aquifer Material: A Microcosm Study," *Environ. Sci. Technol.* 20(10):997–1002 (1986).

Wilson, J. T., J. F. McNabb, D. L. Balkwill, and W. C. Ghiorse "Enumeration and Characterization of Bacteria Indigenous to a Shallow Water-Table Aquifer," *Ground Water* 21(2):134–142 (1983).

Wilson, J. T., G. B. Smith, J. W. Cochran, J. F. Barker, and P. V. Roberts "Field Evaluation of a Simple Microcosm Simulating the Behavior of Volatile Organic Compounds in Subsurface Materials," *Water Resour. Res.* 23(8 August):1547–1553 (1987a).

Wilson, J. T., C. G. Enfield, W. J. Dunlap, R. L. Cosby, D. A. Foster, and L. B. Baskin "Transport and Fate of Selected Organic Pollutants in a Sandy Soil," *J. Environ. Qual.* 10(4):501–506 (1981).

Wilson, J. T., S. Fogel, and P. V. Roberts "Biological Treatment: *In Situ* Treatment of TCE," in *Proceedings of a Symposium on Detecting, Control and Renovation of Contaminated Ground Water*, N. Dee, W. F. McTernan, and K. Kaplan, Eds. (New York: American Society of Civil Engineers, 1987b), pp. 168–178.

Wilson, J. T., M. J. Noonan, and J. F. McNabb "Biodegradation of Contaminants in the Subsurface," in *Ground Water Quality*, C. H. Ward, W. Giger, and P. L. McCarty, Eds. (New York: John Wiley & Sons, Inc., 1985) pp. 483–492.

Wilson, J. L., S. H. Conrad, W. R. Mason, W. Pelinski, and E. Hagan "Laboratory Investigation of Residual Liquid Organics from Spills, Leaks and the Disposal of Hazardous Wastes in Groundwater," EPA/600/56-90/004, U.S. Environmental Protection Agency, Ada, OK (1990).

6

EMPIRICAL MODELS AND VULNERABILITY MAPPING

6.1 INTRODUCTION

The susceptibility of a ground water resource to pollution (e.g., from agricultural chemicals) is related to several local hydrogeological factors. Mapping the aquifer vulnerability to such pollution usually entails the composite consideration of several factors descriptive of the depth and permeability of the unsaturated zone and the area hydrological balance. This chapter will focus on approaches that lead to the development of numerical indices or classifications of geographical areas with regard to the vulnerability of their ground water resources to man-made pollution activities. The primary technical issue addressed in vulnerability mapping is the subsurface transport and fate of potential pollutant chemicals. While not limited to the unsaturated or vadose zone, the general emphasis of most vulnerability mapping techniques is on transport through this media as opposed to transport within the saturated zone.

The first section of this chapter will provide overview information on EPA's Wellhead Protection Plan that requires use of empirical models and/ or vulnerability mapping techniques to delineate protective zones for wells providing drinking water. The second section will address the context of vulnerability relative to ground water resources management. Examples of mapping techniques such as agricultural DRASTIC and soil/aquifer field evaluation (SAFE) will be described. In addition, general information requirements of several other vulnerability mapping techniques will be delineated. Information will also be included on some techniques which give greater emphasis to pesticide characteristics than to hydrogeological factors. Examples include a pesticide index as well as some general qualitative and quantitative information on the state-of-knowledge of pesticide transport in the subsurface environment. One application of vulnerability mapping for developing a monitoring system in the State of Illinois will be described. Potential protection strategies for vulnerable areas will be developed and finally, a discussion of the limitations and needs relative to vulnerability mapping will be presented.

6.2 WELLHEAD PROTECTION (ENVIRONMENTAL MANAGEMENT SUPPORT, INC., 1989)

The Safe Drinking Water Act (SDWA) Ammendments of 1986 established a nationwide program to protect ground water resources used for public water supplies. The program approaches assessment and control of potable ground water quality from a contamination-prevention perspective rather than from the source prioritization or remedial action perspectives of other federal programs. The SDWA places responsibility for developing and implementing ground water protection programs on local governments by calling for the states to establish "Wellhead Protection Plans".

The states have been given considerable flexibility for determining appropriate methods for protecting ground water provided that the individual state wellhead protection programs meet the broad federal guidelines developed by EPA. The SDWA requires EPA to provide technical guidance on the hydrogeologic aspects of wellhead protection and gives EPA the responsibility of managing a program of grants to states who submit approved wellhead protection plans.

The major element of wellhead protection is defining the area around a well within which contaminant sources will be assessed and control measures implemented. These wellhead protection areas are defined in the SDWA as "the surface and subsurface area surrounding a water well or well field, supplying a public water system, through which contaminants are reasonably likely to move toward and reach such water well or well field." In essence, the law requires protection of recharge areas for wells supplying public drinking water.

To develop wellhead protection plans that safeguard public drinking water supplies, states must consider five major areas that require sound scientific evaluations: protection goals, potential contaminant threats, wellhead protection area delineation criteria, criteria thresholds, and delineation area mapping methods. These areas are outlined below.

1.   **Protection Goals** — There are three general protection goals applicable to the delineation of wellhead protection areas:

   a.   Protection based on reaction time provides a remedial action zone to protect wells from unexpected contaminant releases.
   b.   Protection based on the attenuation of contaminants in the subsurface provides a well-field management area that relies on the soil, unsaturated, and saturated subsurface environments to lower the concentrations of specific contaminants to acceptable levels before they reach the wellhead.
   c.   Protection based on the well recharge area (zone of contribution) provides a well-field management area that includes all or a major portion of an aquifer's existing or potential recharge area.

2.   **Wellhead Contamination Threats** — Wellhead area protection can be targeted to the direct introduction of contaminants to the area immediately contiguous to the well through deficient well casings, road runoff, spills, and accidents; to microbial contaminants such as pathogenic bacteria and viruses; or to the broad range of inorganic and organic chemical contaminants. The survival of pathogenic microorganisms in the subsurface has been a key component of drinking water protection for many decades. While a few hundred feet of buffer is usually adequate to safeguard wellheads from microbial threats, many toxic chemicals persist for a long time and may travel great distances in the subsurface. This constitutes the major technical and administrative challenge of the WHP program.

3.   **Delineation Criteria** — Delineation criteria include distance, drawdown, travel time, flow system boundaries, and the capacity of the aquifer to assimilate contaminants (assimilative capacity). These criteria are described below.

   a.   The distance criterion defines the WHPA by a radius measured around a pumping well.
   b.   The drawdown criterion defines the WHPA as the area around the pumping well in which the water table in unconfined aquifers or the potentiometric surface in confined aquifers is lowered by pumping.
   c.   The time of travel criterion bases the WHPA boundary on the time required for contaminants to travel through the subsurface to the water supply.

    d.   The flow boundaries criterion incorporates the known locations of ground water divides and other physical or hydrologic features that control ground water movement.

    e.   The assimilative capacity criterion is based on the subsurface formation capacity to dilute, retard, mineralize, or otherwise attenuate contaminant concentrations to acceptable levels before they reach drinking water wells.

4.   **Criteria Thresholds** — After selection of criteria for WHPA delineation, appropriate threshold values must be chosen to determine the limits above or below which a criterion will not provide the desired degree of protection. Criteria threshold ranges for protection from chemical threats have generally fallen within these ranges:

    a.   Distance — 1000 ft to more than 2 mi

    b.   Drawdown — 0.1 to 1.0 ft

    c.   Time of travel — 5 to 50 years (less than 5 years in high-flow settings)

    d.   Flow boundaries — physical and hydrologic formation

    e.   Assimilative capacity — none available except for a few single constituents

5.   **Delineation Methods** — Delineation methods translate the selected criteria thresholds into a map of the WHPA. Six methods have been identified by EPA as currently in use for WHPA delineations: arbitrary fixed radii, calculated fixed radii, simplified variable shapes, analytical methods, hydrogeologic mapping, and numerical models. The techniques range from simple and inexpensive to highly complex and comprehensive. The methods are described below:

    a.   The arbitrary fixed radius method involves circumscribing a circle around the water supply based on a selected distance. Though simple and inexpensive, this method tends to both overprotect down-gradient from the well by including areas that cannot contribute contaminants to the well and underprotect up-gradient from the well by not including areas that can contribute contaminants. A significant improvement over no delineation, the method is often used for microbial protection and in the early phases of a WHP program for chemical contaminants.

    b.   The calculated fixed radius method applies an analytical equation to calculate the radius of a circular WHPA based on a time of travel criterion. Though still relatively simple and inexpensive to apply, this method provides more accuracy than an arbitrary radius, especially where water tables are relatively flat.

c.   Simplified variable shapes are standard outlines of WHPAs, generated using analytical models and based on a combination of flow boundary and time-of-travel criteria. The shapes are chosen to match or approximate conditions encountered at specific wellheads. This is another inexpensive technique, but is somewhat more accurate in that it reduces over and under protection of the well.

d.   Analytical methods may be used to incorporate ground water flow boundaries and contaminant transport dynamics through the application of empirically derived equations. This is perhaps the most commonly used method where greater precision is needed.

e.   Hydrogeologic mapping can be used to map ground water flow boundaries and to implement other criteria through the use of geological, geomorphic, geophysical, and dye tracing methods. The method is particularly appropriate where aquifers and recharge areas are small.

f.   Numerical models use mathematical approximations of ground water flow and contaminant transport equations that take into account a variety of hydrogeologic and contamination conditions. These models offer the most accurate delineations, although they can be considerably expensive.

The process of wellhead protection utilizes concepts of both vulnerability mapping and empirical models. The delineation of a WHPA is an application of vulnerability mapping. Subsequent to delineation of the WHPA, contaminant sources within the WHPA need to be prioritized. Most prioritization methods are empirically based.

6.3 CONTEXT OF VULNERABILITY MAPPING

Figure 6.1 displays the general conceptual framework which indicates the typical relationship of vulnerability mapping to the overall issue of ground water quality management. Ground water quality management in terms of pollution prevention and control must address the characteristics of the pollution

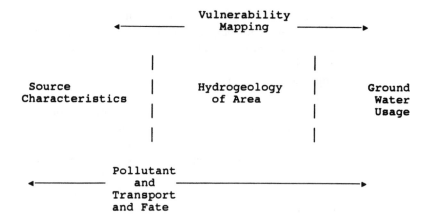

Figure 6.1.   Conceptual framework for vulnerability mapping.

source, the hydrogeological features of the given geographical area, and the usage of the ground water for human consumption and other purposes. An underlying issue relative to ground water quality management is the transport and fate of pollutants in the subsurface environment. Of particular concern relative to the agricultural setting would be the transport and fate of pesticides and fertilizers. While not limited to the hydrogeology of an area, vulnerability mapping is primarily focused on the integrated description of hydrogeologic features.

Vulnerability mapping techniques typically lead to either numerical indices for, or classifications of, geographical areas relative to their susceptibility to ground water contamination. To develop these indices or classifications, the typical approach involves consideration of multiple physical and chemical factors along with their relative importance weighting. In addition, information is typically provided on measurements for the factors and their evaluation, generally through the use of a numerical approach. The final index or classification for a given geographical area will be based upon the summation of the factor scores and/or the summation of the products of the factor scores times their relative importance weights. An important issue to recognize in vulnerability mapping is that there is considerable need for professional judgment to interpret the information. The thing that must be guarded against is becoming enslaved by the numbers or the classification systems without realizing that these are simply tools that can be useful in ground water quality management.

Vulnerability mapping has some particularly important uses or purposes relative to agricultural activities and ground water pollution concerns. One of the most important uses is to identify geographical locations with high susceptibility to ground water pollution, and thus plan monitoring programs in these "hot spots". This is particularly important due to the expenditures necessary for appropriate planning and implementation of monitoring programs. It is more expensive to conduct a ground water monitoring program than a comparable program for surface water resources; accordingly, any tool which can aid in identifying those areas of greatest concern is valuable. Another potential use of vulnerability mapping would be as a basis for permitting various agricultural chemicals for usage in particular geographical locations. For example, it might be desirable to limit the usage of mobile and persistent pesticides in geographical areas that have high susceptibility to ground water pollution. In addition, vulnerability mapping could be the basis for delineating time periods and quantities of chemical usage as well as quantities of irrigation water allowed in particular geographical locations.

There are a number of techniques that have been developed for vulnerability mapping. These techniques are on occasion referred to as empirical assessment approaches, with this term denoting simple approaches for development of numerical indices of the ground water pollution potential of man's activities (Canter, 1985). These empirical assessment methods as well as vulnerability mapping techniques, can be categorized relative to whether they have been developed to address point or nonpoint sources of ground water pollution. Point source techniques or methodologies include the surface impoundment assessment (SIA) approach for pits, ponds, and lagoons; the brine disposal methodology for brine disposal pits for oil and gas well operations; and several examples for landfills and hazardous waste sites, including the landfill site rating (LSR) method, the waste-soil-site interaction matrix (WSSIM), the hazard ranking system (HRS), and the site rating methodology (SRM). Nonpoint sources, including those related to agriculture, can be addressed through the use of the agricultural DRASTIC, the SAFE method, or a pesticide index which incorporates pesticide characteristics and selected characteristics of the local hydrogeological area. The following sections of this paper will provide illustrations of several of these techniques or methodologies.

6.4 AGRICULTURAL DRASTIC INDEX

A numerical rating scheme, called DRASTIC, has been developed for evaluating the potential for ground water pollution at a specific site given its hydrogeological setting (Aller et al., 1985). This rating scheme is based on seven factors chosen by a large number of ground water scientists from throughout the U.S. Information on these factors is presumed to exist for all locations in the U.S. In addition, the scientists also established relative importance weights and a point rating scale for each factor when the method is applied to nonpoint source pollution from pesticides or fertilizers. The acronym DRASTIC is derived from the seven factors in the rating scheme:

D = depth to ground water
R = recharge rate (net)
A = aquifer media
S = soil media
T = topography (slope)
I = impact of the vadose zone
C = conductivity (hydraulic) of the aquifer

Determination of the agricultural DRASTIC index involves multiplying each factor weight by its point rating and summing the total. The higher sum values represent greater potential for ground water pollution, or greater aquifer vulnerability. For a given area being evaluated, each factor is rated on a scale of 1 to 10 indicating the relative pollution potential of the given factor for that area. Once all factors have been assigned a rating, each rating is multiplied by the assigned weight, and the resultant numbers are summed as follows:

$$D_r D_w + R_r R_w + A_r A_w + S_r S_w + T_r T_w + I_r I_w + C_r C_w = \text{Pollution Potential}$$
$$(6.1)$$

where   r = rating for the area being evaluated
         w = importance weight for the parameter

Table 6.1 displays the rating scale for the depth to ground water factor; its importance weight in agricultural DRASTIC is 5 (Aller et al., 1985). Table 6.2 contains the rating information for the net recharge factor, and its agricultural DRASTIC importance weight is 4 (Aller et al., 1985). Figure 6.2 delineates the rating scheme for evaluation of the aquifer media factor. The importance weight for this factor in agricultural DRASTIC is 3 points (Aller et al., 1985). Information for evaluation of the soil media factor is included in Table 6.3, with the importance weight for agricultural DRASTIC being 5 points (Aller et al., 1985). Table 6.4 provides information on evaluation of the topography factor, and the importance weight for this factor in agricultural DRASTIC is 3 points (Aller et al., 1985). The approach for evaluation of the impact of the vadose zone media in agricultural DRASTIC is displayed in Figure 6.3, with the importance weight of this factor being 4 points (Aller et al., 1985). Finally, Table 6.5 summarizes the pertinent information for the hydraulic conductivity of the aquifer factor, with the agricultural DRASTIC importance weight being assigned 2 points (Aller et al., 1985).

The U.S. Environmental Protection Agency Office of Pesticide Programs

**Table 6.1 Evaluation of Depth to Ground Water
Factor in DRASTIC**

Range (ft)	Rating
0 – 5	10
5 – 15	9
15 – 30	7
30 – 50	5
50 – 75	3
75 – 100	2
100+	1

Source: Aller et al., 1985.

**Table 6.2 Evaluation of Net Recharge Factor
in DRASTIC**

Range (in.)	Rating
0 – 2	1
2 – 4	3
4 – 7	6
7 – 10	8
10+	9

Source: Aller et al., 1985.

and Office of Drinking Water are jointly sponsoring a national survey of pesticides in ground water (Alexander and Liddle, 1986). One of the goals of this survey is to assess the relationships between the agricultural usage of pesticides, the measured distribution of pesticide residues in ground water, and the hydrogeologic factors that influence ground water contamination. One aspect of this 2-year survey included a county-level classification of ground water vulnerability for all 3,144 counties in the U.S. The agricultural DRASTIC system was used in a modified form to rank the vulnerability of ground water to pollution for each county. This cursory classification survey was accomplished in a 3-month period; the form used in the survey is in Figure 6.4 (Alexander and Liddle, 1986). As shown in Figure 6.4, information from each county segment was aggregated into a county-level by considering the percentage of the county characterized by the various ranges in the

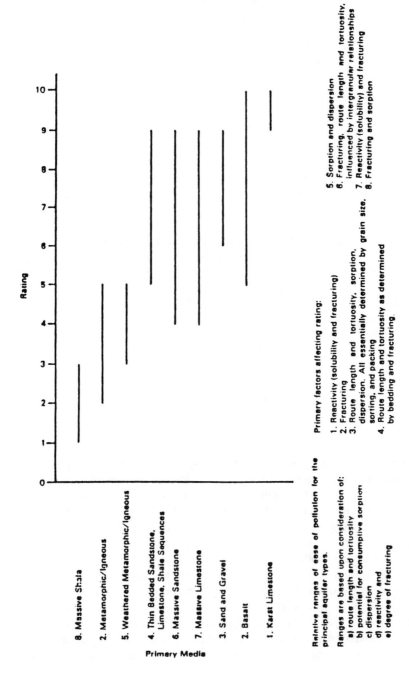

Figure 6.2.   Graph of ranges and ratings for aquifer media in DRASTIC (Aller et al, 1985).

Table 6.3 Evaluation of Soil Media Factor in DRASTIC

Range	Rating
Thin or absent	10
Gravel	10
Sand	9
Shrinking and/or aggregated clay	7
Sandy loam	6
Loam	5
Silty loam	4
Clay loam	3
Nonshrinking and nonaggregated clay	1

Source: Aller et al., 1985.

**Table 6.4 Evaluation of Topography Factor
in DRASTIC**

Range (% Slope)	Rating
0–2	10
2–6	9
6–12	5
12–18	3
18+	1

Source: Aller et al., 1985.

DRASTIC factors. The result of this study was a nationwide categorization of counties into those with high, medium, or low vulnerability to ground water pollution.

6.5 SOIL/AQUIFER FIELD EVALUATION (SAFE)

Another technique which can be used for determining the vulnerability of a ground water system to pesticide contamination is referred to as the soil/aquifer field evaluation (SAFE) methodology (Roux et al., 1986). The technique is somewhat simpler than agricultural DRASTIC, and is based on

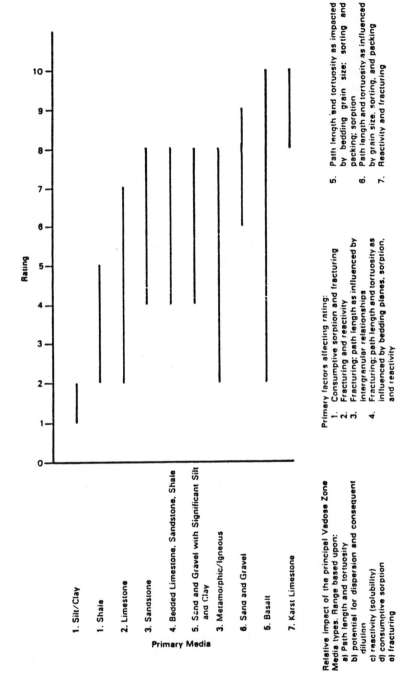

Figure 6.3. Graph of ranges and ratings for impact of the vadose zone in DRASTIC (Aller et al., 1985).

Table 6.5 Evaluation of Hydraulic Conductivity Factor in DRASTIC

Range (GPD/ft^2)	Rating
1–100	1
100–300	2
300–700	4
700–1000	6
1000–2000	8
2000+	10

Source: Aller et al., 1985.

characterization of the soil layer and underlying unsaturated (vadose) and saturated zones in a given geographical area.

To define areas of potentially vulnerable aquifers, subsurface geology is evaluated on a state or county scale using published information. The general flow sequence of the sensitive aquifer determination is displayed in Figure 6.5 (Roux et al., 1986). For those areas where pesticides are used, the first step is to assemble appropriate hydrogeologic data and identify aquifers in terms of whether or not they are present and developable as a major water supply. As suggested in Figure 6.5, if the ground water system is one that yields <50 gal/min, it is eliminated from review. For those ground water systems that are potentially developable as a major water supply (>50 gal/min yield), then the next question is related to the existing water quality in the ground water system. If the ground water quality is not suitable for public supply based on the application of typical water quality criteria, then this system would also be eliminated from further consideration. However, if it is suitable for public supply, then the next question is associated with the degree of aquifer protection. If the aquifer is confined it would not be mapped as a sensitive aquifer in the geographical area. For those aquifer systems that are unconfined or only partially confined, then the next question is related to recharge characteristics. Aquifers can be recharged from precipitation directly on their outcrop/subcrop areas, downward leakage from an overlying aquifer or surface water body, and upward leakage from a deeper, artesian aquifer. Principle routes for surface supplied chemicals such as pesticides to reach an aquifer are precipitation (or irrigation) from the outcrop area and flow from the shallow contaminated water table aquifer into the subcrop. Thus, the outcrop areas (and subcrop areas where they intersect shallow water table aquifers) of major aquifers are mapped in order to define sensitive aquifer areas (Roux et al., 1986).

State County			FIPS Coding		
Ground Water Region ____			Estimated % of County ____		
			____		
			____		
			____		
Hydrogeologic Setting ____ ____			Estimated % of County ____		
____ ____			____		
____ ____			____		
____ ____			____		

Ranges in DRASTIC Factors	Agricultural Weighted Ratings	Estimated % of County	Ranges in DRASTIC Factors	Agricultural Weighted Ratings	Estimated % of County
Depth to Water (ft)			**Topography (% slope)**		
0-5	50	____	0-2	30	____
5-10	45	____	2-6	27	____
10-15	40	____	6-12	15	____
15-30	35	____	12-18	9	____
30-50	25	____	18+	3	____
50-75	15	____	**Impact of Vadose Zone Media**		
75-100	10	____	SI/CL	4	____
100+	5	____	SH	12	____
Net Recharge (in)			LS	24	____
0-2	4	____	SS	24	____
2-4	12	____	LS, SS, SH (bedded)	24	____
4-7	24	____	SA & GVL w/SI & CL	24	____
7-10	32	____	META/IGN	16	____
10+	36	____	SA & GVL	32	____
Aquifer Media			BASALT	36	____
SH (massive)	6	____	LS (karst)	40	____
META/IGN	9	____	**Hydraulic Conductivity (gpd/sq. ft.)**		
META/IGN (wthd)	12	____	1-100	2	____
SS, LS, SH (tn bed)	18	____	100-300	4	____
SS (massive)	18	____	300-700	8	____
LS (massive)	18	____	700-1000	12	____
SA & GVL	24	____	1000-2000	16	____
BASALT	27	____	2000+	20	____
LS (karst)	30	____	**Agricultural DRASTIC Score** _____		
Soil Media					
THIN OR ABSENT	50	____	Index Variability + _____ − _____		
GVL	50	____			
SA	45	____	Data Confidence _____		
CL (shrinking)	35	____	(3 = High, 2 = Medium, 1 = Low)		
LOAM (SA)	30	____			
LOAM	25	____	By _____		
LOAM (SI)	20	____	Firm Date _____ 85		
LOAM (CL)	15	____			
CL (nonshrinking)	5	____	Data Sources or Comments on back →		

Figure 6.4. Data collection form for agricultural DRASTIC (Alexander and Liddle, 1986).

The second emphasis in this technique is related to the evaluation of sensitive (susceptible) soils. The important soil characteristics that may affect pesticide transport through the subsurface environment include permeability, thickness, pH, and temperature (Roux et al., 1986). Permeability, because of the extremes in range, appears to be the most critical factor. Other important factors such as clay content, organic content, and moisture are generally

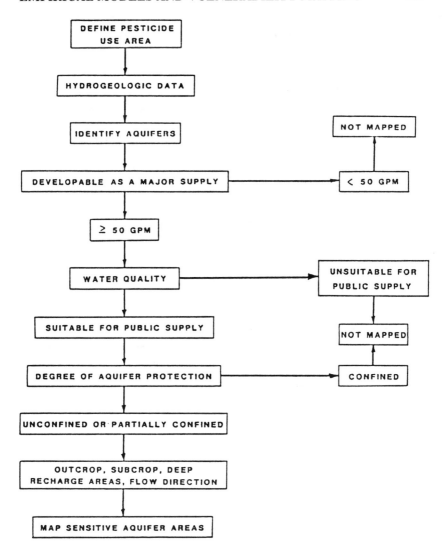

Figure 6.5.   Sensitive aquifer determination in SAFE (Roux et al., 1986).

related to permeability. Since permeability is a factor that can be mapped on a county scale from existing information, it is considered as an appropriate indicator in delineating sensitive soils.

The general approach for this sensitive evaluation is displayed in Figure 6.6 (Roux et al., 1986). The determination of sensitive soils requires a step-wise procedure that is initiated by considering the pesticide use area and the associated soil data. If <20% of the soil associations are farmed, then this would not be considered within this technique. For those soil associations

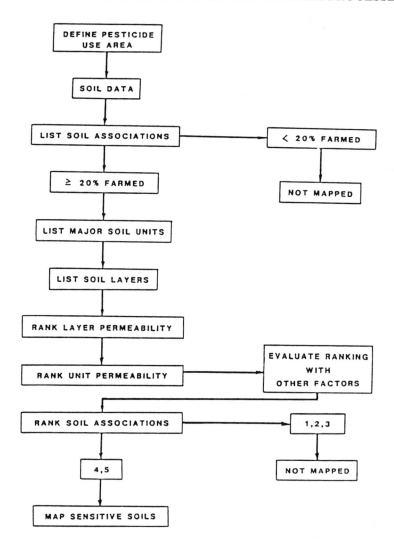

Figure 6.6. Sensitive soils determination in SAFE (Roux et al., 1986).

where >20% is farmed, then major soil units and soil layers are identified along with the associated permeabilities. Sensitive farmed soil associations for pesticide use areas are ranked from 1 (least sensitive) to 5 (most sensitive) according to their comparison with Long Island, New York, soil types. Long Island soil types are used because detectable concentrations of agricultural chemicals have been found in the underlying ground water. Table 6.6 delineates the sensitivity rankings of the soils, with the higher the sensitivity rank, the greater the likelihood of ground water contamination (Roux et al., 1986).

**Table 6.6 Permeability and SAFE Sensitivity Ranks of Long Island
Soils Used as a Standard for This Method**

Soil Unit	Class	Permeability (in./hr)	Sensitivity Rank
Atison	Medium sand	>6.3	5
Carver	Medium coarse sand	>6.3	5
Riverhead	Sandy loam (over sand)	2.0–6.3	4
Walpole	Loamy sand (over sand)	2.0–6.3	4
Bridgehampton	Fine sandy loam	0.63–2.0	3
Haven	Fine sandy loam with clay content <40%	0.63–2.0	3
Montauk variant	Silty sandy loam	0.2–2.0	2
Raynham	Silty sandy loam with clay content >40%	0.2–2.0	2
Whitman	Light loam over fill	<0.63	1
Candice	Silty clayey loam	<0.20	1

Source: Roux et al., 1986.
[a]   A rank of 5 is the most sensitive for pesticide migration to ground water.

Each soil layer within the soil unit is ranked according to its sensitivity as delineated in Table 6.6. When each soil layer rank has been assigned, the sensitivity rank of the overall soil (typically composed of three layers) is calculated. For example, if the permeability ranks of all three layers of soil are 4, then the soil is ranked 4. However, if the individual layers have different sensitivity rankings, then the system takes the thicknesses and relative positions of the various layers into account (Roux et al., 1986).

After each soil layer is ranked and an overall unit permeability is determined based on an analysis of the dominance of the rank assignment and the thickness of the individual layers, then the overall soil associations in the study area are evaluated and one numerical rank is developed. The numerical rank for the overall soil associations is based on a geographically weighted average of the ranks for the individual soil units. At that point, soil associations that have final rankings between 1 and 3 are not considered as sensitive soils. Sensitive soils are those which have a final ranking scale of 4 or 5.

The end results of the aquifer and soil ranking processes are maps showing sensitive areas for the aquifers and sensitive areas for soils in the region being evaluated. The two maps can then be combined and the sensitive or vulnerable areas determined based on the overlap of areas of sensitive aquifers and soil types. This final vulnerability map can then be used to plan monitoring programs for pesticides either by assisting in the placement of new monitoring wells or in the selection of existing wells for monitoring in both vulnerable and nonvulnerable areas (Roux et al., 1986). Other factors such as depth to ground water, vertical hydraulic conductivity within the unsaturated zone, precipitation, irrigation, and so forth, could also be considered in selecting specific monitoring well locations.

6.6 OTHER VULNERABILITY MAPPING TECHNIQUES

As noted earlier, several other techniques for vulnerability mapping, referred to as empirical assessment methods, have been developed. This section briefly delineates the key hydrogeological factors included in some of these methods.

The surface impoundment assessment (SIA) method addresses the permeability and depth of the unsaturated zone, the permeability and depth of the saturated zone, existing ground water quality, and a waste hazard potential factor which focuses on the toxicity and transport characteristics of the waste materials contained in pits, ponds, and lagoons (U.S. Environmental Protection Agency, 1978). The SIA method was used in a nationwide study of pits, ponds, and lagoons by the U.S. Environmental Protection Agency. This study enabled the prioritization of these impoundments based upon their potential for polluting local ground water resources.

The landfill site rating (LSR) methodology incorporates the distance between the landfill and the water supply wells, the depth of the unsaturated zone, the water table gradient from the landfill site, and the permeability of the unsaturated zone (LeGrand and Brown, 1977). This methodology is useful in selecting sites for new landfills and for prioritization of the ground water pollution potential of existing landfills.

Table 6.7 displays the soil and site-related factors in the waste-soil-site interaction matrix used for evaluating existing or new sanitary or chemical landfill sites (Phillips et al., 1977). The factors included in this methodology are the permeability and adsorptive characteristics of the soil, the depth to the water table, the water table gradient, the degree of infiltration, the distance from the landfill site to the nearest point of water use, and the thickness of the porous layer at the site.

Table 6.8 contains a listing of the rating factors associated with the ground water route in the hazard ranking system (HRS) for hazardous waste sites (Caldwell et al., 1981). These factors include route characteristics, waste characteristics, hazardous waste quantity, and targets related to ground water usage. Table 6.9 provides a summary of the rating factors and scales for the site rating methodology (SRM) used for prioritization of Superfund sites (Kufs et al., 1980). These rating factors are divided into three broad categories: receptors; pathways; and waste characteristics.

Table 6.7 Soil-Site Factors in Waste-Soil-Site Interaction Matrix

Group	Factor
Soil	Permeability (NP) — relates to permeability of site materials. Clay is considered to have poor permeability, fine sand moderate permeability, and coarse sand and gravel good permeability. The NP values range from 2.5 (low permeability) to 10 (maximum permeability).
	Sorption (NS) — relates to sorption characteristics of site materials. The NS values range from 1 (high sorption) to 10 (low sorption)
Hydrology	Water table (NWT) — considers the fluctuating boundary free water level and its depth. The zone of aeration occurs above the water table and is important to oxidative degradation and sorption. The NWT values range from 1 (deep water table) to 10 (water table near surface).
	Gradient (NG) — relates to the effect of the hydraulic gradient on both the direction and rate of ground water flow. The NG values range from 1 (gradient away from the disposal site in a desirable direction) to 10 (gradient toward point of water use).
	Infiltration (NI) — relates to the tendency of water to enter the surface of a waste disposal site. Involves consideration of the maximum rate at which a soil can absorb precipitation or water additions. A site with a large amount of infiltration will have greater ground water pollution potential. The NI values range from 1 (minimum infiltration) to 10 (maximum infiltration).
Site	Distance (ND) — relates to the distance from the disposal site to the nearest point of water use. The greater the distance the less chance of contamination, because waste dilution, sorption, and degradation increase with distance. The ND values range from 1 (long distance from disposal site to use site) to 10 (disposal site close to use site).
	Thickness of porous layer (NT) — refers to porous layer at the disposal site. The NT values range from 1 (about 100 ft or more of depth) to 10 (about 10 ft of depth).

Source: Phillips et al., 1977.

Table 6.8 Ground Water Factors in Hazard Ranking System

Hazard Mode	Factor Category	Ground Water Route
Migration	Route characteristics	Depth to aquifer of concern
		Net precipitation
		Permeability of unsaturated zone
	Waste characteristics	Physical state
		Persistence
		Toxicity
	Hazardous waste quantity	Total waste quantity
	Targets	Ground water use
		Distance to nearest well downgradient
		Population served by ground water drawn within 3-mile radius

Source: Caldwell et al., 1981.

The primary point to be made in briefly commenting upon these empirical assessment methodologies which can also be used for vulnerability mapping is to illustrate the similarities of various approaches for evaluating the subsurface environment. While the methodologies include a number of similar factors for consideration, they are certainly not uniform in numbers of factors nor the individual factors included.

6.7 PESTICIDE EVALUATIONS

Another approach related to vulnerability mapping is associated with evaluating the pollutant source, with some techniques also including selected hydrogeological factors. Two illustrations will be presented; one involves a pesticide index, and the other summarizes known information on pesticides in terms of their transport and fate characteristics within the subsurface environment.

Rao et al. (1985) have suggested a simple scheme for ranking the relative potentials of different pesticides to intrude into ground water. This ranking

Table 6.9 Rating Factors and Scales for the Site Rating Methodology for Prioritization of Superfund Sites

Rating Factors	Rating Scale Levels			
	0	1	2	3
Receptors				
Population within 1,000 ft	0	1–25	26–100	>100
Distance to nearest drinking water well	>3 mi	1–3 mi	3,001 ft to 1 mi	0–3,000 ft
Land use/zoning	Completely remote (zoning not applicable)	Agricultural	Commercial or industrial	Residential
Critical environments	Not a critical environment	Pristine natural areas	Wetlands, flood plains, and preserved areas	Major habitat of an endangered or threatened species
Pathways				
Evidence of contamination	No contamination	Indirect evidence	Positive proof from direct observation	Positive proof from laboratory analyses
Level of contamination	No contamination	Low levels, trace levels, or unknown levels	Moderate levels or levels that cannot be sensed during a site visit but which can be confirmed by a laboratory analysis	High levels or levels than can be sensed easily by investigators during a site visit

Table 6.9 (Continued)

Rating Factors	Rating Scale Levels			
	0	**1**	**2**	**3**
Type of contamination	No contamination	Soil contamination only	Biota contamination	Air, water, or food stuff contamination
Distance to nearest surface water	>5 mi	1–5 mi	1,001 ft to 1 mi	0–1,000 ft
Depth to ground water	>100 ft	51–100 ft	21–50 ft	0–20 ft
Net precipitation	<–10 in.	–10–+5 in.	+5–+20 in.	>20 in.
Soil permeability	>50% clay	30–50% clay	15–30% clay	0 to 15% clay
Bedrock permeability	Impermeable	Relatively impermeable	Relatively permeable	Very permeable
Depth to bedrock	>60 ft	31–60 ft	11–30 ft	0–10 ft
Waste Characteristics				
Toxicity	Sax's level 0 or NFPA's level 0	Sax's level 1 or NFPA's level 1	Sax's level 2 or NFPA's level 2	Sax's level 3 or NFPA's levels 3 or 4
Radioactivity	At or below background levels	1 to 3 times background levels	3 to 5 times background levels	Over 5 times background levels

	Easily biodegradable compounds	Straight chain hydrocarbons	Substituted and other ring compounds	Metals, polycyclic compounds, and halogenated hydrocarbons
Persistence	Easily biodegradable compounds	Straight chain hydrocarbons	Substituted and other ring compounds	Metals, polycyclic compounds, and halogenated hydrocarbons
Ignitability	Flash point >200°F or NFPA's level 0	Flash point of 140–200°F, or NFPA's level 1	Flash point of 80–140°F, or NFPA's level 2	Flash point <80°F or NFPA's levels 3 or 4
Reactivity	NFPA's level 0	NFPA's level 1	NFPA's level 2	NFPA's levels 3 or 4
Corrosiveness	pH of 6–9	pH of 5–6 or 9–10	pH of 3–5 or 10–12	pH of 1–3 or 12–14
Solubility	Insoluble	Slightly soluble	Soluble	Very soluble
Volatility	Vapor pressure <0.1 mm Hg	Vapor pressure of 0.1–25 mm Hg	Vapor pressure of 25–78 mm Hg	Vapor pressure >78 mm Hg
Physical state	Solid	Sludge	Liquid	Gas
Waste Management Practices				
Site security	Secure fence	Security guard with lock	Remote location but no fence	No barriers or breachable fence
Hazardous waste quantity	0–250 tons	251–1,000 tons	1,001–2,000 tons	>2,000 tons
Total waste quantity	0–10 ac ft	11–100 ac ft	101–50 ac ft	>250 ac ft

Table 6.9 (Continued)

Rating Factors	Rating Scale Levels			
	0	1	2	3
Waste incompatibility	No incompatible wastes are present	Present, but does not pose a hazard	Present and may pose a future hazard	Present and posing an immediate hazard
Use of liners	Clay or other liner resistant to organic compounds	Synthetic or concrete liner	Asphalt-base liner	No liner used
Use of leachate collection systems	Adequate collection and treatment	Inadequate collection or treatment	Inadequate collection and treatment	No collection or treatment
Use of gas collection systems	Adequate collection and treatment	Collection and controlled flaring	Venting or inadequate treatment	No collection or treatment
Use and condition of containers	Containers are used and appear to be in good condition	Containers are used but a few are leaking	Containers are used but many are leaking	No containers are used

Source: Kufs et al., 1980.

scheme does not require the detailed pesticide and site characteristics information which would be necessary in a complete mathematical model of the subsurface environment. The scheme addresses pesticide transport through the crop root zone and the intermediate vadose zone. The following equations and definitions are used in the ranking scheme (Rao et al., 1985).

$$AF = M_2 / M_o = \exp(-B) \tag{6.2}$$

where AF = Attenuation factor between 0 and 1 = index for pesticide mass emission from the vadose zone
M_2 = Amount of pesticide entering ground water
M_o = Amount of pesticide applied at soil surface

and where

$$B = \frac{0.693 t_r}{t^{\frac{1}{2}}} \tag{6.3}$$

t_r = Time required for pesticide to travel through the root zone and intermediate vadose zone
$t_{1/2}$ = Degradation half-life of the pesticide

and

$$t_r = \frac{(L)(RF)(FC)}{q} \tag{6.4}$$

where L = Distance from the soil surface to ground water
RF = Retardation factor
FC = Volumetric soil-water content at field capacity
q = Net recharge rate

and

$$RF = 1 + \frac{(BD)(OC)(K_{oc})}{FC} + \frac{(AC)(K_h)}{FC} \tag{6.5}$$

where BD = Soil bulk density
OC = Soil organic carbon content
K_{oc} = Sorption coefficient of pesticide on soil
AC = Air-filled porosity of soil
K_h = Henry's constant for pesticide

Table 6.10 Chemical/Physical Properties of Pesticides: Values Which Indicate Potential for Ground Water Contamination

Water solubility	>30 ppm
K_d	<5, usually <1
K_{oc}	<300–500
Henry's Law Constant	<10^{-2} atm-m^{-3} mol
Speciation	negatively charged, fully or partially at ambient pH
Hydrolysis half-life	>25 weeks
Photolysis half-life	>1 week
Field dissipation half-life	>3 weeks

Source: U.S. Environmental Protection Agency, 1986.

Rao et al. (1985) have suggested that the attenuation factor (AF) index can be used by regulatory agencies in the preliminary evaluation of pesticides to be monitored in geographical areas with ground water susceptible to pesticide pollution. Usage of this index is based on the following simplifying assumptions:

1. Vadose zone properties are independent of depth.
2. An average ground water recharge rate can be computed given local rainfall, irrigation, and evapotranspiration data.
3. A K_{oc} value can be estimated for each pesticide, based on the assumption that hydrophobic interactions are dominant.
4. An average $t_{1/2}$ value can be estimated for each pesticide.

The other illustration relative to pesticide characteristics involves summarizing known information about pesticides and whether or not they would represent a threat to ground water. Whether this threat exists or not is dependent upon factors related to the pesticides themselves, the soil, and application factors. Important physical and chemical characteristics of pesticides include water solubility, soil adsorption, volatility, and soil dissipation. Values for these characteristics which indicate increased potential for ground water contamination are summarized in Table 6.10 (U.S. Environmental Protection Agency, 1986). Descriptive summary information for key physical and chemical characteristics of those pesticides that can be characterized as more mobile or leachable are included in Table 6.11 (U.S. Environmental Protection Agency, 1986).

A number of factors related to the physical and chemical characteristics of soils receiving pesticide application also influence the potential for ground

Table 6.11 Descriptive Information on Key Physical and Chemical Characteristics of Pesticides Leachers

Water solubility	The propensity for a pesticide to dissolve in water — the higher a pesticide's water solubility, the greater the amount of pesticide that can be carried in solution to ground water. Water solubility of >30 ppm has been identified as a "flag" for a possible pesticide "leacher."
Soil adsorption	The propensity of a pesticide to "stick" to soil particles — it is defined as the ratio of the pesticide concentration in soil (C_s) to the pesticide concentration in water (K_d, C_s/C_w). There are different mechanisms for pesticide adsorption in soils, with particular important differences occurring in clays as opposed to organic soil matter. A second measure, K_{oc}, is used to help characterize the mechanism of adsorption. K_{oc} is a measure of the pesticide adsorption to the organic part of the soil. The lower a pesticide K_d and K_{oc} values, the more likely these chemicals will not be adsorbed to soil particles but leach to ground water. Of the pesticides found in ground water to date, most have had K_d values of <5, and usually <1.0. These ground water contaminants have also generally been shown to have K_{oc} values of <300.
Volatility	The propensity for a pesticide to disperse into the air — it is primarily a function of the vapor pressure of the chemical and is strongly influenced by environmental conditions (e.g., temperature, wind speed, etc.). Nonvolatile pesticides such as DDT have low vapor pressures, which will increase their persistence on and in the soil. Pesticides with high vapor pressures have not been considered a threat to ground water because they rapidly volatilize from the soil surface. However, the major ground water problems caused by the very volatile pesticides EDB, DBCP, and DCP have dispelled this notion. Contamination by these volatile pesticides has been blamed on their mode of application, which is direct injection into the soil. It has also become apparent that the actual volatility of these pesticides when present in water is critically changed. This aqueous volatility is determined by dividing

Table 6.11 (Continued)

Volatility (continued)	chemical vapor pressure by its solubility; this value is termed Henry's Law Constant (K_h). A compound such as DBCP has a very high vapor pressure but is also very water soluble. High water solubility can cause high vapor pressure chemicals to remain in the soil, particularly when these pesticides are applied just prior to irrigation or rainfall.
Soil Dissipation	A simplified, general measure of pesticide persistence in soil — it is usually measured as the length of time required for dissipation of one-half the concentration of a pesticide and often referred to as pesticide "soil half-life". The soil half-life of a pesticide can be derived from either laboratory or field studies, but care must be taken in recording the conditions of the test, including temperature, type of soil, soil moisture, etc.
	Soil dissipation is dependent on a number of environmental processes, including vaporization and several decomposition processes that cause chemical breakdown, particularly hydrolysis, photolysis, and microbial transformation. Hydrolysis is the reaction of a chemical with water. Photolysis is the breakdown of a chemical from exposure to the energy of the sun. And, microbial transformations result from the metabolic activities of microorganisms within the soil. When a pesticide resists these decomposition processes and does not readily evaporate, it will have a long soil half-life, increasing its potential as a threat to ground water. This is particularly true if the same pesticide is highly soluble and does not readily adsorb to the soil particles. Pesticides with half-lives greater than 2 or 3 weeks should be carefully assessed.

Source: U.S. Environmental Protection Agency, 1986.

water contamination. Table 6.12 provides descriptive information on seven important soil factors (U.S. Environmental Protection Agency, 1986); these include the clay content, organic matter content, soil texture, soil structure, porosity, soil moisture, and depth to ground water. Finally, the methods and conditions of soil application of pesticides can influence the potential for

Table 6.12 Descriptive Information on Key Soil Factors Affecting
Pesticide Transport to Ground Water

Clay content	Refers to the presence of clay minerals — clay minerals contribute to cation exchange capacity, or the ability of the soil to adsorb positively-charged molecules (i.e., cations). Positively-charged pesticides will thus be adsorbed to soil containing negatively-charged clay particles. Clay soils also have a high surface area that further contributes to adsorption capacity. Adsorption onto clay colloids leads to chemical degradation and inactivation of some pesticides, but it inhibits degradation of others.
Organic matter content	Also contributes to adsorption of pesticides in soil — organic matter content affects bioactivity, bioaccumulation, biodegradability, leachability, and pesticide volatility. Soils with high organic content adsorb pesticides and therefore inhibit their movement into ground water. However, pesticides which are highly adsorbed to organic soil will often be applied at higher rates by a farmer to compensate for the adsorbed portion. There is evidence that pesticide residues adsorbed to high organic (humus) soils may eventually be released, intact, to ground water when microbial degradation of the humus occurs.
Soil texture	Also influences pesticide leaching — texture refers to the percent sand, silt, and clay. Leaching is more rapid and deeper in coarse or light-textured sandy soils than in fine or heavy-textured clayey soils.
Soil structure	Refers to the way soil grains are grouped together into larger pieces or aggregates: platy, prismatic, blocky, or granular — structure is affected by texture and percent organic matter. Pesticides and water can seep, unimpeded, through seams between the aggregates.
Porosity	A function of total pore space, pore size, and pore size distribution — it is determined by soil texture, structure, and particle shape. Pesticide transport is more rapid through porous soils.

Table 6.12 (Continued)

Soil moisture	Refers to the presence of water in soil — the water in soil ultimately transports pesticides that are not adsorbed into the water table below. Upward movement may also occur through capillary action and by a process termed evapotranspiration, in which water in the soil is lost to the air.
Depth to ground water	The distance a pesticide must travel through the soil or underlying foundation material to reach ground water — it is, of course, a key determinant in whether contamination will occur at a particular site.

Source: U.S. Environmental Protection Agency, 1986.

pesticide transport to ground water. Table 6.13 contains descriptive information on these key application methods and conditions, including local climatic conditions; rate, timing, and method of application; irrigation and cultivation practices; and whether or not there has been spillage or disposal of pesticides (U.S. Environmental Protection Agency, 1986).

The information in Tables 6.10 to 6.13 is not presented in the context of calculating a specific numerical index for the ground water pollution potential of pesticides in a given geographical location. However, this type of summary information can be used as the basis for development of specific indices or classifications of geographical locations relative to their vulnerability to pesticide contamination of ground water systems.

6.8 EXAMPLE OF VULNERABILITY MAPPING

One example of the use of vulnerability mapping will be described to illustrate a systematic approach for considering hydrogeological factors in ground water quality management. Vulnerability mapping was used in the development of a hazardous substance ground water monitoring program in the State of Illinois (Shafer, 1985). Ground water contamination by hazardous materials is a potential danger to public and environmental health. Since the data for many of these substances in ground water are not available in Illinois, it was determined that a state-wide monitoring program for such substances should be initiated. Table 6.14 contains a list of supporting reasons for the initiation of such a program (Shafer, 1985).

Table 6.13 Descriptive Information on Key Application Methods and Conditions Affecting Pesticide Transport to Ground Water

Local climatic conditions	The degree of pesticide leaching at a particular site — it can be directly dependent on the amount of local rainfall. The temperature of the soil and surrounding air at a site can also greatly affect the processes that result in pesticide movement and degradation in the environment.
Rate of application	How much and how often a pesticide is applied to the soil — it can be the critical determinant in ground water contamination.
Timing of application	When a pesticide is applied — it can be a major factor depending on local environmental conditions and temperature, and rainfall.
Method of application	A pesticide can be applied to crops by aerial spraying, topsoil application (granular, dust, or liquid formulations), soil injection, soil incorporation, or irrigation. Soil injection and incorporation are generally considered to pose the greatest likelihood for ground water contamination problems. The application of pesticides through irrigation, often referred to as chemigation, can also be a significant source of ground water contamination. An irrigation pump may shut down due to a mechanical or electrical failure while the pesticide-adding equipment continues to operate. This malfunction can cause a backflow of pesticides into the well or cause highly concentrated pesticide levels to be applied to a field.
Irrigation practices	Increase the soil moisture content and flow through the soil, raising the potential for chemical leaching — irrigation can decrease the amount of volatilization of some pesticides from the soil. Excess irrigation can also carry pesticides down the well casings of abandoned or poorly constructed wells, directly injecting contaminants into an aquifer. The use of drainage tiles can also lead to direct input of pesticides into ground water regardless of their leaching potential.

Table 6.13 (Continued)

Cultivation practices	Conservation tillage or no-till practices used to decrease soil erosion and pollutant runoff into streams — it will increase water infiltration and hence potential for pesticides to leach. These practices usually require increased use of herbicides that may leach. Other soil conservation practices designed to inhibit runoff may also increase infiltration.
Spillage/disposal	Can result in high concentrations of pesticides in soil — these "slugs" can overwhelm normal decomposition processes and soil adsorption capacity, resulting in high potential for ground water contamination. Spillage, in particular, can be a common problem where pesticide mixing and loading take place.
	Handling of unwanted pesticides and empty containers may also pose problems. Rinse water from the cleaning of spray equipment may also be washed into the soil; the large amounts of contaminated water associated with this practice can increase pesticide leaching.

Source: U.S. Environmental Protection Agency, 1986.

Four categories of criteria were identified for defining targeted areas for hazardous substance ground water monitoring, and they are listed in Table 6.15 (Shafer, 1985). Criterion I was used to evaluate the current use of an aquifer. Usage of this criterion assumes that the population potentially affected by contamination is proportional to water use. Criterion II was used to assess where major future ground water supplies may be located. Criterion III was chosen to evaluate the presence of potential hazardous substance contaminant sources. The assumption is that the density of hazardous substance-related facilities is proportional to the potential for contamination (that is, the higher the density of hazardous substance-related facilities, the greater the risk for contamination of ground water by hazardous substances). Criterion IV was used to focus attention on aquifers which are highly susceptible to contamination. The relative ability of geologic materials, within 50 ft of the surface of the earth, to transmit municipal landfill leachate was a basic factor in evaluation of this criterion.

Geographical areas meeting the criteria outlined in Table 6.15 were initially mapped for the entire state for possible hazardous substance monitoring. Given a rating factor of 1 for each criterion, areas which meet the requirements

Table 6.14 Reasons Basic to a State-Wide Hazardous Substance Ground Water Monitoring Program in Illinois

1. Illinois will continue to be dependent on ground water to furnish supplies for domestic, as well as industrial/agricultural/commercial uses.
2. Large quantities of hazardous substances are generated, transported, and disposed of within the state.
3. Relying on inadvertent detection of water quality problems involves serious uncertainties, since the presence of most contaminants is not readily observable, and once a drinking water source is contaminated, alternative sources of water can be prohibitively expensive.
4. The level of funds and expertise required for sample collection and analysis are not usually at a local level of government.
5. Site-specific monitoring of known potential sites is inadequate for assessing statewide or regional effects.
6. There is a serious lack of data concerning many hazardous substances, particularly synthetic organic compounds, in Illinois ground water.
7. Currently there is no comprehensive program to collect the data necessary to identify regional hazardous substance-related ground water quality problems.
8. The evaluation of the performance of future ground water quality protection programs in Illinois, related to contamination by hazardous substances, will require the collection of baseline information on such substances in Illinois ground water.

Source: Shafer, 1985.

Table 6.15 Criteria Used to Define Targeted Areas for Hazardous Substance Ground Water Monitoring in Illinois

I.   Current ground water withdrawals
     >100,000 gpd withdrawals per township for sand and gravel or shallow bedrock aquifers.
II. Potential ground water withdrawals
     Designation as principal shallow bedrock or sand and gravel aquifer: yields >100,000 gpd/mi^2 and area of more than 50 mi^2.
III. Potential hazardous substance sources
     >2.0 hazardous substance-related facilities per square mile per zip code area.
IV. Aquifers highly susceptible to contamination
     Designation as highly susceptible sand and gravel aquifer or highly susceptible shallow bedrock aquifer.

Source: Shafer, 1985.

of all the criteria would receive a maximum rating of 4 points. With this strategy, the areas which fall into each category were highlighted and the highlighted areas ranked according to the number of points accumulated. This information was then used in planning specific monitoring programs.

6.9 GROUND WATER PROTECTION STRATEGIES IN VULNERABLE AREAS

Aquifer vulnerability maps provide a useful framework within which to designate priorities for the implementation of ground water protection and control measures. Such measures include regulatory and nonregulatory approaches, or a mix of both. Regulatory approaches involve placing a system of legal constraints on land uses, or on particular activities that are potential sources of ground water pollution. Nonregulatory approaches include such activities as public education, voluntary best management practices, governmental coordination, and inspection and training programs.

6.9.1 Regulatory Approaches for Protecting Vulnerable Aquifers

Most ground water contamination results from chemicals or other pollutants deposited on or near the land surface. Controlling the use of land can be effective in limiting contamination. Land use controls are most suitable for highly vulnerable areas or recharge zones. Land use controls that can be used to protect ground water in vulnerable areas include (Henderson et al., 1987):

1. Zoning regulations
2. Siting, development, and construction regulations
3. Public acquisition programs
4. Transferable development rights

Zoning ordinances may be used to limit the uses and forms of development allowed in areas deemed critical for ground water protection. The uses which can be prohibited in a ground water protection zone include the siting of hazardous waste disposal facilities, bulk storage tanks of petroleum and chemical products, and other facilities handling either high concentrations or large volumes of potential ground water contaminants. Some forms of regulation, such as density restrictions and clustering, may be designed to minimize the impacts of certain actions in vulnerable areas (Henderson et al., 1987).

Siting, development and construction regulations establish rules for using land for specified purposes. These regulations specify standards for the division of lands, and the siting and construction of facilities. Existing siting, development, and construction regulations can be modified to minimize ground water contamination. The modifications may include design criteria and permissible technologies for altering surfaces and vegetation, changing natural drainage, and providing artificial drainage (Henderson et al., 1987).

Land acquisition is a way for a state or local government to provide strict protection for selected areas. For ground water protection purposes, this may mean buying the right to restrict uses of the land for waste disposal, industrial purposes, high density residential areas, and specified agricultural practices. The state or local government can restrict all uses of the land while avoiding the possibility of legal challenge (Henderson et al., 1987).

A transferable development rights program is a form of zoning with built-in compensation mechanisms. Owners of land in restricted areas are given development rights which they can sell to others in nonrestricted areas (Henderson et al., 1987).

6.9.2 Best Management Practices (BMPs) for Protecting Vulnerable Aquifers

Agricultural BMPs can be used to protect ground water in vulnerable areas. These include limitations on the types, quantities, and timing of agricultural chemical usage; specifications for the usage of nitrification inhibitors; and the identification of optimal water balances for agricultural production.

Proper siting, density control, design, and construction of on-site wastewater disposal facilities can prevent contamination problems. The use of an alternative or modified disposal may be necessary in geologically vulnerable areas. These systems include alternate dose systems, multiple drain-field systems, mound treatment systems, or waterless toilets (Jaffe and Dinovo, 1987).

Sewer systems are often proposed as a remedial or preventive action to avoid ground water contamination by septic tank systems. The discharges from such systems can also cause localized ground water contamination. Contamination by sewer systems can be avoided by their proper location, design, and construction, and by good maintenance and leakage monitoring practices. Leak proof designs may be necessary in sensitive areas with high water tables and thin soil layers (Jaffe and Dinovo, 1987).

Many business and industries use hazardous materials. The introduction of contaminants to ground water through the infiltration of material spills or by runoff from storage or production areas, is hazardous. The potential for contamination can be reduced by siting industries that use hazardous materials away from sensitive aquifer areas. The risk of contamination can also be reduced through the use of "best management practices". These practices include enclosing storage areas and stockpiles and providing curbs, drains, and sumps to prevent the runoff of contaminants. Management practices can also include process design. Facilities can be designed in such a way to prevent accidental spills or leaks (Jaffe and Dinovo, 1987).

6.9.3 Public Education

It is very important to complement ground water protection with public education. There are several public education approaches that states or local governments could use (U.S. Environmental Protection Agency, 1989):

1. Distributing press releases to newspapers and radio stations
2. Arranging press conferences on ground water protection topics for local radio stations, newspapers, and television stations
3. Distributing ground water protection information in local government newsletters
4. Developing slide shows or video tapes on ground water protection to distribute to local schools and community organizations
5. Establishing voluntary committees to assist local agencies in implementing public education and ground water protection programs
6. Providing speakers on ground water protection to local groups
7. Developing brochures on ground water protection to include in water or tax bills

The purpose of public education programs is to build support for regulatory programs, such as controls on pollution sources in sensitive aquifer areas, and to implement voluntary ground water protection efforts, such as water conservation, and household hazardous waste management.

6.10 LIMITATIONS AND NEEDS RELATIVE TO VULNERABILITY MAPPING

Several limitations as well as advantages and needs of vulnerability mapping techniques can be delineated. Perhaps the primary advantage of vulnerability mapping is as a tool in screening, in a geographical context, those areas that are the most susceptible to ground water contamination. Screening can be used as an aid in planning monitoring programs directed toward determining worst case conditions in a geographical study area. This is particularly important in developing information on ground water quality in large agricultural areas.

One of the primary limitations of vulnerability mapping techniques, at least at this point in time, is the minimal number of field verification studies which demonstrate that those areas anticipated to be most susceptible to ground water pollution are in fact areas with higher ground water concentrations of pesticides and nitrates. Work is needed to provide qualitative relationships, or possibly quantitative relationships based on nonparametric statistical techniques, which could be used as surrogate predictors of ground water concentrations. Problems that might be encountered in field verification studies include variations of concentrations at different times due to different periods of fertilizer or pesticide application and interpreting the results from ground water samples collected at different depths.

Another issue of concern relative to vulnerability mapping is the maximum or minimum geographical area which could be mapped. Should this geographical area be defined based on average hydrogeological conditions for a small area such as several hundred acres, or can geographical areas as large as counties be mapped? For example, the DRASTIC technique was developed as a screening tool for county-size areas down to a resolution of approximately 100 acres (Alexander and Liddle, 1986). Additional consideration needs to be given to geographical limitations in the development of specific numerical indices or classifications.

It is important to realize that vulnerability mapping techniques assume uniform conditions within the subsurface environment, at least for the geographical area to be addressed. However, the subsurface environment is not uniform relative to hydrogeological factors. Pesticides and fertilizers are also not uniformly applied over time and space in a given geographical area. In addition, the subsurface transport and fate of pesticides differs considerably from that for nitrates from fertilizers. The nonuniformity of conditions does not mean that vulnerability mapping should not be used; however, it does suggest that very careful interpretation is needed for developed numerical indices and classification systems.

Another issue is related to the divisions between classifications used to identify areas highly susceptible to ground water pollution in contrast to those that are moderately or minimally susceptible to ground water pollution. Again, it should be recognized that these arbitrary divisions are based on the collective judgment of the individuals developing the technique.

Finally, it should be noted that the most appropriate approach for assessing agricultural sources of ground water pollution should involve the combined consideration of pollutant characteristics, pesticide and fertilizer application rates and practices, hydrogeological factors basic to vulnerability assessment, and the usage of the ground water resource. In other words, vulnerability mapping, while very useful, does not provide a complete assessment of the overall risks related to pesticide and fertilizer applications and their significance as contaminants in ground water resources.

6.11 SUMMARY

Vulnerability mapping represents an emerging tool useful in ground water quality management. A number of techniques for achieving vulnerability mapping have been developed, with these techniques generally involving the composite consideration of several factors descriptive of the hydrogeology in given geographical areas. Vulnerability mapping results can be useful in planning monitoring programs focused on "hot spots"; however, additional work is needed in terms of field studies to verify the basic assumptions of the techniques. Finally, there are a number of management policies that could be derived from vulnerability mapping: limitations on the types, quantities, and timing of agricultural chemical usage; requirements for ground water monitoring in those areas more vulnerable to ground water contamination; specifications for the usage of nitrification inhibitors; and the identification of optimal water balances for agricultural production.

REFERENCES

Alexander, W. J. and S. K. Liddle "Ground Water Vulnerability Assessment in Support of the First Stage of the National Pesticide Survey," in *Proceedings of the Conference on Agricultural Impacts on Ground Water*, (Dublin, OH: National Water Well Association, 1986), pp. 77–87.

Aller, L., T. Bennett, J. H. Lehr, R. J. Petty, and G. Hackett "DRASTIC: A Standard System for Evaluating Ground Water Pollution Potential Using Hydrogeologic Settings," EPA/600/2-85/018, U.S. Environmental Protection Agency, Robert S. Kerr Environmental Research Laboratory, Ada, OK.

Caldwell, S., K. W. Barrett, and S. S. Chang "Ranking System for Releases of Hazardous Substances," in *Proceedings of the National Conference on Management of Uncontrolled Hazardous Waste Sites,* (Silver Spring, MD: Hazardous Materials Control Research Institute, 1981), pp. 14–20.

Canter, L. W. "Methods for Assessment of Ground Water Pollution Potential," in *Ground Water Quality,* C. H. Ward, W. Giger, and P. L. McCarty, Eds. (New York: John Wiley & Sons, Inc., 1985), pp. 270–306.

Environmental Management Support, Inc. "Draft: Wellhead Protection Five Year Research Plan," U.S. Environmental Protection Agency, Robert S. Kerr Environmental Research Laboratory, Ada, OK (1989).

Henderson, T. R., J. Trauberman, and T. Gallagher *Ground Water Strategies for State Action,* 3rd ed. (Washington D.C.: Environmental Law Institute, 1987), pp. 117–127.

Jaffe, M. and F. Dinovo *Local Ground Water Protection,* (Washington, D.C.: American Planning Association, 1987), pp. 141–160, 225–250.

Kufs, C., D. Twedell, S. Paige, R. Wetzel, P. Spooner, R. Colonna, and M. Kilpatrick "Rating of the Hazard Potential of Waste Disposal Facilities," in *Proceedings of the National Conference on Management of Uncontrolled Hazardous Waste Sites,* (Silver Spring, MD: Hazardous Materials Control Research Institute, 1980), pp. 30–41.

LeGrand, H. E. and H. S. Brown "Evaluation of Ground Water Contamination Potential from Waste Disposal Sources," Office of Water and Hazardous Materials, U.S. Environmental Protection Agency, Washington, D.C. (1977).

Phillips, C. R., J. D. Nathwani, and H. Mooij "Development of a Soil-Waste Interaction Matrix for Assessing Land Disposal of Industrial Wastes," *Water Res.* 11(November):859–868 (1977).

Rao, P. S., A. G. Hornsby, and R. E. Jessup "Indices for Ranking the Potential for Pesticide Contamination of Groundwater," *Proc. Soil Crop Sci. Soc. Fla.* 44:1–8 (1985).

Roux, P., J. DeMartinis, and G. Dickson "Sensitivity Analysis for Pesticide Application on a Regional Scale," in *Proceeding of the Conference on Agricultural Impacts on Ground Water,* (Dublin, OH: National Water Well Association, 1986), pp. 145–158.

Shafer, J. M. "An Assessment of Ground-Water Quality and Hazardous Substance Activities in Illinois with Recommendations for a Statewide MonitoringStrategy," Illinois Department of Energy and National Resources, Champaign, IL (1985), pp. 79–90.

U.S. Environmental Protection Agency "A Manual for Evaluating Contamination Potential of Surface Impoundments," EPA/570/9-78-003, Office of Drinking Water, Washington D.C. (1978).

U.S. Environmental Protection Agency "Pesticides in Ground Water: Background Document," Office of Ground Water Protection, U.S. Environmental Protection Agency, Washington, D.C. (1986).

7

MODELING SUBSURFACE TRANSPORT AND FATE PROCESSES

7.1 INTRODUCTION

An area of interest within the field of subsurface transport and fate processes receiving much attention is the area of modeling. Ironically, modeling of subsurface transport and fate processes is truly in its infancy. Early ground water models focused on determining head distributions and flow patterns in aquifer systems. The first ground water quality models focused on solute transport, including the effects of advection and dispersion of non-reactive contaminants. Recently, modeling efforts have focused on incorporating the effects of other subsurface processes including adsorption and biological reactions. Currently, much interest is being shown in developing models for depicting the behavior of multiple phase flow systems involving aqueous and non-aqueous phases in the subsurface.

This chapter presents an overview of ground water modeling processes and how the effects of subsurface transport and fate processes can be incorporated into the models. The topics to be addressed include the purposes of model usage, basic information on models, limitations of models, and selection of an appropriate model. Also included is a section describing several overview studies on models. The final section will highlight several conclusions drawn from this overview.

7.2 PURPOSES FOR MODEL USAGE

Within the last several years there has been a growing interest in the use of ground water models, particularly for solute (contaminant) transport or quality models. Several factors have contributed to this expanding interest including

1. The regulatory emphasis being given to ground water quality management and the prevention or control of contamination from various man-made sources
2. An increasing realization of society's dependency on ground water for economic development
3. The expansion in computational capabilities, including the mushrooming of available personal computers and software
4. The improvements in numerical analysis techniques so as to enable the use of more sophisticated models.

Mathematical models can be used for several purposes in ground water quality management. Examples include (U.S. Environmental Protection Agency, 1987): (1) appraising the physical extent, and chemical and biological quality, of ground water reservoirs (e.g., for planning purposes); (2) assessing the potential impact of domestic, agricultural and industrial practices (e.g., for permit issuance); (3) evaluating the probable outcome of remedial actions at waste sites and aquifer restoration techniques generally; and (4) providing health-effects exposure estimates. In addition, modeling can also be used in planning ground water monitoring networks, evaluating the contributions of different potential sources of ground water pollution at a specific site, risk assessment studies, development of wellhead protection policies, and analyzing the effects of salt water intrusion, artificial recharge, and land-use policies.

7.3 GOVERNING EQUATIONS

Ground water modeling can encompass several different aspects of the behavior of subterranean water systems. Five processes of potential relevance include ground water flow, solute transport, heat transport, deformation, and multiphase flow. Ground water flow modeling studies are usually undertaken to determine the responses of an aquifer to pumping, injection, or recharge stresses. Mass transport modeling studies are usually concerned with the movement within an aquifer system of a solute or contaminant. Heat transport models are usually focused on developing geothermal energy resources, and deformation studies are employed to analyze the effects of ground water removal on land subsidence. An area of intense current research is multiphase flow modeling which involves movement of fluids released to the subsurface that are immiscible with ground water.

The need for ground water models stems from the fact that the fundamental equations governing the behavior of the five processes listed earlier are all complex second order partial differential equations which are not amenable to direct analytical solution without significant simplifying assumptions. Ground water models try to circumvent these difficulties by either: (1) simulating the behavior of the aquifer system on a small scale or with coarse grids or (2) using simplifying assumptions or numerical approximations to the governing equations. The basic equations to be addressed herein include the Laplace and Poisson equations, and the equations for transient flow, unconfined aquifers, and solute transport (Canter et al., 1987).

Described below are some of the equations governing flow in ground water formations under confined conditions. Traditionally the development of the equations utilizes a mass balance approach for an infinitesimal volume within an aquifer (Figure 7.1). With this approach, the saturated thickness of the volume (Δz) is constant which is characteristic of confined formations. The unconfined equations are developed with the saturated thickness being variable. In addition, the manipulation of equations oftentimes results in the combination of hydraulic conductivity times saturated thickness. This product is referred to as the transmissivity.

7.3.1 Laplace Equation

If steady state flow of ground water through an infinitesimal volume is assumed, then by combining Darcy's Law and the equation of continuity the following equation can be developed:

$$\frac{\partial}{\partial x}\left(-K_x \frac{\partial h}{\partial x}\right) + \frac{\partial}{\partial y}\left(-K_y \frac{\partial h}{\partial y}\right) + \frac{\partial}{\partial z}\left(-K_z \frac{\partial h}{\partial z}\right) = 0 \qquad (7.1)$$

where   K = hydraulic conductivity of aquifer in the three directions (L/t)

h = hydraulic head (L)

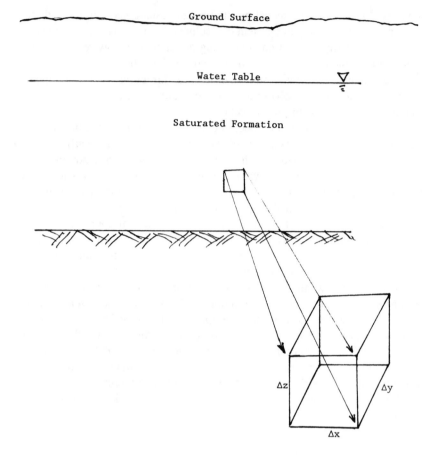

Figure 7.1. Elemental volume for equation development.

If K is assumed to be constant throughout the aquifer and independent of direction, that is, the aquifer is assumed to be homogeneous and isotropic, then Equation 7.1 becomes:

$$\frac{\partial^2 h}{\partial x^2} + \frac{\partial^2 h}{\partial y^2} + \frac{\partial^2 h}{\partial z^2} = 0 \tag{7.2}$$

Equation 7.2 is the well-known Laplace equation. The Laplace equation, as applied to hydraulic head in an aquifer, can be used to describe steady state flow through an isotropic, homogeneous aquifer with no sources of recharge or sinks of removal.

7.3.2 Poisson Equation

If sources of recharge (such as direct precipitation) or sinks of removal (such as extraction wells) need to be addressed, the Laplace equation must be modified. By considering a volumetric water balance at steady-state conditions, that is, outflow equals inflow, for an infinitesimal volume of aquifer the following two-dimensional equation can be identified:

$$\frac{\partial^2 h}{\partial x^2} + \frac{\partial^2 h}{\partial y^2} = \frac{-R(x,y)}{T}$$ (7.3)

where $R(x,y)$ = volume of water added per unit time per unit aquifer area (L/t)

T = transmissivity of the aquifer = hydraulic conductivity × the saturated thickness (L^2/t)

Equation 7.3 is known as Poisson's equation, and it can be used to predict two-dimensional steady state flow in an isotropic, homogeneous aquifer with possible sources and/or sinks included.

7.3.3 Transient Flow Equation

Equations 7.2 and 7.3 both govern steady-state flow conditions. The more common case encountered is transient flow conditions where the position of the water table (or piezometric surface) is changing with time. By setting the volume outflow rate in an infinitesimal aquifer equal to the volume inflow rate plus the rate of release of water from storage, the following equation is developed:

$$\frac{\partial^2 h}{\partial x^2} + \frac{\partial^2 h}{\partial y^2} = \frac{S}{T}\frac{\partial h}{\partial t} + \frac{R(x,y)}{T}$$ (7.4)

where t = time

S = storage coefficient of aquifer = volume of water released per unit area of aquifer per unit drop in head (dimensionless)

Equation 7.4 is a two-dimensional equation for addressing transient flow in an isotropic, homogeneous aquifer.

7.3.4 Unconfined Aquifers

Equations 7.2, 7.3, and 7.4 are most often applied to confined aquifers. Equations governing the behavior of unconfined aquifers must take into consideration the fact that the saturated thickness is variable. The Dupuit-Forchheimer assumptions of horizontal flow and hydraulic gradient equal to the slope of the water table are needed for development of these equations. The equation for two-dimensional steady state flow in an unconfined aquifer becomes:

$$\frac{\partial^2 h^2}{\partial x^2} + \frac{\partial^2 h^2}{\partial y^2} = -2\frac{R(x,y)}{K} \tag{7.5}$$

The equation for two-dimensional transient flow in an unconfined aquifer is:

$$\frac{K}{2}\left(\frac{\partial^2 h^2}{\partial x^2} + \frac{\partial^2 h^2}{\partial y^2}\right) = S_y\frac{\partial h}{\partial t} - R(x,y,t) \tag{7.6}$$

where S_y = specific yield of aquifer = volume of water released per unit area per unit drop in water table (dimensionless).

7.3.5 Solute Transport Equations

Mass (solute) transport is more difficult than flow modeling due to the mathematical character of the governing equations and the potential multiplicity of subsurface processes that can affect the concentration of a contaminant at a point in time in the subsurface environment. The many natural processes that can affect transport and fate in the subsurface can be divided into physical, chemical and biological categories as shown in Table 7.1 (National Research Council, 1990).

The equation governing the movement of a dissolved species in ground water can be developed by utilizing a conservation of mass approach and assuming Fick's Law to govern dispersion as was done in Chapter 2. The general form of the solute transport equation in statements is

net rate of change of mass of solute within the element	=	flux of solute out of the element	−	flux of solute into the element	±	loss or gain of solute mass due to reactions

**Table 7.1 A Summary of the Processes Important in Dissolved
Contaminant Transport and Their Impact on
Contaminant Spreading**

Process	Definition	Impact on Transport
Mass transport		
1. Advection	Movement of mass as a consequence of ground water flow	Most important way of transporting mass away from source
2. Diffusion	Mass spreading due to molecular diffusion in response to concentration gradients	An attenuation mechanism of second order in most flow systems where advection and dispersion dominate
3. Dispersion	Fluid mixing due to effects of unresolved heterogeneities in the permeability distribution	An attenuation mechanism that reduces contaminant concentration in the plume; however, spreads to a greater extent than predicted by advection alone
Chemical mass transfer		
4. Radioactive decay	Irreversible declined in the activity of a radionuclide through a nuclear reaction	An important mechanism for contaminant attenuation when the half-life for decay is comparable to or less than the residence time of the flow system, also adds complexity in production of daughter products
5. Sorption	Partitioning of a contaminant between the ground water and mineral or organic solids in the aquifer	An important mechanism that reduces the rate at which the contaminants are apparently moving; makes it more difficult to remove contamination at a site
6. Dissolution/ precipitation	The process of adding contaminants to, or removing them from, solution by reactions dissolving or creating various solids	An important mechanism that can control the concentration of contaminant in solution; solution concentration is mainly controlled either at the source or at a reaction front

Table 7.1 (Continued)

Process	Definition	Impact on Transport
7. Acid/base reactions	Reaction involving a transfer of protons (H$^+$)	Mainly an indirect control on contaminant transport by controlling the pH of ground water
8. Complexation	Combination of cations and anions to form a more complex ion	An important mechanism resulting in increased solubility if metals in ground water, if adsorption is not enhanced; major ion complexation will increase the quantity of a solid dissolved in solution
9. Hydrolysis/ substitution	Reaction of a halogenated organic compound with water or a component ion of water (hydrolysis) or with another anion (substitution)	Often hydrolysis/ substitution reactions make an organic compound more susceptible to biodegradation and more soluble
10. Redox reactions (biodegradation)	Reactions that involve a transfer of electrons and include elements with more than one oxidation state	An extremely important family of reactions in retarding contaminant spread through the precipitation of metals
Biologically mediated mass transfer		
11. Biological	Reaction involving the degradation of organic compounds, whose rate is controlled by the abundance of the microorganisms and redox conditions	Important mechanism for contaminant reduction, but can lead to undesirable daughter products

The solute transport equation can take on different forms depending on the assumptions made or reactions considered. The one-dimensional form of the equation for a nonreactive dissolved constituent in a homogeneous, isotropic aquifer under steady-state, uniform flow is

$$D_\ell \frac{\partial^2 C}{\partial \ell^2} - V_\ell \frac{\partial C}{\partial \ell} = \frac{\partial C}{\partial t} \tag{7.7}$$

where ℓ = curvilinear coordinate direction taken along the flowline
D_ℓ = coefficient of hydrodynamic dispersion in the flow path direction
V_ℓ = average linear ground water velocity or seepage velocity
C = solute concentration

The two-dimensional equation for a nonreactive dissolved chemical species in flowing ground water can be written as follows (Konikow and Bredehoeft, 1978):

$$\frac{\partial(Cb)}{\partial t} = \frac{\partial}{\partial x_i}\left(bD_{ij}C_j \frac{\partial C}{\partial x_j}\right) - \frac{\partial}{\partial x_i}(bCV_i) - \frac{C'W}{\eta} \tag{7.8}$$

$$i, j = 1, 2$$

where C = the concentration of the dissolved chemical species (M/L^3)
V_i = the seepage velocity in the direction of x_i (L/t)
D_{ij} = the coefficient of hydrodynamic dispersion (a second-order tensor) (L^2/t)
C' = the concentration of the dissolved chemical in a source or sink fluid (M/L^3)
W = the volume flux per unit area (L/t)
b = the saturated thickness of the aquifer (L)

The first term on the right side of Equation 7.8 represents the change in concentration due to hydrodynamic dispersion. The second term describes the effects of convective transport, while the third term represents a fluid source or sink (Konikow and Bredehoeft, 1978).

Javendel et al. (1984) have presented a generalized two-dimensional equation describing solute transport and including reactions as:

$$\frac{\partial}{\partial x_i}\left(D_{ij}\frac{\partial C}{\partial x_j}\right) - \frac{\partial}{\partial x_i}(CV_i) - \frac{C'W^*}{\eta} + \sum_{K=1}^{N} R_K = \frac{\partial C}{\partial t} \tag{7.9}$$

$$i, j = 1, 2$$

$$V_i = \frac{-K_{ij}}{\eta} \frac{\partial h}{\partial x_j}$$

(7.10)

$$i, j = 1, 2$$

and C = solute concentration
 V_i = seepage or average pore water velocity in the direction x_i
 D_{ij} = dispersion coefficient tensor
 C' = solute concentration in the source or sink fluid
 $W*$ = volume flow rate per unit volume of the source or sink
 η = effective porosity
 h = hydraulic head
 K_{ij} = hydraulic conductivity tensor
 R_k = rate of solute production in reaction K of N different reactions
 x_i = Cartesian coordinate

If equilibrium-controlled ion exchange reactions are considered, the summation in Equation 7.9 may be set equal to:

$$\sum_{K=1}^{N} R_K = \frac{-\rho_b}{\eta} \frac{\partial \overline{C}}{\partial t}$$

(7.11)

where ρ_b is the bulk density of the solid and $\overline{C}$ is the concentration of species adsorbed on the solid.

To incorporate Equation 7.11 into Equation 7.9, an expression relating the adsorbed concentration $\overline{C}$ to the solute concentration C is required. Considering equilibrium transport and assuming that the adsorption isotherm can be described with a linear and reversible equation, then

$$\overline{C} = K_d C$$

(7.12)

where K_d is called the distribution coefficient. Now, by incorporating Equations 7.11 and 7.12 into Equation 7.9, the following is obtained:

$$\frac{\partial}{\partial x_i}\left(D_{ij}\frac{\partial C}{\partial x_j}\right) - \frac{\partial}{\partial x_i}(CV_i) + \frac{C'W^*}{\eta} = R\frac{\partial C}{\partial t}$$

(7.13)

where

$$R = \left(1 + \frac{\rho_b K_d}{\eta}\right) \tag{7.14}$$

The parameter R is called the retardation factor. Retardation can be used to account for various chemical processes in the subsurface such as adsorption and ion exchange.

7.3.6 Multiphase Flow Equations

Several authors have developed mathematical descriptions for the behavior of multiphase flow systems. For further development and description of the various models the reader is referred to Faust (1985); Abriola and Pinder (1985); Osborne and Sykes 1986); Parker et al. (1986); Reible et al. (1986); Corapcioglu and Hossain (1986); and references therein. Each of these references contains a review of multiphase flow theory.

The development of the governing equations for multiphase flow is based on a mass balance for each of the phases about an elemental volume. As noted in Reible et al. (1986), single fluid flow equations are written for each phase and then coupled using capillary pressures and saturations of each phase. The equations are also implicitly coupled by their relative permeabilities (see Figures 2.11 and 2.12). Although simple in concept, this procedure is difficult to implement. Parker et al. (1986) point out that a major difficulty in modeling multiphase transport has been the lack of information on the relationships between fluid pressures, saturations, and permeabilities of the coexisting phases.

One way of presenting the multiphase transport equation is to start with the comprehensive development and then work back to the simpler relationships by outlining the simplifying assumptions. Abriola and Pinder (1985) developed the macroscopic mass balance law for a species "i" in a phase "α" as

$$\frac{\partial}{\partial t}\left(\rho^\alpha \varepsilon^\alpha w_i^\alpha\right) + \nabla \bullet \left(\rho^\alpha \varepsilon^\alpha V^\alpha w_i^\alpha\right) - \nabla \bullet J_i^\alpha - \rho^\alpha \varepsilon^\alpha f_i^\alpha$$

$$= \rho^\alpha \varepsilon^\alpha \left[e^\alpha \left(\rho w_i\right) + I_i^\alpha\right] \tag{7.15}$$

where ρ^α = α phase mass density
ε^α = fraction of voids occupied by the α phase
V^α = α phase velocity
w_i^α = mass fraction of species "i" in α phase
J_i^α = average nonadvective flux vector of species "i" in the α phase
f_i^α = external supply of species "i" in the α phase
$e^\alpha(\rho w_i)$ = exchange of mass of species "i" due to phase changes
I_i^α = exchange of mass species "i" due to interphase diffusion

Equation 7.15 must be written for each phase (e.g., aqueous, nonaqueous, gaseous) in which species "i" can exist. These equations (one for each α) are then summed to produce a total mass balance for the species. The above procedure must be done for each species.

One of the most common simplifying assumptions is that the species are non-reacting. Equation 7.15 then becomes

$$\frac{\partial}{\partial t}\left(\rho^\alpha \varepsilon^\alpha w_i^\alpha\right) + \nabla \bullet \left(\rho^\alpha \varepsilon^\alpha V^\alpha w_i^\alpha\right) - \nabla \bullet J_i^\alpha - \rho^\alpha \varepsilon^\alpha f_i^\alpha = 0 \qquad (7.16)$$

Equation 7.16 would be written and summed for all "α" phases containing constituent "i". This could include a dissolved aqueous phase, a nonaqueous phase, and a gaseous phase. It is important to note that Equation 7.16 is simply a mass balance equation. The first term reflects the change in mass (of species "i" in the "α" phase) with respect to time. This time rate of change is due to three processes: advective flow (second term), nonadvective or diffusive and dispersive flow (third term), and the influence of reactive sources or sinks (fourth term).

A second simplifying assumption is to ignore the nonadvective flow term in Equation 7.16. Reible et al. (1986) note that for water immiscible chemicals, the infiltrating chemical phase will maintain a sharp wetting front. For water miscible chemicals, the infiltrating chemical phase will be preceded by a dissolution zone. Hence, the applicability of the assumption of nonadvective flow being negligible is dependent on the chemical (species) in question.

By assuming negligible nonadvective flow and by applying Darcy's Law to the advective term, Equation 7.16 can be rewritten as (Faust, 1985).

$$\frac{\Sigma}{\text{all phases } \alpha} \nabla \bullet \frac{K\rho_\alpha k_{ra}}{\mu_\alpha}\left(\nabla \bullet P_\alpha - \rho_\alpha g \nabla \bullet D\right) + f_\alpha = \frac{\partial\left(\eta \rho_\alpha S_\alpha\right)}{\partial t} \qquad (7.17)$$

where D = depth of the phase
P_α = pressure of the phase
μ_α = viscosity of the phase

$k_{r\alpha}$ = relative permeability of the phase
K = intrinsic permeability of the medium
η = porosity of the medium
S_α = saturation of the phase

Note that for simplification, consideration of different species is dropped from Equation 7.17, the subscripts now refer to the various phases. As was noted earlier, the simplifying assumption made most often is that the phases behave separately. (Consideration of phase transport only is a significant limitation to the usefulness of the resulting models. Significant species transport can occur in the aqueous and gaseous phases for species (chemicals) contained in the nonaqueous phase). This results in the consideration of three phases: an aqueous or water phase (w), a nonaqueous or immiscible phase (n), and an air phase (a). Applying this assumption to Equation 7.17 results in

$$\nabla \bullet \frac{K\rho_w k_{rw}}{\mu_\alpha}\left(\nabla \bullet P_w - \rho_w g \nabla \bullet D\right) + f_w = \frac{\partial\left(\eta\rho_w S_w\right)}{\partial t} \qquad (7.18a)$$

$$\nabla \bullet \frac{K\rho_n k_{rn}}{\mu_n}\left(\nabla \bullet P_n - \rho_n g \nabla \bullet D\right) + f_n = \frac{\partial\left(\eta\rho_n S_n\right)}{\partial t} \qquad (7.18b)$$

$$\nabla \bullet \frac{K\rho_a k_{ra}}{\mu_a}\left(\nabla \bullet P_a - \rho_a g \nabla \bullet D\right) + f_n = \frac{\partial\left(\eta\rho_a S_a\right)}{\partial t} \qquad (7.18c)$$

Equations 7.18a, 7.18b, and 7.18c have 16 dependent variables. In order to solve this system, thirteen additional relationships are needed. The most common relationships are, as outlined by Faust (1985),

1. Saturations sum to one ($S_w + S_n + S_a = 1$) (1).
2. Densities and viscosities are functions of phase pressure (6).
3. Relative permeabilities are functions of the saturations (3).
4. Capillary pressures are functions of saturations (2).
5. Porosity is a function of pressure (1).

Often times, the densities and viscosities are assumed to be independent of the pressure. The most sweeping assumption is to assume that the pressure of the gaseous phase is atmospheric. This allows Equation 7.18c to be ignored and the problem is essentially approximated as a two-phase (aqueous-immiscible) fluid flow problem.

7.4 TYPES OF MODELS

Ground water models can be divided into two broad groups; physical models and numerical models. The physical models can be subdivided into scale models and analog models. The numerical models can be subdivided into analytical models, stochastic models, and numerical techniques models.

7.4.1 Physical Models

The earliest attempts at modeling ground water were of the physical type. Physical models can be divided into two classes; scale models and analog models (Canter et al., 1987). Scale models are actual physical replicas of an aquifer that have been "scaled down" for study in the laboratory. The most common scale models are the soil column (one-dimensional model), the Hele-Shaw apparatus (two-dimensional model) and the sand-tank (three-dimensional model). In these models, media is placed in such a way that it parallels the soil structure of the aquifer of concern. The models are then subjected to certain stresses such as water removal or injection, or contaminated recharge. The response of the models is obtained through direct measurements of selected variables. The behavior of the prototype aquifer to real life stresses can then be predicted by using scale relationships.

Analog models are based on the fact that a direct analogy can be made between ground water flow and some easily measurable phenomenon from a different field of study. One type of analog model is the electric analog. In this type of model, the properties of an aquifer (permeability, storage coefficient, etc.) are simulated by various electronic components (resistors, capacitors, etc.), and the voltage across these components is analogous to the potential (or head) of water in the aquifer. For example, Shamberger and Domenico (1965) described an electrical analog model for the ground water system in the Las Vegas Valley, Nevada. The model was used to predict aquifer response to pumping by measuring appropriate voltages and currents. The analog model was considered to be a useful tool in analyzing ground water management options.

With the advent of high-speed computational capabilities, the use of physical models for predicting ground water flows has decreased. Physical models are also limited by space, time, cost, and accuracy deficiencies. However, scale models or microcosms are currently being used in laboratory studies of the transport and fate of selected ground water contaminants (Canter, 1988). Analog models are finding decreasing applications, especially in the area of ground water quality modeling.

7.4.2 Mathematical Models

Mathematical models are currently the most popular types of ground water models. Mathematical models can be grouped into three classes: (1) analytical models, (2) stochastic models, and (3) numerical technique models (Canter, et al., 1987). Analytical models are solutions to the equations governing subsurface processes. Analytical models are usually developed by considering highly idealized conditions or using significant simplifying assumptions to obtain a solution to the governing equations. The utility of analytical models is their ability to screen and test numerical models.

Frequently cited examples of analytical solutions are the steady-state flow of ground water under either confined or unconfined conditions using the Dupuit-Forchheimer (DF) assumptions. The DF assumptions are (1) purely horizontal flow and (2) the hydraulic gradient is equal to the slope of the potentiometric surface and invariant with depth.

Utilizing the DF assumptions and employing Darcy's Law with the continuity equation gives rise to two of the more famous analytical models or equations. The equations describing the steady-state flow of ground water to fully penetrating wells under the DF assumptions are known as the Theim equations and are shown below:

$$Q = \frac{2\Pi KD\left(h_2 - h_1\right)}{\ln\left(r_2 / r_1\right)}$$ (7.19 confined)

or where   Q = volumetric flow rate (L^3/t)
       h = hydraulic head (L)

$$Q = \frac{\Pi K\left(h_2^2 - h_1^2\right)}{\ln\left(r_2 / r_1\right)}$$ (7.20 unconfined)

where   K = hydraulic conductivity (L/t)
      D = depth of aquifer (L)
      r = radius from well (L)

Equations 7.19 and 7.20 are overly simplistic for all but the most general studies. Conversely, many other analytical solutions have been developed for extremely idealized or specific situations. Tables 7.2 and 7.3 are comprehensive listings of analytical solutions (models) for ground water flow under both steady-state and transient conditions.

Analytical solutions for solute transport problems are difficult to develop owing to the multiple processes involved. Bear (1979) discusses several analytical solutions to relatively simple, one-dimensional solute transport problems. However, even simple solutions tend to get overwhelmed with advanced mathematics. As an example, consider the one-dimensional flow of a solute through the soil column shown in Figure 2.2(a). The boundary conditions represented by the step function input are described mathematically as

$$
\begin{array}{lll}
C(\ell,0) & = & 0 & \ell > 0 \\
C(0,t) & = & C_0 & t > 0 \\
C(\infty,t) & = & 0 & t > 0
\end{array}
$$

For these boundary conditions the solution to Equation 7.7 for a saturated, homogeneous porous medium is

$$
\frac{C}{C_o} = \frac{1}{2}\left[erfc\,\frac{\left(\ell - \bar{v}_\ell t\right)}{2\sqrt{D_\ell t}} + \exp\frac{\left(\bar{v}_\ell \cdot \ell\right)}{D_\ell}\,erfc\,\frac{\left(\ell + \bar{v}_\ell t\right)}{2\sqrt{D_\ell t}} \right] \qquad (7.21)
$$

where erfc represents the complementary error function, ℓ is the distance along the flow path, and $\bar{v}_\ell$ is the average linear water velocity. For conditions in which the dispersivity (D_ℓ) of the porous medium is large or when ℓ or t is large, the second term on the right-hand side of Equation 7.21 is negligible. Equation 7.21 can be used to compute the shapes of the breakthrough curves and concentration profiles illustrated in Figure 2.2(c) and 2.2(d).

Analytical models represent an attractive alternative to both physical and numerical models in terms of decreased complexity and input data requirements. However, analytical models are often only feasible when based on significant simplifying assumptions, and these assumptions may not allow the model to accurately reflect the conditions of interest. Additionally, even some of the simplest analytical models tend to involve complex mathematics. A comprehensive compilation of solute transport analytical models can be found in van Genuchten and Alves (1982).

Stochastic models are generally considered to be those models that incorporate the uncertain nature of certain subsurface properties within the model. In simple terms, the basic information governing ground water, such as hydraulic conductivity values, is considered to be variable with space and/or time and is best represented by stochastic processes. The results of the modeling process, therefore, include some sort of mean value accompanied by the variance to account for uncertainty. A more complete description of stochastic models, along with applications and problems associated with stochastic models can be found in van der Heijde et al. (1988).

The most popular approach for modeling ground water behavior is through the use of numerical models involving computer-based solutions. Numerical

Table 7.2 Analytical Solutions of Steady Flow Applicable to Corrective Actions (Cohen and Miller, 1983)

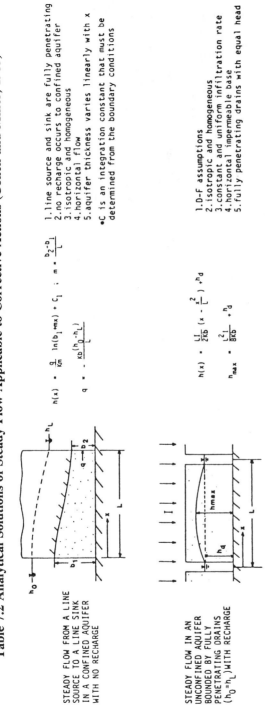

STEADY FLOW FROM A LINE SOURCE TO A LINE SINK IN A CONFINED AQUIFER WITH NO RECHARGE

$$h(x) = \frac{q}{Km} \ln(b_1 + mx) + C_1 \quad ; \quad m = \frac{b_2 - b_1}{L}$$

$$q = -\frac{KD(h_0 - h_L)}{L}$$

1. line source and sink are fully penetrating
2. no recharge occurs to confined aquifer
3. isotropic and homogeneous
4. horizontal flow
5. aquifer thickness varies linearly with x

• C is an integration constant that must be determined from the boundary conditions

STEADY FLOW IN AN UNCONFINED AQUIFER BOUNDED BY FULLY PENETRATING DRAINS ($h_0 = h_L$) WITH RECHARGE

$$h(x) = \frac{LI}{2Kb}(x - \frac{x^2}{L}) + h_d$$

$$h_{max} = \frac{L^2 I}{8Kb} + h_d$$

1. D-F assumptions
2. isotropic and homogeneous
3. constant and uniform infiltration rate
4. horizontal impermeable base
5. fully penetrating drains with equal head

Table 7.2 (continued)

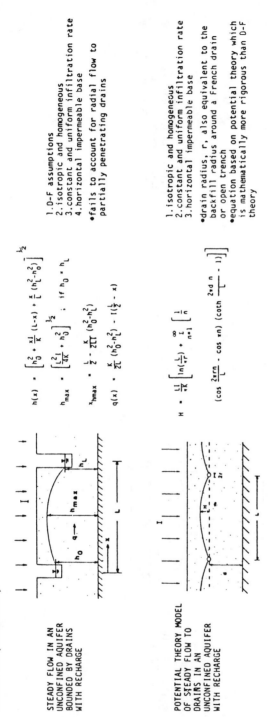

STEADY FLOW IN AN UNCONFINED AQUIFER BOUNDED BY DRAINS WITH RECHARGE

$$h(x) = \left[h_0^2 + \frac{xI}{K}(L-x) + \frac{x}{L}(h_L^2 - h_0^2) \right]^{1/2}$$

$$h_{max} = \left[\frac{L^2 I}{4K} + h_0^2 \right]^{1/2} \; ; \quad \text{if } h_0 = h_L$$

$$x_{hmax} = \frac{L}{2} - \frac{K}{2IL}(h_0^2 - h_L^2)$$

$$q(x) = \frac{K}{2L}(h_0^2 - h_L^2) - I\left(\frac{L}{2} - x\right)$$

1. D-F assumptions
2. isotropic and homogeneous
3. constant and uniform infiltration rate
4. horizontal impermeable base

• fails to account for radial flow to partially penetrating drains

POTENTIAL THEORY MODEL OF STEADY FLOW TO DRAINS IN AN UNCONFINED AQUIFER WITH RECHARGE

$$H = \frac{LI}{\pi K}\left[\ln\left(\frac{L}{\pi r}\right) + \sum_{n=1}^{\infty} \frac{1}{n} \left[\left(\cos\frac{2\pi rn}{L} - \cos \pi n\right)\left(\coth\frac{2\pi d n}{L} - 1\right)\right]\right]$$

1. isotropic and homogeneous
2. constant and uniform infiltration rate
3. horizontal impermeable base

• drain radius, r, also equivalent to the backfill radius around a French drain or open trench

• equation based on potential theory which is mathematically more rigorous than D-F theory

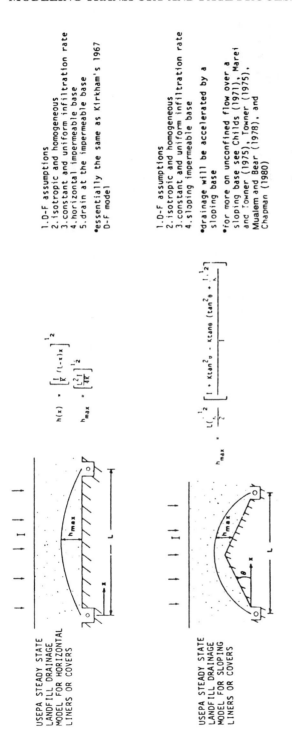

USEPA STEADY STATE
LANDFILL DRAINAGE
MODEL FOR HORIZONTAL
LINERS OR COVERS

1. D-F assumptions
2. isotropic and homogeneous
3. constant and uniform infiltration rate
4. horizontal impermeable base
5. drain at the impermeable base

• essentially the same as Kirkham's 1967
  D-F model

$$h(x) = \left[\frac{1}{K}'^{(L-x)x} \right]^{\frac{1}{2}}$$

$$h_{max} = \left[\frac{L^2 I}{4K} \right]^{\frac{1}{2}}$$

USEPA STEADY STATE
LANDFILL DRAINAGE
MODEL FOR SLOPING
LINERS OR COVERS

1. D-F assumptions
2. isotropic and homogeneous
3. constant and uniform infiltration rate
4. sloping impermeable base

• drainage will be accelerated by a
  sloping base

• for more on unconfined flow over a
  sloping base see Childs (1971), Marei
  and Towner (1975), Towner (1975),
  Mualem and Bear (1978), and
  Chapman (1980)

$$h_{max} = L(\frac{I}{K})^{\frac{1}{2}} \left[\frac{1 + Ktan^2\theta - Ktan\theta (tan^2\theta + \frac{I}{K})^{\frac{1}{2}}}{1} \right]$$

Table 7.2 (continued)

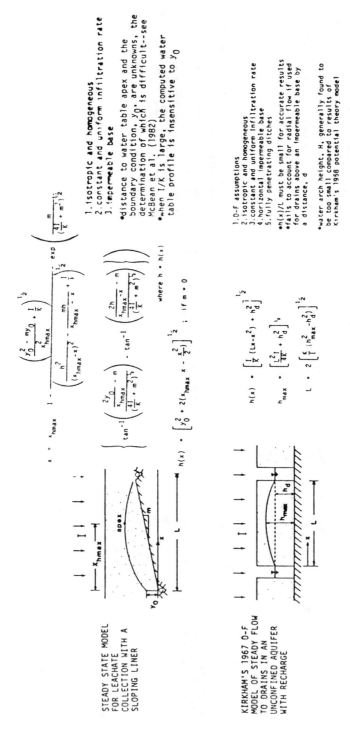

STEADY STATE MODEL FOR LEACHATE COLLECTION WITH A SLOPING LINER

$$h(x) = \left[y_0^2 + 2(x_{hmax} x - \frac{x^2}{2}) \right]^{\frac{1}{2}}; \quad \text{if } m = 0$$

$$\left\{ \tan^{-1} \frac{\frac{2y_0}{x_{hmax}} - m}{(\frac{4I}{K} + m^2)^{\frac{1}{2}}} - \tan^{-1} \frac{\frac{2h}{x_{hmax} - x}}{(\frac{4I}{K} + m^2)^{\frac{1}{2}}} \right\}$$

$$x = x_{hmax} \left\{ 1 - \frac{\left[\frac{h^2}{(x_{hmax} - x)^2} - \frac{mh}{x_{hmax} - x} + \frac{I}{K} \right]^{\frac{1}{2}}}{\left(\frac{y_0^2 - my_0}{x_{hmax}^2} + \frac{I}{K} \right)^{\frac{1}{2}}} \exp \left(\frac{4I}{K} + m^2 \right)^{\frac{1}{2}} \right\}$$

where $h = h(x)$

1. isotropic and homogeneous
2. constant and uniform infiltration rate
3. impermeable base

• distance to water table apex and the boundary condition, y_0, are unknowns, the determination of which is difficult--see McBean et al. (1982)
• when I/K is large, the computed water table profile is insensitive to y_0

KIRKHAM'S 1967 0-F MODEL OF STEADY FLOW TO DRAINS IN AN UNCONFINED AQUIFER WITH RECHARGE

$$h(x) = \left[\frac{I}{K} (Lx - x^2) + h_d^2 \right]^{\frac{1}{2}}$$

$$h_{max} = \left[\frac{2I}{4K} + h_d^2 \right]^{\frac{1}{2}}$$

$$L = 2 \left[\frac{K}{I} (h_{max}^2 - h_d^2) \right]^{\frac{1}{2}}$$

1. 0-F assumptions
2. isotropic and homogeneous
3. constant and uniform infiltration rate
4. horizontal impermeable base
5. fully penetrating ditches

• $h(x)/L$ must be small for accurate results
• fails to account for radial flow if used for drains above an impermeable base by a distance, d

• water arch height, H, generally found to be too small compared to results of Kirkham's 1958 potential theory model

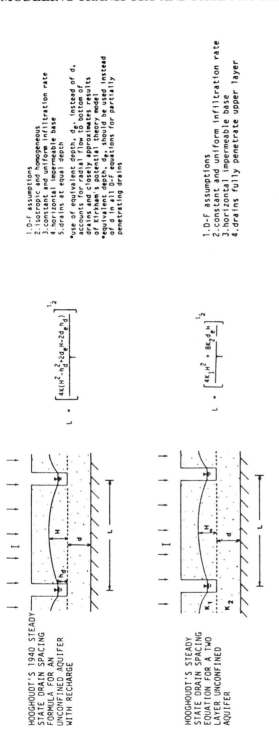

HOOGHOUDT'S 1940 STEADY
STATE DRAIN SPACING
FORMULA FOR AN
UNCONFINED AQUIFER
WITH RECHARGE

$$L = \left[\frac{4K(H^2 - h_d^2 + 2d_e H - 2d_e h_d)}{} \right]^{\frac{1}{2}}$$

1. D-F assumptions
2. isotropic and homogeneous
3. constant and uniform infiltration rate
4. horizontal impermeable base
5. drains at equal depth

• use of equivalent depth, d_e, instead of d, accounts for radial flow to bottom of drains and closely approximates results of Kirkham's potential theory model
• equivalent depth, d_e, should be used instead of d in all D-F equations for partially penetrating drains

HOOGHOUDT'S STEADY
STATE DRAIN SPACING
EQUATION FOR A TWO
LAYER UNCONFINED
AQUIFER

$$L = \left[\frac{4K_1 H^2 + 8K_2 d_e H}{} \right]^{\frac{1}{2}}$$

1. D-F assumptions
2. constant and uniform infiltration rate
3. horizontal impermeable base
4. drains fully penetrate upper layer

Table 7.2 (continued)

YOUNGS' STEADY STATE INEQUALITY DRAIN SPACING MODEL

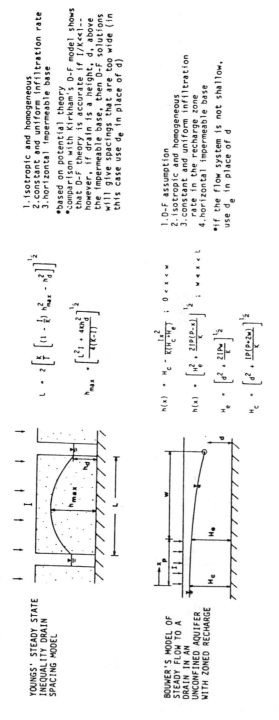

$$L = 2\left[\frac{K}{I}\left[\left(1-\frac{I}{K}\right)h_{max}^2 - h_d^2\right]\right]^{1/2}$$

$$h_{max} = \left[\frac{L^2 I + 4Kh_d^2}{4(K-I)}\right]^{1/2}$$

1. isotropic and homogeneous
2. constant and uniform infiltration rate
3. horizontal impermeable base

• based on potential theory
• comparison with Kirkham's D-F model shows that D-F theory is accurate if I/K<<1-- however, if drain is a height, d, above the impermeable base, then D-F solutions will give spacings that are too wide (in this case use d_e in place of d)

BOUWER'S MODEL OF STEADY FLOW TO A DRAIN IN AN UNCONFINED AQUIFER WITH ZONED RECHARGE

$$h(x) = H_c - \frac{Ix^2}{K(H_c+H_e)} \; ; \; 0 < x < w$$

$$h(x) = \left[H_e^2 + \frac{2IP(P-x)}{K}\right]^{1/2} \; ; \; w < x < L$$

$$H_e = \left[d^2 + \frac{2IPw}{K}\right]^{1/2}$$

$$H_c = \left[d^2 + \frac{IP(P+2w)}{K}\right]^{1/2}$$

1. D-F assumption
2. isotropic and homogeneous
3. constant and uniform infiltration rate in the recharge zone
4. horizontal impermeable base

• if the flow system is not shallow, use d_e in place of d

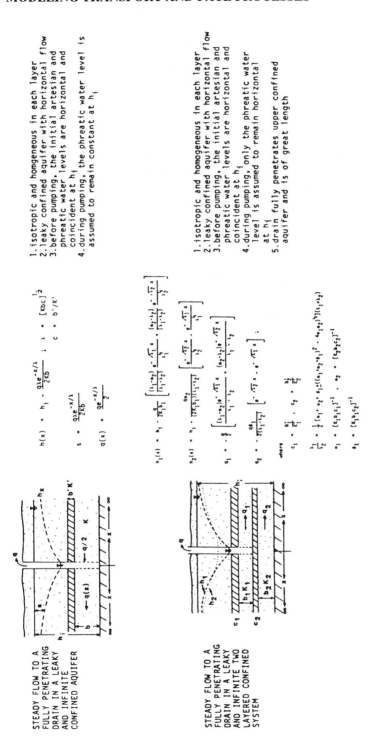

STEADY FLOW TO A FULLY PENETRATING DRAIN IN A LEAKY AND INFINITE CONFINED AQUIFER

1. isotropic and homgeneous in each layer
2. leaky confined aquifer with horizontal flow
3. before pumping, the initial artesian and phreatic water levels are horizontal and coincident at h_i
4. during pumping, the phreatic water level is assumed to remain constant at h_i

$$h(x) = h_i - \frac{q_1 e^{-x/\lambda}}{2Kb}; \quad \lambda = [Kbc]^{\frac{1}{2}}$$
$$c = b'/K'$$

$$s = \frac{q_1 e^{-x/\lambda}}{2Kb}$$

$$q(x) = \frac{q e^{-x/\lambda}}{2}$$

STEADY FLOW TO A FULLY PENETRATING DRAIN IN A LEAKY AND INFINITE TWO LAYERED CONFINED SYSTEM

1. isotropic and homogeneous in each layer
2. leaky confined aquifer with horizontal flow
3. before pumping, the initial artesian and phreatic water levels are horizontal and coincident at h_i
4. during pumping, only the phreatic water level is assumed to remain horizontal at h_i
5. drain fully penetrates upper confined aquifer and is of great length

Table 7.2 (continued)

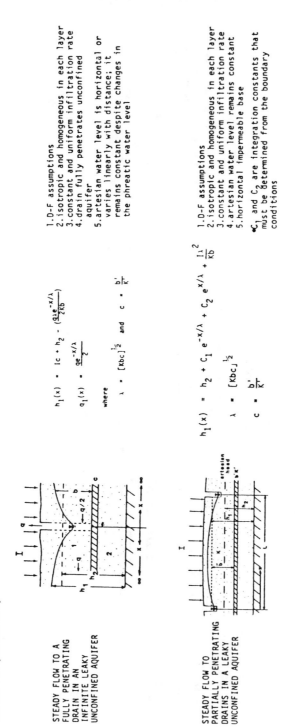

STEADY FLOW TO A FULLY PENETRATING DRAIN IN AN INFINITE LEAKY UNCONFINED AQUIFER

$$h_1(x) = Ic + h_2 - \left(\frac{qe^{-x/\lambda}}{2Kb}\right)$$

$$q_1(x) = \frac{qe^{-x/\lambda}}{2}$$

where

$$\lambda = [Kbc]^{\frac{1}{2}} \quad \text{and} \quad c = \frac{b'}{K'}$$

1. D-F assumptions
2. isotropic and homogeneous in each layer
3. constant and uniform infiltration rate
4. drain fully penetrates unconfined aquifer
5. artesian water level is horizontal or varies linearly with distance; it remains constant despite changes in the phreatic water level

STEADY FLOW TO PARTIALLY PENETRATING DRAINS IN A LEAKY UNCONFINED AQUIFER

$$h_1(x) = h_2 + C_1 e^{-x/\lambda} + C_2 e^{x/\lambda} + \frac{I\lambda^2}{Kb}$$

$$\lambda = [Kbc]^{\frac{1}{2}}$$

$$c = \frac{b'}{K'}$$

1. D-F assumptions
2. isotropic and homogeneous in each layer
3. constant and uniform infiltration rate
4. artesian water level remains constant
5. horizontal impermeable base

C_1 and C_2 are integration constants that must be determined from the boundary conditions

$$h_1(x) = C_1 e^{-x/\lambda} + C_2 e^{x/\lambda} - \frac{Icx^2}{2(\lambda_1^2+\lambda_2^2)} + C_3 x + \frac{Ic}{(1+a)}z + C_4$$

$$h_2(x) = -aC_1 e^{-x/\lambda} - aC_2 e^{x/\lambda} - \frac{Icx^2}{2(\lambda_1^2+\lambda_2^2)} + C_3 x - \frac{aIc}{(1-a)}z + C_4$$

where

$$a = \frac{K_1 b_1}{K_2 b_2}$$

$$\lambda_1 = [K_1 b_1 c]^{\frac{1}{2}} \quad , \quad \lambda_2 = [K_2 b_2 c]^{\frac{1}{2}}$$

$$\lambda_2^2 = \frac{\lambda_1^2}{1+a}$$

$$c = \frac{b'}{K'}$$

1. horizontal flow in both aquifers
2. isotropic and homogeneous in each layer
3. constant and uniform infiltration rate
4. horizontal impermeable base to the confined aquifer

• $C_1 - C_4$ are integration constants that must be determined from boundary conditions

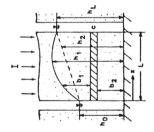

STEADY FLOW TO A
FULLY PENETRATING
DRAIN IN A LEAKY
CONFINED-UNCONFINED
AQUIFER SYSTEM

Table 7.3 Analytical Solutions of Transient Flow Applicable to Corrective Solutions (Cohen and Miller, 1983)

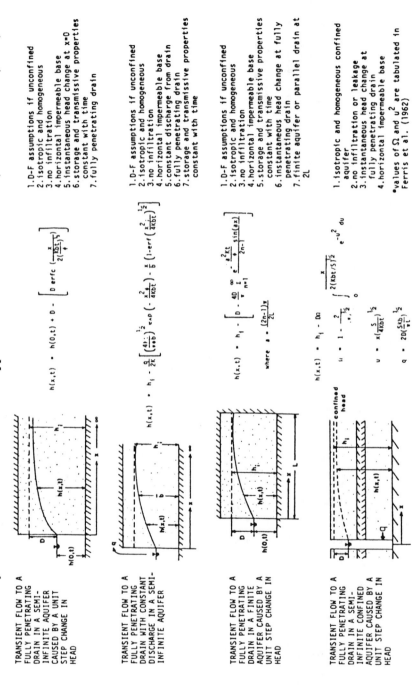

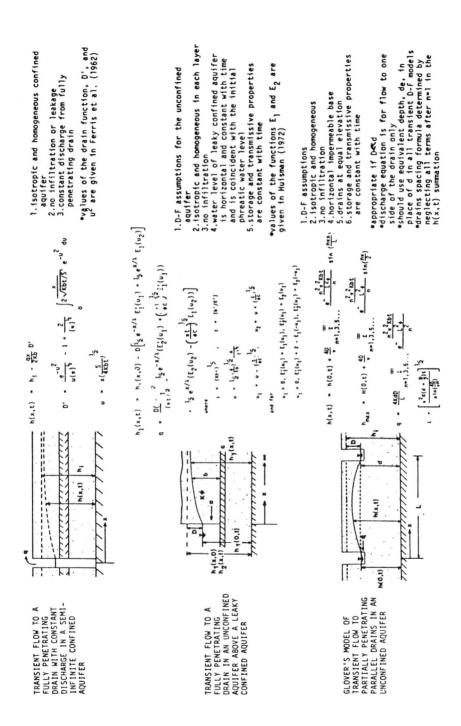

TRANSIENT FLOW TO A FULLY PENETRATING DRAIN WITH CONSTANT DISCHARGE IN A SEMI-INFINITE CONFINED AQUIFER

1. isotropic and homogeneous confined aquifer
2. no infiltration or leakage
3. constant discharge from fully penetrating drain

• values of the drain function, D', and u^2 are given in Ferris et al. (1962)

$$h(x,t) = h_i - \frac{Qx}{2Kb} D'$$

$$D' = \frac{e^{-u^2}}{u(\pi)^{\frac{1}{2}}} - 1 + \frac{2}{(\pi)^{\frac{1}{2}}} \int_0^{\frac{x}{2\sqrt{Kbt/S}}} e^{-u^2}\, du$$

$$u = x\frac{S^{\frac{1}{2}}}{4Kbt}$$

TRANSIENT FLOW TO A FULLY PENETRATING DRAIN IN AN UNCONFINED AQUIFER ABOVE A LEAKY CONFINED AQUIFER

1. D-F assumptions for the unconfined aquifer
2. isotropic and homogeneous in each layer
3. no infiltration
4. water level of leaky confined aquifer is horizontal and constant with time and is coincident with the initial phreatic water level
5. storage and transmissive properties are constant with time

• values of the functions E_1 and E_2 are given in Huisman (1972)

$$h_1(x,t) = h_1(x,0) - D\left[\tfrac{1}{2}e^{-x/\lambda} E_1'(u_1) + \tfrac{1}{2}e^{x/\lambda} E_1'(u_2)\right]$$

$$q = \frac{DL}{(\pi t)^{\frac{1}{2}}} \cdot \tfrac{1}{2}e^{x/\lambda}\left[E_2'(u_2) - \left(\frac{\pi t}{\phi c}\right)^{\frac{1}{2}} E_1'(u_2)\right]$$

where

$$\lambda = (\tau c)^{\frac{1}{2}} \qquad c = (b'/k')$$

$$u = \tfrac{1}{2}\left(\tfrac{1}{\phi c}\right)^{\frac{1}{2}} - \frac{x}{L}$$

$$u_1 = \tfrac{1}{2}\left(\tfrac{1}{\phi c}\right)^{\frac{1}{2}} \qquad u_2 = u - \left(\tfrac{1}{\phi c}\right)^{\frac{1}{2}}$$

and for

$$u_1 > 0, \ E_1'(u_1) = E_1(u_1),\ E_1'(u_1),\ E_1'(u_1)$$
$$u_1 < 0, \ E_1'(u_1) = 2 - E_1(-u_1),\ E_1'(-u_1),\ E_1'(-u_1)$$

GLOVER'S MODEL OF TRANSIENT FLOW TO PARTIALLY PENETRATING PARALLEL DRAINS IN AN UNCONFINED AQUIFER

1. D-F assumptions
2. isotropic and homogeneous
3. no infiltration
4. horizontal impermeable base
5. drains at equal elevation
6. storage and transmissive properties are constant with time

• appropriate if D<d
• discharge equation is for flow to one side of the drain only
• should use equivalent depth, de, in place of d in all transient D-F models
• drains spacing formula determined by neglecting all terms after n=1 in the h(x,t) summation

$$h(x,t) = h(0,t) + \frac{4Q}{\pi}\sum_{n=1,3,5,...}^{\infty} \frac{e^{-\frac{n^2 \pi^2 Kbt}{L^2}}}{n}\sin\left(\frac{n\pi x}{L}\right)$$

$$h_{max} = H(0,t) + \frac{4Q}{\pi}\sum_{n=1,3,5,...}^{\infty} \frac{e^{-\frac{n^2 \pi^2 Kbt}{L^2}}}{n}\sin\left(\frac{n\pi}{2}\right)$$

$$q = \frac{4Kd Q}{L^2}$$

$$L = \left[\frac{2Kt(d + \frac{D}{2})t}{S \ln\left(\frac{4D}{\pi h}\right)}\right]^{\frac{1}{2}}$$

Table 7.3 (continued)

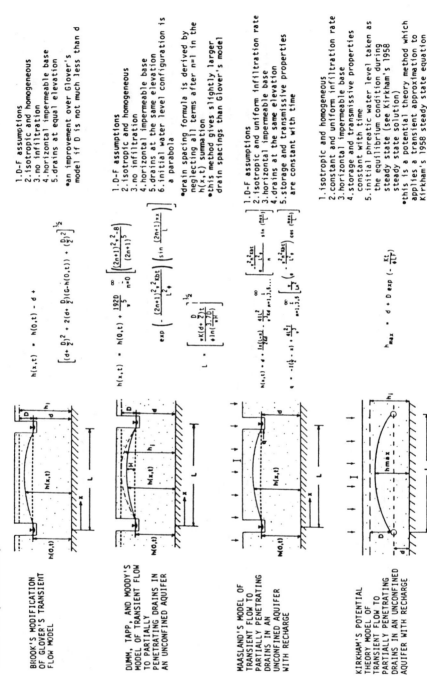

BROOK'S MODIFICATION OF GLOVER'S TRANSIENT FLOW MODEL

$$h(x,t) = h(0,t) - d +$$

$$\left[\left(d+\tfrac{D}{2}\right)^2 + 2\left(d+\tfrac{D}{2}\right)(G-h(0,t)) + \left(\tfrac{D}{2}\right)^2 \right]^{\frac{1}{2}}$$

1. D-F assumptions
2. isotropic and homogeneous
3. no infiltration
4. horizontal impermeable base
5. drains at equal elevation

• an improvement over Glover's model if D is not much less than d

DUMM, TAPP, AND MOODY'S MODEL OF TRANSIENT FLOW TO PARTIALLY PENETRATING DRAINS IN AN UNCONFINED AQUIFER

$$h(x,t) = h(0,t) + \frac{1920}{\pi^5} \sum_{n=0}^{8} \left[\frac{(2n+1)^2 \pi^2 - 8}{(2n+1)^5} \right]$$

$$\exp\left(-\frac{(2n+1)^2 \pi^2}{L^2} \alpha t \right) \left(\sin \frac{(2n+1)\pi x}{L} \right)$$

$$L = \left[\frac{-K(d+\tfrac{D}{2})t}{\phi \ln\left(\tfrac{3.7D}{\pi h}\right)} \right]^{\frac{1}{2}}$$

1. D-F assumptions
2. isotropic and homogeneous
3. no infiltration
4. horizontal impermeable base
5. drains at the same elevation
6. initial water level configuration is a parabola

• drain spacing formula is derived by neglecting all terms after n=1 in the h(x,t) summation
• this method gives slightly larger drain spacings than Glover's model

MAASLAND'S MODEL OF TRANSIENT FLOW TO PARTIALLY PENETRATING DRAINS IN AN UNCONFINED AQUIFER WITH RECHARGE

$$h(x,t) = d + \frac{i(d+x)}{K} + \frac{4iL^2}{K\pi^3} \sum_{n=1,3,5...} \left[\frac{1}{n^3} \left(e^{-\frac{n^2\pi^2\alpha t}{L^2}} \right) \sin\left(\frac{n\pi x}{L}\right) \right]$$

$$q = \pi i \left(\tfrac{L}{2} - x\right) + \frac{4iL}{\pi^2} \sum_{n=1,3,5} \left[\frac{1}{n^2} \left(e^{-\frac{n^2\pi^2\alpha t}{L^2}} \right) \cos\left(\frac{n\pi x}{L}\right) \right]$$

1. D-F assumptions
2. isotropic and uniform infiltration rate
3. horizontal impermeable base
4. drains at the same elevation
5. storage and transmissive properties are constant with time

KIRKHAM'S POTENTIAL THEORY MODEL OF TRANSIENT FLOW TO PARTIALLY PENETRATING DRAINS IN AN UNCONFINED AQUIFER WITH RECHARGE

$$h_{max} = d + D \exp\left(-\frac{Kt}{\phi L^2} \right)$$

1. isotropic and homogeneous
2. constant and uniform infiltration rate
3. horizontal impermeable base
4. storage and transmissive properties constant with time
5. initial phreatic water level taken as the equilibrium condition during steady state (see Kirkham's 1958 steady state solution)

• this is a potential theory method which applies a transient approximation to Kirkham's 1958 steady state equation

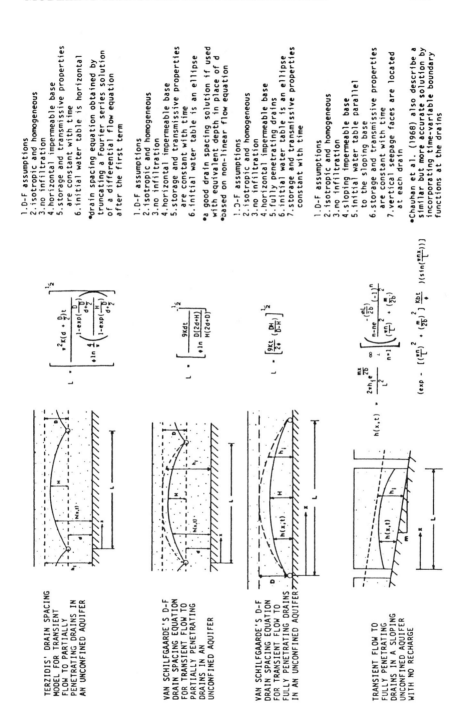

TERZIDIS' DRAIN SPACING MODEL FOR TRANSIENT FLOW TO PARTIALLY PENETRATING DRAINS IN AN UNCONFINED AQUIFER

1. D-F assumptions
2. isotropic and homogeneous
3. no infiltration
4. horizontal impermeable base
5. storage and transmissive properties are constant with time
6. initial water table is horizontal

• drain spacing equation obtained by truncating a Fourier series solution of a differential flow equation after the first term

VAN SCHILFGAARDE'S D-F DRAIN SPACING EQUATION FOR TRANSIENT FLOW TO PARTIALLY PENETRATING DRAINS IN AN UNCONFINED AQUIFER

1. D-F assumptions
2. isotropic and homogeneous
3. no infiltration
4. horizontal impermeable base
5. storage and transmissive properties are constant with time
6. initial water table is an ellipse

• a good drain spacing solution if used with equivalent depth in place of d
• based on non-linear flow equation

VAN SCHILFGAARDE'S D-F DRAIN SPACING EQUATION FOR TRANSIENT FLOW TO FULLY PENETRATING DRAINS IN AN UNCONFINED AQUIFER

1. D-F assumptions
2. isotropic and homogeneous
3. no infiltration
4. horizontal impermeable base
5. fully penetrating drains
6. initial water table is an ellipse
7. storage and transmissive properties constant with time

TRANSIENT FLOW TO FULLY PENETRATING DRAINS IN A SLOPING UNCONFINED AQUIFER WITH NO RECHARGE

1. D-F assumptions
2. isotropic and homogeneous
3. no infiltration
4. sloping impermeable base
5. initial water table parallel to the sloping base
6. storage and transmissive properties are constant with time
7. vertical seepage faces are located at each drain

• Chauhan et al. (1968) also describe a similar but more accurate solution by incorporating time-variable boundary functions at the drains

Table 7.3 (continued)

FOUR TRANSIENT MODELS:
a. CONSTANT HEAD BOUNDARIES AND CONSTANT RATES OF RECHARGE AND DISCHARGE
b. CONSTANT HEAD BOUNDARIES AND VARIABLE RATES OF RECHARGE AND DISCHARGE
c. VARIABLE HEAD BOUNDARIES AND CONSTANT RATES OF RECHARGE AND DISCHARGE
d. VARIABLE HEAD BOUNDARIES AND VARIABLE RATES OF RECHARGE AND DISCHARGE

refer to Singh and Jacob (1977)

HANTUSH'S 1967 D-F MODEL OF THE GROWTH AND DECAY OF WATER TABLE MOUNDS IN RESPONSE TO UNIFORM PERCOLATION

refer to Singh and Jacob (1977)

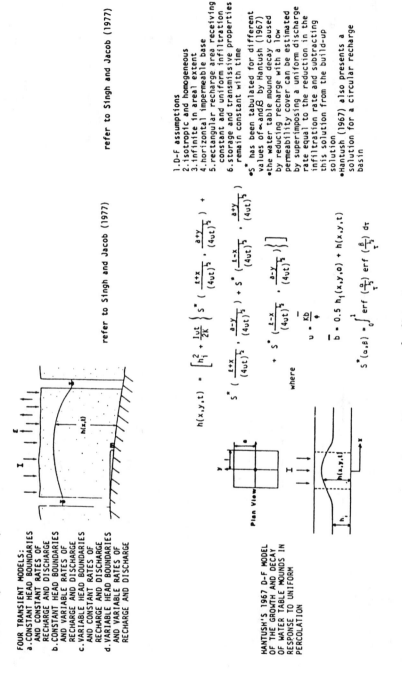

Plan View

$$h(x,y,t) = \left[h_i^2 + \frac{Iut}{2K} \left\{ S^* \left(\frac{l+x}{(4ut)^{\frac{1}{2}}}, \frac{a-y}{(4ut)^{\frac{1}{2}}} \right) + S^* \left(\frac{l-x}{(4ut)^{\frac{1}{2}}}, \frac{a+y}{(4ut)^{\frac{1}{2}}} \right) \right. \right.$$
$$\left. \left. + S^* \left(\frac{l+x}{(4ut)^{\frac{1}{2}}}, \frac{a-y}{(4ut)^{\frac{1}{2}}} \right) + S^* \left(\frac{l-x}{(4ut)^{\frac{1}{2}}}, \frac{a-y}{(4ut)^{\frac{1}{2}}} \right) \right\} \right]$$

where

$$u = \frac{Kb}{\phi}$$

$$\bar{b} = 0.5 \, h_1(x,y,o) + h(x,y,t)$$

$$S^*(\alpha,\beta) = \int_0^1 erf\left(\frac{\alpha}{\tau^{\frac{1}{2}}}\right) erf\left(\frac{\beta}{\tau^{\frac{1}{2}}}\right) d\tau$$

$$h_{max} = h_i^2 + \frac{2Iut}{K} S^* \left(\frac{l}{(4ut)^{\frac{1}{2}}}, \frac{a}{(4ut)^{\frac{1}{2}}} \right)$$

1. D-F assumptions
2. isotropic and homogeneous
3. infinite in areal extent
4. horizontal impermeable base
5. rectangular recharge area receiving constant and uniform infiltration
6. storage and transmissive properties remain constant with time

● S^* has been tabulated for different values of α and β by Hantush (1967)

● the water table mound decay caused by reducing recharge with a low permeability cover can be estimated by superimposing a uniform discharge rate equal to the reduction in the infiltration rate and subtracting this solution from the build-up solution

● Hantush (1967) also presents a solution for a circular recharge basin

models utilize techniques that generate matrices that are so large they require the use of computers, capable of multiple irerations, to converge on a solution. Computer models can be grouped into broad classifications based on their spatial approach to the aquifer system: (1) lumped models and (2) distributed models.

Lumped models attempt to predict the behavior of the aquifer as a whole unit. The approach used is to estimate the total change in a given parameter as the difference between total input and output. For example, a simple water balance equation for a stream-connected phreatic aquifer system can be represented by (McLin and Gelhar, 1979):

$$\eta \frac{dh}{dt} = q_n + q_a - q_o - q_p \pm q_\ell \tag{7.22}$$

where   h = average thickness of the ground water zone
            t = time
           η = average effective porosity
          q_n = natural recharge rate per unit surface area
          q_a = artificial recharge rate per unit surface area
          q_o = natural aquifer outflow rate per unit surface area
          q_p = aquifer pumping rate per unit surface area
          q_l = river leakage rate per unit surface area

The corresponding mass balance equation for the stream-connected phreatic aquifer system would be (McLin and Gelhar, 1979):

$$\eta \frac{d(hc)}{dt} = q_n c_n + q_a c_a - q_o c - q_p c \pm q_\ell c + \eta hr \tag{7.23}$$

where   r = volumetric source-sink term that accounts for contaminant additions or degradation within the flow zone
         c = average aquifer concentration
        c_n = concentration in natural recharge
        c_a = concentration in artificial recharge

Hence, an aquifer-wide accounting procedure for the parameters listed in Equations 7.22 and 7.23 will provide the data necessary for a predictive model.

The second and most widely used classification of computer models are the distributed models. Distributed models use numerical techniques to describe the behavior of an aquifer at selected points (nodes). Several different numerical techniques exist for solving the relevant partial differential equations. However, the three most popular techniques for ground water studies are finite difference methods, finite element methods, and the method of characteristics.

The finite difference method gets its name from the way the method approximates the governing equations. The ground water basin is divided into a mesh of polygons, usually rectangular. Nodes are then defined either in the center of the blocks formed by the grid lines as shown in Figure 7.2(a), or at the intersections of the lines as shown in Figure 7.2(b) (American Society of Civil Engineers, 1987). The finite difference technique is based on approximating the partial differential terms in the governing equations by their truncated Taylor's series expansion. Consider, for example, the distribution of head in one dimension as shown in Figure 7.3. The finite difference approximation for the rate of change of head at some point "i" could be written as:

$$\frac{\partial h}{\partial x} \approx \frac{h(i+1) - h(i)}{\Delta x} \tag{7.24}$$

This is called the first forward approximation in that it involves the "i+1" term. Similarly, the first backward approximation would be written as

$$\frac{\partial h}{\partial x} \cong \frac{h(i) - h(i-1)}{\Delta x} \tag{7.25}$$

The central difference approximation is independent of the head at node "i", and can be shown as

$$\frac{\partial h}{\partial x} \cong \frac{h(i+1) - h(i-1)}{2\Delta x} \tag{7.26}$$

Similarly, the approximations for second order or higher derivatives can be derived. For example, the second difference at node "i" could be written as

$$\frac{\partial^2 h}{\partial x^2} \cong \frac{h(i-1) - 2h(i) + h(i+1)}{(\Delta x)^2} \tag{7.27}$$

The equation for transient flow with no sources/sinks would be

$$\frac{\partial^2 h}{\partial x^2} + \frac{\partial^2 h}{\partial y^2} = \frac{S}{T} \frac{\partial h}{\partial t} \tag{7.28}$$

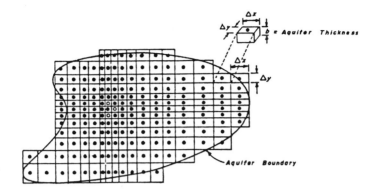

(a.) FINITE DIFFERENCE GRID WITH BLOCK CENTERED NODES.

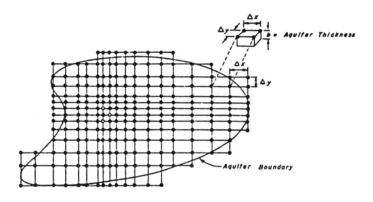

(b.) FINITE DIFFERENCE GRID WITH MESH CENTERED NODES.

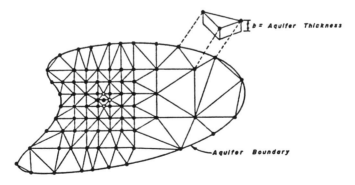

(c.) FINITE ELEMENT MESH WITH TRIANGULAR ELEMENTS.

Figure 7.2. Finite difference-finite element configurations for aquifer study (American Society of Civil Engineers, 1987).

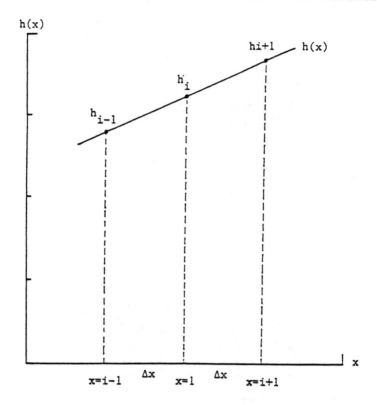

Figure 7.3. **Distribution of head in one dimension.**

Using a central difference approximation for the spatial derivatives and a forward difference approximation for temporal derivatives yields the following:

$$\frac{h_{i-1,j}^k - 2h_{i,j}^k + h_{i+1,j}^k}{(\Delta x)^2} + \frac{h_{i,j-1}^k - 2h_{i,j}^k + h_{i,j+1}^k}{(\Delta y)^2} =$$

$$\frac{S}{T}\left(\frac{h_{i,j}^{k+1} - h_{i,j}^k}{\Delta t}\right) \tag{7.29}$$

where   i = denotes the position in x direction
          j = denotes the position in y direction
         k = elapsed time

In Equation 7.29 all values of h are known at the "k" time step and the value of $h_{i,j}$ at the "k+l" time step can be solved for directly. This is called an explicit scheme. If Equation 7.28 is written with the backward difference approximation for the time derivative the following is obtained:

$$\frac{h_{i-1,j}^{k+1} - 2h_{i,j}^{k+1} + h_{i+1,j}^{k+1}}{(\Delta x)^2} + \frac{h_{i,j-1}^{k+1} - 2h_{i,j}^{k+1} + h_{i,j+1}^{k+1}}{(\Delta y)^2} =$$

$$\frac{S}{T}\left(\frac{h_{i,j}^{k+1} - h_{i,j}^k}{\Delta t}\right) \quad (7.30)$$

Equation 7.30 has five unknowns; however, it can be written for each node in the grid resulting in a set of simultaneous equations. These equations can be solved to give a new value of "h" at each node for the "k+l" time increment. This is called an implicit scheme.

Explicit schemes are simple to formulate but have severe restrictions on the grid spacing and time increments. Implicit methods are more complicated, but more versatile. They require greater computer storage capacity but use less running time than explicit methods. Implicit methods are superior in that they also permit the use of larger time values.

In summary, finite difference models are probably the most popular method of ground water simulation. This is because they are relatively easy to understand and code into usable computer programs (American Society of Civil Engineers, 1987). A very simple finite difference model may require only two pages of computer coding and can be applied to large-scale simulations.

Whereas the finite difference techniques approximate the partial differential equations by a differential approach, the finite element method approximates the equations by the integral approach. The method proceeds by dividing a two-dimensional ground water system into a mesh of triangular or quadrilateral shapes as shown in Figure 7.2(c), each of which is called an element. The ground water surface (head) in each of these elements is some shape that is determined by the values of the heads at the nodes (corners) of the elements (American Society of Civil Engineers, 1987).

The finite element technique traditionally applied to ground water systems is called the method of weighted residuals. The approach of the method is to first assume a trial solution to the governing equation. The degree to which this solution does not satisfy the governing equation is referred to as the residual. The objective of the method is to drive the cumulative residuals over the domain to zero.

As a simple example consider Laplace's equation (Equation 7.2) in two dimensions. A trial solution involving the summation of head values at each node (h_L) and an associated nodal basis function $N_L(x,y)$ can be written as

$$\hat{h}(x,y) = \sum_{L=1}^{NN} h_L N_L(x,y) \quad (7.31)$$

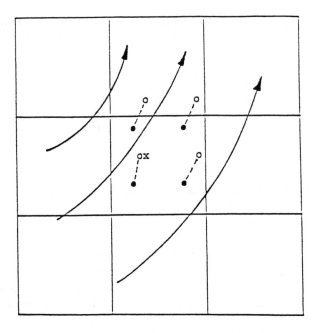

x finite-difference node
● reference particle
o new location
⬈ flow line

Figure 7.4.   Finite difference grid showing reference particles.

The subscript L indicates the nodal number and NN is the total number of nodes in the grid. Substituting the trial solution into Laplace's equation results in a residual value. The residual values are then weighted using the nodal basis functions, hence the term method of weighted residuals. Summing the weighted residual values over the problem domain (D) and equating them to zero results in

$$\int_D \int \left(\frac{\partial^2 \hat{h}}{\partial x^2} + \frac{\partial^2 \hat{h}}{\partial y^2} \right) N_L(x,y) dx dy = 0 \qquad (7.32)$$

To obtain a numerical solution to Equation 7.32, the domain is divided into finite elements as shown in Figure 7.2(c). The order of Equation 7.32 is reduced by applying integration by parts. The resulting equations are then applied to the individual elements with the elemental values then summed for the entire grid. This results in a set of simultaneous equations which can be solved to give nodal head values. More detailed information on finite element analysis can be found in Wang and Anderson (1982).

In general, the finite element method requires more complicated computer programming and more computer run time than the finite difference method. Its advantage is that its triangular and quadrilateral elements can be used to accurately emulate irregular geographic and geologic features, such as re-charging rivers, faults, aquifer boundaries, and water agency boundaries (American Society of Civil Engineers, 1987). The finite element model can be very detailed in areas of interest and relatively general in other areas where ground water hydrology and geology are more uniform. Because of this, a well designed finite element model may have significantly fewer elements than a comparable finite difference model. This advantage makes the finite element method potentially more efficient in basins where geographic patterns of ground water hydrology and geology are irregular.

The finite difference and finite element techniques have been applied to both ground water flow and solute transport studies. The method of characteristics is a technique used specifically for solute transport problems, especially those situations where convective transport dominates. The approach is not to solve the transport equation directly, but rather to solve an equivalent system of ordinary differential equations. The ordinary differential equations are obtained by rewriting the transport equation using the fluid particles as the point of reference. That is, instead of observing how the concentration changes with time at a fixed position in space, changing concentrations associated with fluid movement are noted. Therefore, the velocity distribution represents necessary information. In two dimensions, the end result is three equations for x-velocity, y-velocity and concentration; the solution of which are called the characteristic curves, hence the name, method of characteristics (Geo Trans, Inc., 1982).

This method of characteristics is accomplished numerically by introducing a set of moving points (or reference particles) that can be traced within the stationary coordinates of a finite difference grid. Points are placed in each finite difference block and then allowed to move a distance proportional to the length of the time increment and the velocity at that point as shown in Figure 7.4. The moving points effectively simulate convective transport because the concentrations at each node vary as different points having different concentrations enter and leave the area of that block. Once the convective effect is determined, the remaining parts of the transport equation are solved using finite difference approximations and matrix methods (Geo Trans, Inc., 1982). Such a procedure is used in the two-dimensional solute transport model developed by Konikow and Bredehoeft (1978).

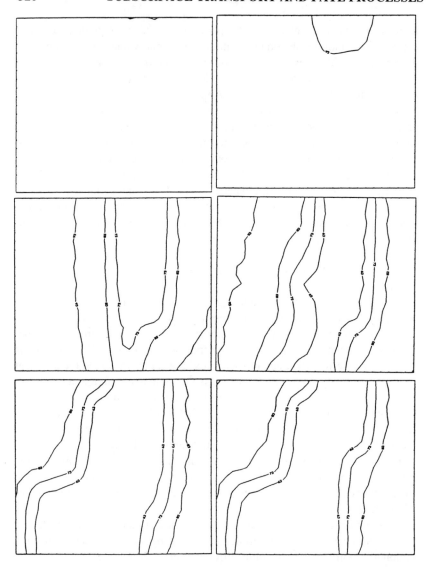

Figure 7.5. Water saturation profiles for μ_w/μ_{nw} = 1.72 and ρ_{nw} = 2.00 g/cm^3 with continuous injection at various times (Corapcioglu and Hossain, 1986).

7.5 APPLICATIONS

Biological degradation in the subsurface environment can also be addressed via solute transport modeling. For example, Rifai et al. (1988, 1989) have developed a numerical code for modeling of biodegradation and sorption in contaminant transport analysis. The model, called BIOPLUME II, simulates the transport of dissolved hydrocarbons in two dimensions in a saturated porous media under the influence of oxygen limited biodegradation. The model is based on the U.S. Geological Survey 2-D solute transport model (Konikow and Bredehoeft, 1978). The BIOPLUME II model accounts for (1) advective and dispersive transport of hydrocarbons and oxygen, (2) instantaneous reactions between hydrocarbons and oxygen at every step to simulate aerobic biodegradation, (3) adsorption of hydrocarbons, (4) mass balances for oxygen and hydrocarbons, (5) first order decay to represent anaerobic biodegradation, and (6) injection wells as oxygen sources. The model uses equations describing contaminant transport, oxygen transport, and microbial growth and decay. The governing two-dimensional equations are shown in Table 7.4. Newell et al. (1990) describe a decision support system software package called OASIS that enables modelers to apply the BIOPLUME model totally by computer through use of extensive onscreen tutorials.

Most of the solute transport models developed to date have addressed processes occurring in the saturated zone. However, Nofziger (1988a) has summarized some of the key transport and fate processes in the unsaturated zone. He noted that the transport and fate of chemicals moving vertically from the soil surface is highly dependent upon water infiltration, redistribution, and evaporation. The depth of water penetration, the water content at different places in the soil, and the initial condition of the soil, combined with the properties of the chemical, determine the depth of chemical movement and hence the time required to reach a critical depth or the water table. These results, along with the degradation rate of the chemical, determine the quantity of chemical passing some critical depth or reaching the water table. Nofziger (1988a) also noted that volatilization can result in the loss of a portion of the chemical from the soil environment to the atmosphere, especially when the chemical is near the surface. Volatilization also results in the gaseous form of the chemical which can diffuse through the soil. This process depends upon the wetness of the soil and hence the movement of water in the unsaturated soil.

Short et al. (1987) developed the Regulatory and Investigative Treatment Zone (RITZ) model that incorporates the processes discussed above. Nofziger et al. (1988b) describe a user's guide for the RITZ model. In an extension of their previous work, Nofziger et al. (1989) developed a one-dimensional model for water and chemical movement in unsaturated soils (CHEMFLO) that also incorporates that transport process discussed above. Stevens et al. (1989) evaluated the effects of sensitive soil and modeling parameters for hazardous wastes in the vadose zone.

Attempts at modeling two-phase flow systems have met with increasing success. Intuitively, the migration of dense, immiscible hydrocarbons through a saturated formation would be gravity dominated (i.e., the migration would be predominantly downward despite the lateral flow of ground water). Corapcioglu and Hossain (1986) modeled both continuous and pulse injection

Table 7.4 Two-Dimensional Transport Equations with Reaction Terms

$$\frac{\partial H}{\partial t} = \frac{\nabla(D\nabla H - vH)}{R_h} - \frac{M_t k}{R_h} \cdot \frac{H}{k_h + H} \cdot \frac{O}{k_o + O} \tag{1}$$

$$\frac{\partial O}{\partial t} = \nabla(D\nabla O - vO) - M_t \cdot k \cdot F \cdot \frac{H}{k_h + H} \cdot \frac{O}{k_o + O} \tag{2}$$

$$\frac{\partial M_s}{\partial t} = \frac{\nabla(D\nabla M_s) - vM_s}{R_m} + M_s \cdot k \cdot Y \cdot \frac{H}{k_h + H} \cdot \frac{O}{k_o + O}$$
$$+ \frac{K_c \cdot Y \cdot OC}{R_m} - bM_s \tag{3}$$

where,

D	=	Dispersion tensor
v	=	Ground water velocity vector
O	=	Concentration of dissolved oxygen
H	=	Concentration of contaminant
R_h	=	Retardation factor for contaminant
M_s	=	Concentration of microbes in solution
R_m	=	Microbial retardation factor
M_t	=	$R_m \cdot M_s$
K_h	=	Contaminant half saturated constant
K_o	=	Oxygen half saturated constant
K_c	=	First order decay of natural carbon
OC	=	Natural organic carbon concentration
b	=	Microbial decay rate
F	=	Ratio of oxygen to contaminant consumed
Y	=	Microbial yield coefficient (g cells/g contamiant)
k	=	Maximum contaminant utilization rate per unit mass microorganism

Instantaneous reaction assumption

$$H(t+1) = H(t) - O(t)/F; \quad O(t+1) = 0$$
$$\text{where } H(t) > O(t)/F \tag{4}$$

Table 7.4 (Continued)

$$0(t+1) = 0(t) - H(t) \bullet F; \quad H(t+1) = 0$$
$$\text{where } 0(t) > H(t) \bullet F \tag{5}$$

where H (t), H (t+1), 0(t) and 0 (t+1) are the concentrations of contaminant oxygen at time t and t + 1, respectively.

Source: Rifai et al., 1989.

of high density immiscible hydrocarbons into a ground water formation. The resulting saturation curves (Figures 7.5 and 7.6) show the dominance of gravity flow. Similarly, Parker et al. (1986) modeled the movement of TCE through the vadose and saturated zones with constant ground water flow. As shown in Figure 7.7, their model also showed gravity flow to dominate until the plume encounters an impermeable formation, at which point it tends to spread laterally.

In spite of all the simplifications used to develop the two-phase models, they remain disturbingly complex. As noted by Reible et al. (1986) the sophistication of these models generally far exceeds that of the available database. In their study, Reible et al. (1986) applied single capillary tube flow principles to develop a one-dimensional model for oil infiltrating the unsaturated zone (Figure 7.8). The simple model assumes that the immiscible fluid flows through the pore channels that are partially occupied by air and by residual water saturation. The immiscible phase displaces the air, but not the residual water, then leaves a residual oil saturation. Both the wetting and drainage fronts are assumed to remain sharp, which allows the assumption of a linear variation in capillary pressure with depth in the medium. The model was found to be relatively accurate and is promoted for use in making crude exposure assessment type estimates of free phase organic transport.

Abdul (1988) describes a one-dimensional soil column experiment involving oil and water movement monitored with tensiometers. The overall results of the study showed that oil infiltrating a water-wet soil will move rather freely through the pendular zone, but its mobility in the funicular and saturated zones depends on its ability to displace the pore water present in these zones. Increasing resistance to vertical movement of the oil below the water table results in the lateral spreading of the oil on top of the saturated zone. These results are identical to the typical conditions encountered in the field.

Metcalfe and Zukovs (1986) present a simple assessment model that addresses the nonaqueous fluid (immiscible with water) phase only. The conceptual basis of their model is shown in Figure 7.9. In essence, the model assumes that spilled immiscible fluids travel downward as saturated slugs through the subsurface in accordance with Darcy's Law. Migration continues until the spill volume is exhausted by residual saturation. The model is ap-

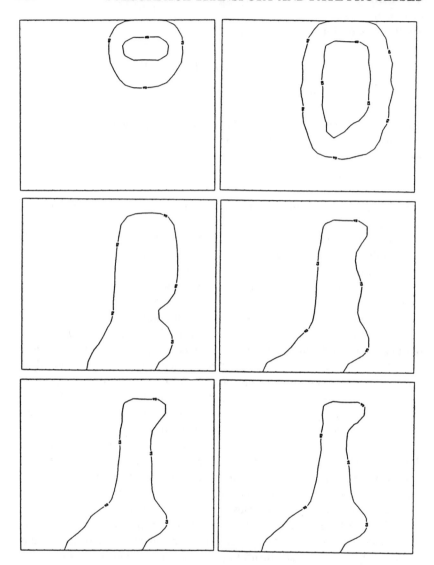

Figure 7.6. Water saturation profiles for $\mu_w/\mu_{nw} = 1.72$ and $\rho_{nw} = 2.00$ g/cm^3 with initial pulse injection at various times (Corapcioglu and Hossain, 1986).

plicable to immiscible fluids both lighter and heavier than water. In addition, the solution algorithm is simplified by providing a series of nomograms.

Charbeneau et al. (1989) note that most numerical multiphase flow models require comprehensive data sets and are computationally expensive. In addition, numerical and analytical models are poorly suited for studying the multiphase flow equations. They discuss the development of their simplified

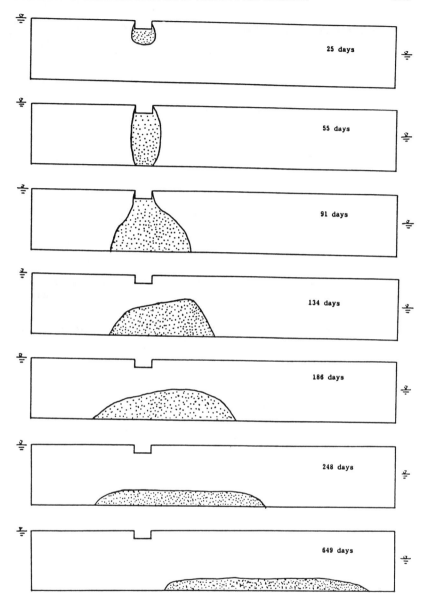

Figure 7.7.   Finite element model predicted TCE plume at selected times
after initiation of 100 d TCE leakage event from storage
tank under constant TCE head with ground water flow
under fixed gradient (Parker et al., 1986).

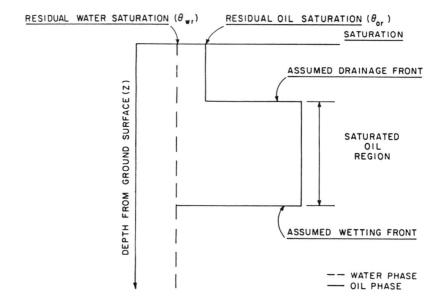

Figure 7.8.   Conceptual model of one-dimensional oil infiltration (Reible et al., 1986).

physically based model that uses kinematic theory of multiphase flow. Two prototype models are developed for analyzing problems involving oily pollutant transport.

The growth in subsurface modeling studies has also reached into areas related to porous media flow of fluids. For example, Yates and Yates (1988) discuss the biological, chemical, and physical factors that are known to influence virus and bacterial survival and transport in the subsurface. Models available to predict the fate of microorganisms are presented, but the models generally do not explicitly address these factors.

An area that is receiving much attention is modeling water flow and solute transport through fractured media. Because of the heterogeneous and anisotropic nature of fractures, the entire modeling process is more complex and difficult to interpret than for unconsolidated media. Schmelling and Ross (1989) give an excellent overview discussion of contaminant transport in fractured media and available models.

7.6 OVERVIEW STUDIES OF MODELS

Numerous flow and/or solute transport models for ground water have been developed within the last several years. Several overview studies of the status

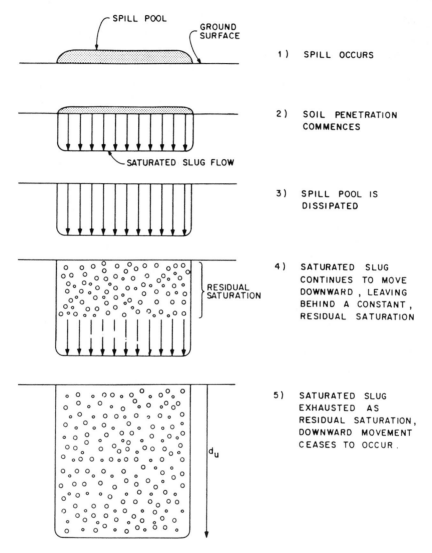

Figure 7.9. Idealized representation of the subsurface penetration and movement of fluids immiscible with water (Metcalf and Zukovs, 1986).

of model development and usage have been published (Anderson, 1979; Appel and Bredehoeft, 1976; Bachmat et al., 1978; and Mercer and Faust, 1980a, 1980b), and several books have been published (Wang and Anderson, 1982; Bear and Verruijt, 1987; and Willis and Yeh, 1987). Anderson (1979) discussed the use of models to simulate the movement of contaminants through ground water flow systems, and Mercer and Faust (1980a, 1980b) discussed various types of models and their applications.

Appel and Bredehoeft (1976) identified the following types of problems for which models were being developed: ground water flow in saturated or partially unsaturated material; land subsidence resulting from ground water extraction; flow in coupled ground water-stream systems; coupling of rainfall-runoff basin models with soil moisture accounting aquifer flow models; interaction of economic and hydrologic considerations; transport of contaminants in an aquifer; and effects of proposed development schemes for geothermal systems. The status of the modeling activity for various models was reported by Appel and Bredehoeft (1976) as being in a development, verification, operational, or continued improvement phase.

Bachmat et al. (1978) assessed the status of 250 models used internationally as tools for ground water related water resources management. In a follow-on to the Bachmat et al. (1978) study, van der Heijde et al. (1985) defined three types of numerical ground water models: (1) prediction models, which simulate the behavior of the ground water system and its response to stress; (2) resource management models, which integrate hydrologic predictions with explicit management decision procedures; and (3) identification models, which determine input parameters for both of the above. Most of the models produced to date are prediction models focused on flow, solute transport, heat transport, and deformation (subsidence). A comprehensive review of 399 models from 19 countries was completed by van der Heijde et al. (1985). A summary of the types of models and countries of the reports is in Table 7.5 (van der Heijde et al., 1985). A detailed taxomony of the surveyed models is in Figure 7.10 (van der Heijde et al., 1985). The focus of this discussion will be on the findings relative to flow and mass (solute) transport prediction models.

Table 7.6 summarizes the tasks that the surveyed distributed prediction models can address and relevant typical outputs (van der Heijde et al. 1985). The most common numerical methods employed are finite differences (FD) and finite elements (FE). Other less common numerical methods include the method of characteristics (MOC) and random walk, which are used for mass transport models, and boundary integral elements (BIEM) and integrated or integral finite differences (IFDM).

Almost all flow models reviewed during the survey are based on Darcy's Law (van der Heijde et al., 1985). Those distributed flow models which are also based on the Dupuit approximations are termed hydraulic flow models. To this group belong one-dimensional and two-dimensional models of horizontal flow in a single or multiple leaky or nonleaky aquifer system, with a vertical leakage component. Saturated-unsaturated flow models with horizontal flow in the saturated zone and vertical flow in the unsaturated zone are also considered hydraulic flow models. Distributed flow models which do not use the Dupuit approximations are termed hydrodynamic. Models with two-dimensional flow in the vertical plane or three-dimensional flow belong to this group.

Mass transport models are more complex than flow models in that they

Table 7.5 Summary of the Number of Reviewed Model Reports

Country	Total	Flow	Mass Transport	Heat Transport	Defor-mation	Multi-Purpose	Manage-ment	Identifi-cation
				Prediction Models				
U.S.	219	94	50	15	7	14	24	15
France	38	25	6	4	—	0	2	1
Israel	23	7	10	—	—	—	4	2
United Kingdom	27	20	4	—	1	—	—	3
Netherlands	26	21	1	1	—	1	—	1
Canada	16	6	8	—	—	1	—	1
West Germany	10	7	2	—	—	—	—	1
Japan	11	7	—	—	4	—	—	—
Australia	8	7	—	—	—	—	—	0
Argentina	0	0	—	—	—	—	—	0
Spain	1	1	—	—	—	—	—	—
Belgium	2	1	—	1	—	—	—	—
New Zealand	2	2	—	—	—	—	—	—
India	5	2	1	—	—	—	2	—
Switzerland	2	2	2	—	—	—	—	—
UN/Argentina	4	3	—	—	—	—	—	1
U.S./Sweden	1	—	—	1	—	—	—	—
U.S./Canada	1	—	—	—	—	—	—	1

Table 7.5 Summary of the Number of Reviewed Model Reports

Country	Total	Flow	Prediction Models					
			Mass Transport	Heat Transport	Defor- mation	Multi- Purpose	Manage- ment	Identifi- cation
U.S.S.R.	1	—	—	—	—	—	1	1
Italy	1	—	—	—	—	—	—	1
U.S./Israel/ Vennezuela	1	—	—	—	—	—	1	—
Total		399	203	84	22	12	16	33
29								
Percentage	100	51	21	6	3	4	8	7

Source: van der Heijde et al., 1985.

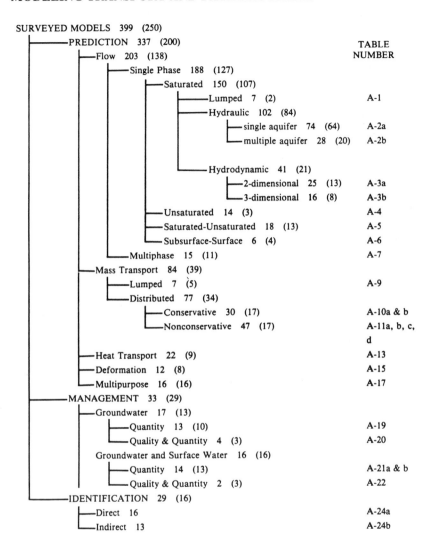

Figure 7.10.   Taxonomy of surveyed models (van der Heijde et al., 1985).

consider quality in conjunction with quantity. In principle, a mass transport model contains a flow submodel which computes the fluid flow velocity and then utilizes these velocities in a quality submodel which transports the contaminant in the flow field, allowing for additional spreading (dispersion) and transformations by reactions (van der Heijde et al., 1985). Under certain circumstances, such as low concentrations of contaminants, flow and quality submodels can operate independently. In other cases, however, their mutual effects cannot be uncoupled. Thus, relatively high contaminant concentra-

Table 7.6 Mathematical Methods Used in Surveyed Distributed Prediction Models

Purpose and Output	Total	Semi-Analytical	Finite Difference	Finite Element	Finite Difference or Finite Element	Integrated Finite Difference
General						
Head	79	3	39	28	—	1
Head and flow rates	46	—	27	16	—	1
Concentrations	69	6	27	24	2	—
Temperature	13	1	7	5	—	—
Multi-purpose	16	—	3	7	—	5
Steam-aquifer relation	25	0	21	4	—	—
Flow pattern	9	5	3	1	—	—
Interface						
Position	14	2	9	3	—	—
Dispersion zone	8	—	3	4	—	—
Free surface	3	—	1	2	0	—
Multiphase flow	5	1	3	1	—	—
Wells (drawdown)	13	—	9	3	—	1
Geothermal reservoirs	6	—	3	3	0	—
Subsidence	12	—	5	6	—	1
Frost propagation	3	—	3	—	—	—
Prediction and calibration	6	—	6	—	—	—
Response coefficient	3	—	3	—	—	—
Total	330	18	172	107	2	9

Purpose and Output	Finite Difference and Method of Characteristics	Finite Difference and Randon Walk	Boundary Element	Hybrid	Other
General					
Head	—	—	2	1	5
Head and flow rates	—	—	1	—	1
Concentrations	6	2	—	1	1
Temperature	1	—	—	—	—
Multi-purpose	—	—	—	—	—
Steam-aquifer relation	—	—	—	—	—
Flow pattern	—	—	—	—	0
Interface					
Position	1	—	—	—	0
Dispersion zone	—	—	—	—	—
Free surface	—	—	—	—	—
Multiphase flow	—	—	—	—	—
Wells (drawndown)	—	—	—	—	—
Geothermal reservoirs	—	—	—	—	—
Subsidence	—	—	—	—	—
Frost propagation	—	—	—	—	—
Prediction and calibration	—	—	—	—	—
Response coefficient	—	—	—	—	—
Total	8	2	3	2	7

Source: van der Heijde et al., 1985.

tions in wastewater or salt water can alter densities and affect the flow pattern of the ground water, which, in turn, affects the movement and spreading of the contaminants; this requires interactive coupling between the flow and quality submodels which account for the effects of varying density upon the flow field. There are fewer mass transport models which consider contaminant transport by flow only (conservative mass transport) than those which consider reactions as well (nonconservative mass transport). The latter include primarily adsorption and radioactive decay. Some of the multiconstituent models also consider chemical reactions under simplified assumptions. However, only a few of the surveyed models handle the more complicated biochemical transformations of nitrogen and other compounds which are important in modeling the effects of contamination by organic waste products.

A common feature of all the prediction models surveyed is that their forecasts are deterministic (one value) rather than probabilistic (a range of values of varying probability). In most cases, the predictions are also unconditional, implying that the models do not contain any restrictions on acceptable values or operations.

The review of models conducted by van der Heijde et al. (1985) also gave some attention to model documentation, availability, and applications. Table 7.7 summarizes the collected information with regard to these three issues (van der Heijde et al., 1985). Of the surveyed models, 53% are documented, including a description of the model's theoretical framework, underlying assumptions and limitations and operational characteristics, a set of user's instructions, and a series of example problems to test computer implementation and to familiarize the user with the model's operations. The percentage of documented models is higher for the more recently developed complex models than for earlier models. A model is defined as available if the program code associated with it can be obtained or accessed by potential users with or without restrictions. On the other hand, proprietary models are available only with restrictions and under various conditions, such as licensed use with limited distribution, a lease that includes maintenance and update services, and royalty-based use for which a fee must be paid to the model manager each time the model is used (van der Heijde et al., 1985). Approximately 68% of the surveyed models have been applied to a field case at least once. Models developed and operated by consulting firms as well as federal and state agencies have usually been applied many times. A model can be defined as usable if it is fully documented, available, and applied once or more in the field; 52% of all surveyed models fall in this category (van der Heijde et al. 1985).

van der Heijde et al. (1985) also developed a series of recommendations and identified some key issues associated with the use of the ground water models. The four major problem areas, ranked in order of importance, were noted by van der Heijde et al. (1985) as follows:

Table 7.7 State of Usability of the Surveyed Models

Model Category	Total	Documented[1]	Available			Usable[4]	
			Total[2]	Public Domain	Applied[3]	Total	Supported[5]
Prediction							
Flow	203	120	172	117	155	118	44
Mass transport	84	39	69	45	49	39	22
Heat transport	22	15	18	15	13	15	9
Deformation	12	8	11	10	7	8	5
Multipurpose	16	8	14	5	8	8	8
Management	33	8	18	13	19	7	1
Identification	29	13	16	15	21	11	1
Total	399	211	318	220	272	206	90
Percentage	100	53	80	55	68	52	23

Source: van der Heijde et al., 1985.

[1] Including theoretical framework, instructions, and example problems.
[2] Available as public domain or proprietary model through sale, leasing, licensing, etc.
[3] At least once.
[4] Documented and available.
[5] Usable models for which active support by code-developing or code-maintaining team is provided.

1.   Accessibility of models to users — top priority should be given to making existing models more accessible to potential users as the most necessary immediate improvement. Increasing the accessibility of models consists not only in improving the quality of information about models and making this information and the models more available, but also in improving the training of those persons who use models.

2.   Communications between managers and technical personnel — measures must be taken to improve the links between management and those who provide technical services employing models. This will involve designing model outputs to be more responsive to management needs as well as to involve more interactive participation by managers and technical personnel in problem definition and model applications.

3.   Inadequacies of data — solutions to problems of data will require increased attention to the identification of those data critical to the solution of ground water management problems. Improved methods of data collection, storage, and retrieval are needed.

4.   Inadequacies in modeling — in certain areas, models still do not exist or are considered inadequate. The development or improvement of these models should be encouraged; in many instances, however, models will have to be preceded by improved scientific understanding of the fundamental processes that the models are to describe.

Table 7.8 delineates some specific concerns associated with the four major problem areas. Documentation is related to the accessibility problem, with proper documentation including (van der Heijde et al., 1985) (1) a brief description of the model, providing information identifying the model, the author (or the person who provides model support), the organization where the model was developed, the date of completion of the first documentation, the version number or updates (if any), the programming language, the organization where the model may be obtained, and an abstract or concise description of the model; (2) engineering documentation, including a description of the types of problems solved by the model, the basic theory and the method of solution, and its limitations and underlying assumptions; (3) program documentation describing program capabilities and limitations, different options available to users, and lists of input and output variables; (4) system documentation containing information on the structure of the program, data structures of external files and the core storage required, a list of variables and subroutines, references to code listing, and a description of required computer hardware and communication equipment; and (5) sample runs including both input and output files.

Table 7.8 Key Concerns Associated with Ground Water Modeling

Problem Area	Issues of Concern
Accessibility of models to users	Lack of adequate documentation of existing models
	A well-documented model needs to be publicized and distributed to prospective users
	Model users must be trained for the development of an understanding of the model structure, to verify underlying assumptions, and to respect its limitations
	Need for development and/or expansion of programs designed to teach technical personnel how to better select and use models for addressing field problems as well as the capabilities and limits of newly developed modeling techniques
Communications between managers and technical personnel	Managers may lack confidence in the use of models as decision-making tools
	Overselling of models wherein modelers claim that their models can perform more than is actually possible
	Form of model output can either aid or hinder communication of resultant communication information
	Need for usage of nontechnical jargon between technical personnel and managers
	There may be difficulties in precisely formulating the management problem to be addressed via modeling
	Potential usability of management models needs to be more thoroughly explored

Table 7.8 (Continued)

Problem Area	Issues of Concern
Communications between managers and technical personnel (continued)	Incompatibility between the use of models in their present state and the nature of the evidence required by a court of law
Inadequacies of data	Acquisition of data is expensive, and questions often arise as the the necessary extent and accuracy of such data
	Need for input data to calibrate and validate a model prior to its application to a particular field problem
	Need for data manipulation coeds where large amounts of data already exist and require efficient organization and processing to prepare model inputs
	Need for compatible data storage and retrieval systems among governmental agencies and private groups collecting ground water data
	Matching the more intense data needs for complex models with the lesser data requirements for simpler models which may not be as powerful in analyzing ground water problems
Inadequacies of modeling	Transferability of models with different codes to other computer systems
	Need to improve existing codes in terms of simplicity and structure and use; flexibility and visual illustration of outputs; stability of outputs; adaptability to other computers and codes; and reduction of computer storage and time requirements

Table 7.8 (Continued)

Problem Area	Issues of Concern
Inadequacies of modeling (continued)	Need for models which fully integrate subsurface flow, surface flow, and flow through the unsaturated zone
	Need for models which describe flow and mass transport through fractured media or secondary permeability
	Additional contaminant transport models that include chemical and biological interactions between ground water constituents and the rock or soil matrix and among the constituents themselves are needed
	Need for ground water models describing the flow of two immiscible fluids
	Should the variability of ground water parameters and system inputs be addressed stochastically or deterministically

Source: van der Heijde et al., 1985.

The inadequacies in the modeling portion of Table 7.8 identifies several needed types of models. Fundamental research may be necessary in order to develop such models. Six key research areas and needs identified by van der Heijde et al. (1985) include (1) adequate descriptions of the kinetics of chemical and biological processes; (2) movement of pollutants through the unsaturated zone and fractured rock; (3) quantification of management objectives; (4) better parameter identification methodologies; (5) effects of scale and heterogeneity on transport phenomena; and (6) characterization of spatial and temporal variability in system parameters and inputs.

van der Heijde et al. (1989) describe the establishment of the Groundwater Research Data Center. The Center provides information on public datasets from modeling studies and distributes datasets for testing and validation of models for flow and contaminant transport in the subsurface. The Center also analyzes and evaluates information from field and laboratory experiments.

Table 7.9 Problems with Numerical Models

Data requirements
    Head over time
    Concentration over time
    Hydraulic conductivity
    Aquifer dispersivities
Boundary conditions
    Constant head
    Constant flow
Accuracy
    Model verification
    Input data dependence
    Dynamic flow
    Numerical problems
    Sensitivity analysis
Complicated, expensive, and time consuming
Must deal with uncertainty
May be misused
Incomplete understanding of system
Lack of quality assurance/quality control program

7.7 LIMITATIONS OF MODELS

Because of their increased popularity and wide availability, it is important to note the limitations of ground water models. Table 7.9 contains a listing of some example problems and limitations. One limitation of computer models is that they can have significant data requirements. Models can require a variety of input data, and this data may be required for several years. Additionally, some available data may not be useful. For example, values for hydraulic conductivities may be available from a series of pump tests. However, when this information is used in the model, the results may not accurately predict the behavior of the aquifer because the pump test may reflect local trends. These values must be changed until the model does accurately reflect the aquifer conditions. Hence, even though the values obtained may be correct, they do not reflect the overall hydraulic conductivity of the aquifer and, as such, are of lesser value in modeling studies. Dispersivity values are usually calibrated rather than measured. Field measurements of dispersivity can be difficult to make and questioned based on their representativeness (Anderson, 1984).

The second limitation associated with computer models is their required boundary conditions. Because each model has a finite size associated with it, certain constraints or conditions must be applied at the outermost nodes. These are most often constant head or constant flow conditions. These mandated numerical conditions may not be truly reflective of the actual field conditions at the points represented by the nodes.

Numerical models can be very precise in their predictions, but these predictions are not always accurate. The accuracy of the output is highly dependent on the accuracy of the input. Accordingly, in order to use a ground water model for future predictions, evidence must be developed to demonstrate its ability to simulate plausibly the historic past. The general procedure in verification is to estimate the range of values of both ground water flow and solute transport parameters and then test the model, using first the "best guess" values of the basin parameters to obtain system response to external stresses over time. The result is computed values of the system state variables. System response may be expressed in terms of changes in water levels or salinity or contaminant profiles. Then a comparison is made between the computed values and the known histories of the basin. If a good job has been done in developing the data, a fairly close match may result in the initial run. However, some adjustment of the parameters of storativity, transmissivity, or net deep percolation almost always is required (American Society of Civil Engineers, 1987).

Sensitivity analyses may be desirable in aquifer studies. For example, Aquado et al. (1977) presented a method for determining in which areas detailed knowledge of aquifer characteristics and conditions is most critical to the success of a management plan. These questions are answered by using sensitivity analysis to determine how variations in parameters and input data affect the optimal solution of a linear programming management model. The model uses finite element or finite difference approximations of the ground water equations as constraints. The optimal locations and discharge rates of wells have been determined for dewatering a rectangular area to a specified level while minimizing the steady-state total pumping rate and maintaining hydraulic heads in the dewatered area at or below the specified value. The area is in a small aquifer having constant head boundaries. Sensitivity analysis has shown that the optimal steady-state solution is most sensitive to hydraulic conductivity at and near the aquifer boundaries parallel to the length of the dewatered area. Thus, field exploration and testing should be concentrated on determination of hydraulic conductivity in those areas.

Another problem associated with some numerical models is that they can be quite complicated from a mathematical perspective. This is especially true of the finite element models, and also the comprehensive models that are written for three dimensions. For these models, it is imperative that their availability be accompanied by extensive documentation as to the workings of the model. In addition, numerical modeling studies can be expensive. This is due to the typically large computer storage and time requirements of most models. Numerical modeling can also be a time-consuming venture. This is especially true if insufficient data is available. In fact, even with sufficient data available, models can sometimes require months to be calibrated.

Uncertainty relative to model assumptions and usability should be recognized. For example, McLaughlin (1979) addressed the role of uncertainty in the Rockwell Hanford Operations ground water model development and application program at the Hanford Site in Washington. Methods of applying statistical probability theory in quantifying the propagation of uncertainty from field measurements to model predictions were discussed. It was shown that measures of model accuracy or uncertainty provided by a statistical analysis can be useful in guiding model development and sampling network design.

Another limitation or problem listed in Table 7.9 for models is that they have, in the past, been misused. Misuse in this sense means that models have been applied to cases where they were not appropriate. More importantly, misuse of models has come in the form of blind faith in the results of model applications. The output from any modeling study must be interpreted. It is imperative that the output be accurate and realistic. The interpretation of accuracy and reality can only be made by professionals trained in ground water hydrology and quality management.

Finally, van der Heijde (1986) identified two additional concerns related to ground water quality modeling: (1) inadequate or incomplete understanding of subsurface systems and (2) the absence of a quality assurance/quality control (QA/QC) program. One of the key problems in inadequate modeling is that the model is based on an incorrect conceptual description of the system being studied. To be able to conceptualize a complex ground water system correctly, computer modeling, data collection, and interpretation should go hand in hand.

For each modeling project an adequate quality assurance plan should be developed and followed (van der Heijde, 1986). Quality assurance (QA) in ground water modeling is the procedural and operational framework put in place by the organization managing the modeling study to assure technically and scientifically adequate execution of all project tasks included in the study. The two major elements of QA are quality control (QC) and quality assessment. Quality control refers to the procedures that ensure the quality of the final product. Appropriate quality control for ground water models has been suggested to encompass the following (U.S. Environmental Protection Agency, 1987):

1. Validation of the model mathematics by comparison of its output with known analytical solutions to specific problems
2. Verification of the general framework of the model by successful simulation of observed field data
3. Benchmarking of the model efficiency in solving problems by comparison with other models

4.   Critical review of the problem conceptualization to ensure that the modeling effort considers all physical, chemical, and biological processes that may affect the problem
5.   Evaluation of the specifics of the application, e.g., appropriateness of the boundary conditions, grid design, time steps, etc.

Developing and implementing an appropriate QA/QC plan for a particular modeling study does not guarantee a perfect or optimally useful product. However, careful planning eliminates many of the common problems inherent in ground water modeling and the interpretation of model outputs.

7.8 SELECTION OF APPROPRIATE MODEL

Selection of an appropriate model to meet a given need typically involves consideration of both the technical capabilities of models as well as managerial issues. Technical comparisons of a number of models are contained elsewhere (van der Heijde et al., 1985). Managerial issues include economic considerations, necessary training and experience of model users, and information communication. Summary information on these managerial issues will be addressed in this section.

The nominal costs of the support staff, computing facilities, and specialized graphics production equipment associated with numerical modeling efforts can be high (U.S. Environmental Protection Agency, 1987). As a general rule, costs are greatest for personnel, moderate for hardware, and minimal for software. The exception to this ordering relates to the combination of software and hardware purchased. An optimally outfitted computer may cost about $100,000, but it can rapidly pay for itself in terms of dramatically increased speed and computational power. A well complemented personal computer may cost $10,000, but the significantly slower speed and limited computational power may infer hidden costs in terms of the inability to perform specific tasks. For example, highly desirable statistical packages like SAS and SPSS are unavailable or available only with reduced capabilities for personal computers; many of the most sophisticated mathematical models are available in their fully capable form only on larger computers.

Figure 7.11 gives a brief comparison of typical costs for software for different levels of computing power (U.S. Environmental Protection Agency, 1987). Obviously, the software for less capable computers is cheaper, but the programs are not equivalent; so managers need to thoroughly think through what level is appropriate. If the decisions to be made are to be based on very little data, it may not make sense to insist on the most elegant software and

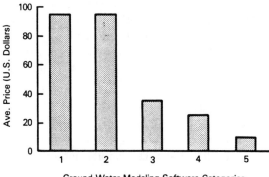

Ground-Water Modeling Software Categories

Categories
  1 Mainframe / business computer models
  2 Personal computer versions of mainframe models
  3 Original IBM-PC and compatibles' models
  4. Handheld microcomputer models (e.g., Sharp
     PC1500)
  5 Programmable calculator models (e.g., HP41-CV)
Prices include software and all available
documentation, reports, etc.

Figure 7.11.   Average price per category for ground water models from the International Ground Water Modeling Center (U.S. Environmental Protection Agency, 1987).

hardware. If the intended use involves substantial amounts of data and sophisticated analyses are necessary, it may be unwise to opt for the least expensive combination.

Based on experience and observation, there does seem to be a trend away from both ends of the spectrum and toward the middle; that is, the use of powerful personal computers is increasing rapidly. Most importantly for ground water managers, many of the mathematical and data packages have been "down-sized" from mainframe computers to personal computers; many more are being written directly for this market. Figure 7.12 has been prepared to provide some idea of the costs of available software and hardware for personal computers (U.S. Environmental Protection Agency, 1987).

A fair degree of specialized training and experience are necessary to develop and apply mathematical models, and relatively few technical support staff can be expected to have such skills presently (van der Heijde et al., 1985). This is due in part to the need for familiarity with a number of scientific disciplines, so that the model may be structured to faithfully simulate real-world problems. In practice, this means that ground water modelers have a great need to become involved in continuing education efforts (U.S. Environmental Protection Agency, 1987).

Information communication is also necessary in the selection and use of ground water models. Communication is needed between managers, model-

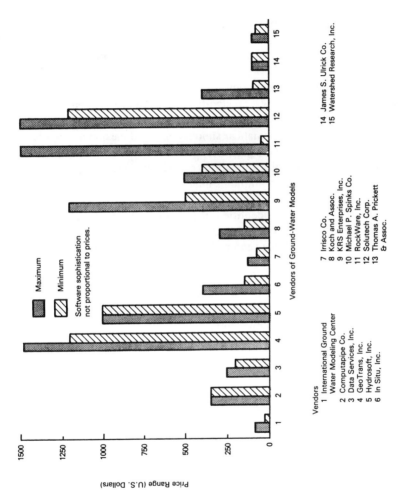

Figure 7.12. Price range for IBM-PC ground water models available from various sources (U.S. Environmental Protection Agency, 1987).

**Table 7.10 Screening-Level Questions for Mathematical
Modeling Efforts**

General problem definition
    What are the key issues: quantity, quality, or both?
    What are the controlling geologic, hydrologic, chemical, and biological
        features?
    Are there reliable data (proper field scale, quality control, etc.) for
        preliminary assessments?
    Do we have the model (s) needed for appropriate simulations?
Initial responses needed
    What is the time-frame for action (imminent or long-term)?
    What actions, if taken now, can significantly delay the projected
        impacts?
    To what degree can mathematical simulations yield meaningful results
        for the action alternatives, given available data?
    What other techniques or information (generic models, past experience,
        etc.) would be useful for initial estimates?
Strategies for further study
    Are the critical data gaps identified; if not, how well can simulations
        determine the specific data needs?
    What are the trade-offs between additional data and increased certainty
        of the simulations?
    How much additional manpower and resources are necessary for further
        modeling efforts?
    How long will it take to produce useful simulation, including quality
        control and error-estimation efforts?

Source: U.S. Environmental Protection Agency, 1987.

ers, and interested publics. Some of the questions managers should ask technical support staff, and vice versa, to ensure that the solution being developed is appropriate to the actual problems are listed in Tables 7.10 through 7.12. Table 7.10 consists of "screening level" questions. Table 7.11 addresses the need for correct conceptualizations, and Table 7.12 is comprised of sociopolitical concerns (U.S. Environmental Protection Agency, 1987).

7.9 CONCLUSIONS

As an update to their previous overview studies, van der Heijde et al. (1988) compiled a status report on ground water modeling. The report in-

Table 7.11 Conceptualization Questions for Mathematical Modeling Efforts

Assumptions and limitations

What are the assumptions made, and do they cast doubt on the model projections for this problem?

What are the model limitations regarding the natural processes controlling the problem; can the full spectrum of probable conditions be addressed?

How far in space and time can the results of the model simulations be extrapolated?

Where are the weak spots in the application, and can these be further minimized or eliminated?

Input parameters and boundary conditions

How reliable are the estimates of the input parameters; are they quantified within acceptable statistical bounds.

What are the boundary conditions, and why are they appropriate to this problem?

Have the initial conditions with which the model is calibrated been checked for accuracy and internal consistency?

Are the spatial grid design(s) and time-steps of the model optimized for this problem?

Quality control and error estimation

Have these models been mathematically validated against other solutions to this kind of problem?

Has anyone field verified these models before by direct applications or simulation of controlled experiments?

How do these models compare with others in terms of computational efficiency and ease of use or modification?

What special measures are being taken to estimate the overall errors of the simulations?

Source: U.S. Environmental Protection Agency, 1987.

cludes an extensive set of appendices categorizing the currently available models according to their applications. The report also includes an overview discussion of the availability and usability of existing flow and solute transport models as outlined below.

Several new developments are emerging in ground water modeling. It is evident that recently ground water flow models have evolved to a point where a wide range of flow characterizations are possible. The newer models may include options for various types of boundary conditions, as well as the ability to handle a wide variety of hydrologic processes such as evapotranspiration, stream-aquifer exchanges, spatial and temporal variations in recharge, and the more complex characterization necessary for simulating unsaturated

Table 7.12 Sociopolitical Questions for Mathematical Modeling Efforts

Demographic considerations
   Is there a larger population endangered by the problem, than we are able
   to provide sufficient responses to?
   Is it possible to present the model results in both nontechnical and
   technical formats to reach all audiences?
   What role can modeling play in public information efforts?
   How prepared are we to respond to criticism of the model (s)?
Political constraints
   Are there nontechnical barriers to using this model, such as "tainted by
   association" with a controversy elsewhere?
   Do we have the cooperation of all involved parties in obtaining the
   necessary data and implementing the solution?
   Are similar technical efforts for this problem being undertaken by friend
   or foe?
   Can the results of the model simulations be turned against us; are the
   results ambiguous or equivocal?
Legal concerns
   Will the present schedule allow all regulatory requirements to be met in
   a timely manner?
   If we are dependent on others for key inputs to the model (s), how do
   we recoup losses stemming from their nonperformance?
   What liabilities are incurred for projections which later turn out to be
   misinterpretations originating in the model?
   Do any of the issues relying on the application of the model (s) require
   the advice of attorneys?

Source: U.S. Environmental Protection Agency, 1987.

flow. Similarly, recently developed models for simulation of solute transport
for new versions of earlier models often include increased flexibility in describing the solute source and simulating transport and fate processes such as
radioactive decay or chemical transformation and effects of both equilibrium
and nonequilibrium adsorption. In some instances, these transport models are
coupled with existing geochemical models to provide a more complete analysis of the solute chemistry. Such a development is also noticeable with respect to the simulation of biodegradation, e.g., for the analysis of
bioremediation schemes (Borden and Bedient, 1986; Borden et al., 1986).
Furthermore, important developments have occurred in the modeling of flow
and transport in fractured rock systems. Here, both improved site characterization and stochastic analysis of fracture geometry, together with an improved capability to describe the interactions of chemicals between the active
and passive fluid phases and the rock matrix, have facilitated increasingly
realistic simulation of real-world fractured rock systems. Multiphase flow

models have become increasingly available, especially those designed for studying the movement of immiscible fluids such as NAPLs. Also, new approaches have been developed for parameter identification and are increasingly used in practical applications (Yeh, 1986). Finally, optimization-based management models have been applied to a growing variety of decision problems, especially in the area of ground water protection (Wagner and Gorelick, 1987).

REFERENCES

Abdul, A. S. "Migration of Petroleum Products Through Sandy Hydrogeologic Systems," *Ground Water Monit. Rev.* VIII(4 Fall):73–81 (1988)

Abriola, L. M. and G. F. Pinder "A Multiphase Approach to the Modeling of Porous Media Contamination by Organic Compounds 1. Equation Development," *Water Resour. Res.* 21(1 January):11–18 (1985).

"Ground Water Management," *ASCE Manual No. 40,* 3rd ed. (New York: American Society of Civil Engineers, 1987), pp. 138–152.

Anderson, M. P. "Movement of Contaminants in Groundwater: Groundwater Transport—Advection and Dispersion," in *Groundwater Contamination* (Washington, D.C.: National Academy Press, 1984), pp. 37–45.

Anderson, M. P. "Using Models to Simulate the Movement of Contaminants through Ground Water Flow Systems," *CRC Crit. Rev. Environ. Control* 9(2 November):97–156 (1979).

Appel, C. A., and J. D. Bredehoeft "Status of Ground Water Modeling in the U.S. Geological Survey," Circular No. 737, U.S. Geological Survey, Washington, D.C. (1976).

Aquado, E., N. Sitar, and I. Remson "Sensitivity Analysis in Aquifer Studies," *Water Resour. Res.* 13(4 August):733–737 (1977).

Bachmat, Y., B. Andrews, D. Holtz, and S. Sebastian "Utilization of Numerical Groundwater Models for Water Resources Management," EPA-600/8-78/012, U.S. Environmental Protection Agency, Ada, OK (1978).

Bear, J. and A. Verruijt *Modeling Groundwater Flow and Pollution,* (Dordrecht, The Netherlands: D. Reidel Publishing Company, 1987).

Bear, J. "Groundwater Quality Problem (Hydrodynamic Dispersion)," in *Hydraulics of Groundwater* (New York: McGraw-Hill Book Company, Inc., 1979), pp. 263–275.

Borden, R. C. and P. B. Bedient "Transport of Dissolved Hydrocarbons Influenced by Reaeration and Oxygen-Limited Biodegradation, 1. Theoretical Development," *Water Resour. Res.* 22(13):1973–1982 (1986).

Borden, R. C., P. B. Bedient, M. D. Lee, C. H. Ward, and J. T. Wilson "Transport of Dissolved Hydrocarbons Influenced by Reaeration and

Oxygen-Limited Biodegradation, 2. Field Application", *Water Resour. Res.* 22(13):1983–1990 (1986).

Canter, L. W., R. C. Knox, and D. M. Fairchild *Ground Water Quality Protection* (Chelsea, MI: Lewis Publishers, Inc., 1987) pp. 209–276.

Canter, L. W. "Use of Microcosms for Subsurface Transport and Fate Studies," presented at the J. K. G. Silvey Conference on Land and Water Resource Challenges in North Central Texas, University of North Texas, Denton, TX, September 22–23, 1988.

Charbeneau, R. J., J. W. Weaver, and V. J. Smith "Kinematic Modeling of Multiphase Solute Transport in the Vadose Zone," EPA/600/52-89/035, U.S. Environmental Protection Agency, Ada, OK (1989).

Cohen, R. M. and W. J. Miller "Use of Analytical Models for Evaluating Corrective Actions at Hazardous Waste Disposal Facilities," in *Proceedings of the Third National Symposium on Aquifer Restoration and Ground Water Monitoring,* (Worthington, OH: National Water Well Association, 1983), pp. 85–97.

Corapcioglu, M. Y. and M. A. Hossain "Migration of Chlorinated Hydrocarbons in Groundwater," in *Hydrocarbons in Ground Water: Prevention, Detection and Restoration* (Worthington, OH: National Water Well Association, 1986), pp. 33–52.

Faust, C. R. "Transport of Immiscible Fluids Within and Below the Unsaturated Zone: A Numerical Model," *Water Resour. Res.* 21(4 April):587–596 (1985).

Geo Trans, Inc. "Notes for Ground Water Models Workshop," Holcomb Research Institute Modeling Workshop, Butler University, Indianapolis, IN (1982).

Javandel, I., C. Doughty, and C. F. Tsang "Groundwater Transport: Handbook of Mathematical Models," Water Resources Monograph Series 10, American Geophysical Union, Washington, D.C. (1984), pp. 9–13.

Konikow, L. F. and J. D. Bredehoeft "Computer Model of Two-Dimensional Solute Transport and Dispersion in Ground Water," Techniques of Water-Resources Investigations of the U.S. Geological Survey, Washington, D.C. (1978)

McLaughlin, D. C. "Hanford Groundwater Modeling: Statistical Methods for Evaluating Uncertainty and Assessing Sampling Effectiveness," Report No. RMA-8310, National Technical Information Service, U.S. Department of Commerce, Springfield, VA (1979).

McLin, S. G. and L. W. Gelhar "A Field Comparison Between the USBR-EPA Hydrosalinity and Generalized Lumped Parameter Models,"in *Proceedings of the Canberra Symposium* (IAHS-AISH Pub. No. 128, 1979), pp. 339–348.

Mercer, J. W. and C. R. Faust "Ground Water Modeling: An Overview," *Ground Water* 18(2 March-April):108–115 (1980a).

Mercer, J. W. and C. R. Faust "Ground Water Modeling: Mathematical Models," *Ground Water* 18(3 May-June):212–227 (1980b).

Metcalf, D. E. and G. Zukovs "A Rapid Assessment Model for Spills on Soil of Oily Fluids That Are Immiscible with Water," in *Hydrocarbons in Ground Water: Prevention, Detection and Restoration* (Worthington, OH: National Water Well Association, 1986), pp. 128–148.

National Research Council *Ground Water Models-Scientific and Regulatory Applications* (Washington, D.C.: National Academy Press, 1990).

Newell, C. J., J. F. Haasbeek, L. P. Hopkins, S. E. Alder-Schaller, H. S. Rifai, P. B. Bedient, and G. A. Gorry "OASIS: Parameter Estimation System for Aquifer Restoration Models — User's Manual Version 2.0," EPA/ 600/58-90/039, U.S. Environmental Protection Agency, Ada. OK (1990).

Nofziger, D. L., J. R. Williams, and T. E. Short "Interactive Simulation of the Fate of Hazardous Chemicals During Land Treatment of Oily Wastes: RITZ User's Guide," EPA/600/58-88/001, U.S. Environmental Protection Agency, Ada, OK (1988).

Nofziger, D. L. "The Role of the Unsaturated Zone in Ground Water Modeling," in *Ground Water Modeling Newsletter,* International Ground Water Modeling Center, Butler University, Indianapolis, IN VII(1 February):1, 9–10 (1988).

Nofziger, D. L., K. Rajender, S. K. Nayudu, and P. Y. Su "CHEMFLO: One-Dimensional Water and Chemical Movement in Unsaturated Soils," EPA/600/58-89/076, U.S. Environmental Protection Agency, Ada, OK (1989).

Osborne, M. and J. Sykes "Numerical Modeling of Immiscible Organic Transport at the Hyde Park," *Water Resour. Res.* 22(1):25–33 (1986).

Parker, J. C., Lenhard, R. J., and T. Kuppusamy "Modeling Multiphase Contaminant Transport in Ground Water and Vadose Zones," in *Hydrocarbons in Ground Water: Prevention, Detection and Restoration* (Worthington, OH: National Water Well Association, 1986), pp. 189–200.

Reible, D. D., T. H. Illangasekare, D. V. Doshi, and A. F. Ayonb "Development and Experimental Verification of a Model for Transport of Concentrated Organics in the Unsaturated Zone," in *Hydrocarbons in Ground Water: Prevention, Detection and Restoration* (Worthington, OH: National Water Well Association, 1986), pp. 107–126.

Rifai, H. S. et al. *Computer Model of Two-Dimensional Transport Under the Influence of Oxygen Limited Biodegradation in Ground Water* (Houston, TX: National Center for Ground Water Research, Rice University, 1988).

Rifai, H. S., P. B. Bedient, and J. T. Wilson "BIOPLUME Models for Contaminant Transport Affected by Oxygen Limited Biodegradation," Environmental Research Brief, EPA/600/M-89/019, U.S. Environmental Protection Agency, Ada, OK (1989).

Schmelling, S. G. and R. R. Ross "Contaminant Transport in Fractured Media: Models for Decision Makers," EPA/540/4-89/004, U.S. Environmental Protection Agency, Ada, OK (1989).

Shamberger, H. A. and P. A. Domenico "Application of Analog Techniques to Water Management," presented at A.S.C.E. Fourteenth Hydraulics Division Conference, University of Arizona, Tucson, AZ (1965), pp. 1–22.

Short, T. E., W. J. Grenney, C. L. Caupp, and R. C. Sims "Mathematical Model for the Fate of Hazardous Substances in Soil: Model Description and Experimental Results," EPA/600/J-87/204, U.S. Environmental Protection Agency, Ada, OK, (1987).

Stevens, D. K., W. J. Grenney, Z. Yan, and R. C. Sims "Sensitive Parameter Evaluation for a Vadose Zone Fate and Transport Model," EPA/600/S2-89/039 U.S. Environmental Protection Agency, Ada, OK (1989).

U.S. Environmental Protection Agency "Handbook: Ground Water," U.S. EPA Report-625/6-87/016, Center for Environmental Research Information, Cincinnati, OH (1987).

van der Heijde, P. K. M., Y. Bachmat, J. Bredehoeft, B. Andrews, D. Holtz, and S. Sebastion "Groundwater Management: The Use of Numerical Models," *Water Resources Monograph 5*, 2nd ed. (Washington, D.C.: American Geophysical Union, 1985), pp. 13–51.

van der Heijde, P. "Why is the Usefulness of Models Contested?" *Ground Water Modeling Newsletter*, International Ground Water Modeling Center, Butler University, Indianapolis, IN V(3 October) (1986).

van der Heijde, P., A. I. El-Kadi, and S. A. Williams "Groundwater Modeling: An Overview and Status Report," EPA/600/2-89/028, Ada, OK (1988).

van der Heijde, P. K. M., W. I. M. Elderhorst, R. A. Miller, and M. F. Trehan "The Establishment of a Groundwater Research Data Center for Validation of Subsurface Flow and Transport Models", Project Summary, EPA/600/S2-89/040, U. S. Environmental Protection Agency, Ada, OK (1989).

van Genuchten, M. Th. and W. J. Alves "Analytical Solutions of the One-Dimensional Convective-Dispersive Solute Transport Equation," Technical Bulletin No. 1661, U.S. Department of Agriculture (1982).

Wagner, B. J. and S. M. Gorelick "Optimal Groundwater Quality Management under Parameter Uncertainty," *Water Resour. Res.* 23(7):1162–1174 (1987).

Wang, H. F. and M. P. Anderson *Introduction to Groundwater Modeling — Finite Difference and Finite Element Methods* (San Francisco: W.H. Freeman Book Company, Inc., 1982).

Willis, R. and W. E. Yeh *Groundwater Systems Planning and Management* (Englewood Cliffs, NJ: Prentice-Hall, Inc., 1987).

Yates, M. V. and S. R. Yates "Modeling Microbial Fate in the Subsurface Environment," EPA/600/J-88/022, U.S. Environmental Protection Agency, Ada, OK (1988).

Yeh, G.T. and D. D. Huff "FEWA: A Finite Element Model of Water Flow Through Aquifers," ORNL-5976, Oak Ridge National Laboratory, Oak Ridge, TN (1986).

8

APPLICATIONS OF TRANSPORT
AND FATE INFORMATION

8.1 INTRODUCTION

If one were to consider only the most basic information regarding the predominant transport and fate processes in the subsurface, one could easily be overcome by the multitude of potential reactions for just a single species. If one further considers that these potential reactions can occur simultaneously and/or sequentially, the complexity of the problem of assessing the transport and ultimate fate of this single species grows dramatically. Coupling the numerous potential reactions and interactions with the non-uniformity of the real world flow regime and the variability associated with all measurements, the assessment of subsurface transport and fate would seem to be an exercise in futility. The complexity of the problem can be attributed mainly to the nature of the environment in which the materials are moving, i.e., the porous medium.

Porous media flow always involves two phases: a solid phase and a liquid phase. In unsaturated conditions, the gases present in the porous matrix represent a third phase. The gaseous phase can be significant when considering volatile contaminants. However, it is the mere presence of the solid matrix within the fluid flow regime that dramatically increases the complexity of the subsurface transport and fate system. In addition to the properties and characteristics of the fluid and chemical species of concern, one must now consider the characteristics of the solid matrix, and interactions between the chemical in the fluid and the matrix.

In contrast to the perspective outlined above, it should be noted that the complexity of a given situation involving subsurface transport and fate of a material can be simplified to a workable system. By obtaining basic information regarding the soil, fluid and chemical characteristics and by applying the principles governing the predominant processes of subsurface transport and fate for the given situation, one can develop a general (preliminary) impression of the behavior of the material in the subsurface. Hence, a broad-based but thorough working knowledge of subsurface transport and fate phenomena is a prerequisite for any assessment of the behavior of materials in subsurface soils.

The basic objective of this chapter is to demonstrate how to use the abundant information presented in the previous chapters for developing and completing a comprehensive and cohesive assessment of the transport and fate of a contaminant in the subsurface. In that sense, this chapter represents a capstone summary that integrates all the previous information and allows us to extend our analysis beyond simply understanding the problem. The goal is to utilize the available analytical information in order to make projections as to future behavior or perhaps to design measures that will compensate for or counteract the natural movement of the contaminant.

To achieve the stated objective, this chapter will focus on real world applications for which subsurface transport and fate information is needed. The first part of the chapter will give an overview discussion of the influences of various transport and fate processes on ground water tracer studies. Tracer studies are often used to assess the subsurface environment and some of the transport and fate processes operating in that environment. The next section will focus on identifying some of the implications of certain transport and fate processes on ground water monitoring and remediation technologies. Ground water monitoring and remediation schemes are similar, especially for remediation schemes involving pump-and-treat technologies. The primary objective of both systems is to situate and operate well networks in such a manner that they intercept any contaminants of concern. The discussion of the behavior of a contaminant being targeted by an extraction well is also applicable for ground water monitoring situations. This section will also include overview information on some of the innovative measures for remediating subsurface contamination, especially those measures based on consideration of the principles of subsurface transport and fate processes. The final two sections will include case studies, one involving inorganic contaminants and one involving organic contaminants. The case studies will draw on the information from the previous chapters to explain the observed migration and changes of the contaminant plumes.

8.2 TRACER STUDIES

Tracer studies are engineered investigations designed to determine transport characteristics of the subsurface including: hydraulic conductivity, porosity, dispersivity, chemical distribution coefficients, and other properties. Ground water tracer studies can involve variations in all aspects of the study. Variations can include the type of flow regime, the type of tracer, the number of dimensions of the problem, and the scale of the investigation. This discussion will focus on field scale tracer studies. Related information on smaller scale studies can be found in Chapter 5.

Davis et al. (1985) give an exhaustive discussion of the design, construction, operation, and analysis of ground water tracer tests. This discussion covers variations in the type of test utilized (single vs multiple wells), the myriad of different types of tracers, and different means of analyzing for the various tracers. The manual gives an excellent discussion of both the advantages and limitations of the various options for completing field scale ground water tracer tests. A second manual on designing and conducting field scale tracer tests is found in Molz et al. (1986). This document discusses the design of both single-well and multiple-well tracer tests including an in-depth discussion of the design and construction of multilevel monitoring wells. The manual also presents the results and analyses of the two types of tests conducted at a field site in Mobile, Alabama. The site has since been extensively studied and reported in the literature (Guven et al., 1985; Guven et al., 1986; Molz et al., 1985; and Molz et al., 1986).

An understanding of the processes affecting the transport and fate of a given tracer is essential in developing accurate analytical information from the test. Because advection is the dominant transport process in most tracer studies, the three-dimensional velocity distribution will have the most profound impact on the transport of the tracer. Spatial (planar) variations in hydraulic gradient and hydraulic conductivity are provided by monitoring wells. Determination of the vertical distribution of those parameters represents a formidable challenge that involves the use of multilevel sampling systems such as those shown in Figure 8.1.

The vertical variation in horizontal hydraulic conductivity and, hence, the flow velocity of a tracer as depicted schematically in Figure 8.2 results in a vertical distribution of contaminants at any one point in the flow field. Most often, the single points in a flow field are assessed by means of a monitoring well with a finite screened interval. Samples from these monitoring wells are assumed to represent the vertically-averaged concentration of the tracer. The hypothetical movement of tracers using vertically-averaged concentrations is depicted in Figure 8.3.

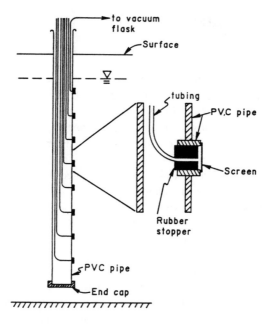

Figure 8.1. Multilevel sampling system (Molz et al., 1986).

Owing to the inherent three-dimensional nature of the flow regime, recent investigations have focused on assessing the temporal, spatial, and vertical distribution of tracers. One means of analyzing spatial and temporal variability is through use of spatial moments. Freyberg (1986) proposed the use of spatial moments for developing estimates of total mass and center of mass of contaminant plumes using vertically averaged point concentration values. The spatial moments equation can be written as:

$$M_{i,j,k}(t) = \int_{-\infty}^{\infty} \int_{-\infty}^{\infty} \int_{-\infty}^{\infty} C(x,y,z,t) x^i y^j z^k \, dx\,dy\,dz$$

$$i,j,k = 0,1,2$$

(8.1)

Where $C(x,y,z,t)$ represents a variable characteristic of the domain (subsurface). The coordinates of the center of mass of the aquifer characteristic can be calculated from

$$X_c = M_{100} / M_{000}, \quad Y_c = M_{010} / M_{000}, \quad Z_c = M_{001} / M_{000} \quad (8.2)$$

Freyberg (1986) describes the second moment about the center of mass as the

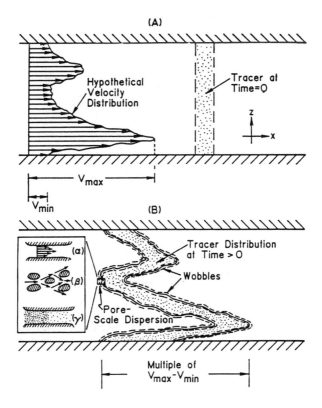

Figure 8.2. Vertical velocity and tracer distribution (Molz et al., 1986).

spatial covariance tensor. The components of the covariance tensor are indicative of the variation of the characteristic about the center of mass. The covariance equations are

$$\sigma_{xx} = \frac{M_{200}}{M_{000}} - X_c^2, \quad \sigma_{yy} = \frac{M_{020}}{M_{000}} - Y_c^2, \quad \sigma_{zz} = \frac{M_{002}}{M_{000}} - Z_c^2 \qquad (8.3a)$$

and

$$\sigma_{xy} = \sigma_{yx} = \frac{M_{110}}{M_{000}} - X_c Y_c, \quad \sigma_{xz} = \sigma_{zx}$$

$$= \frac{M_{101}}{M_{000}} - X_c Z_c, \quad \sigma_{yz} = \sigma_{zy} = \frac{M_{011}}{M_{000}} - Y_c Z_c \qquad (8.3b)$$

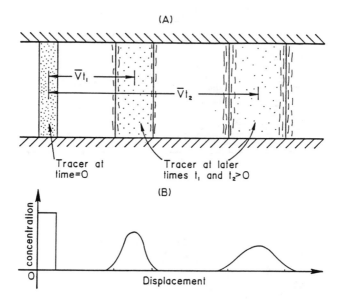

Figure 8.3. **Tracer transport for vertically averaged conditions (Molz et al., 1986).**

Analogous to the multivariate normal distribution, the components of the covariance tensor can be described in terms of an ellipsoid. The values of σ_{xx} and σ_{yy} represent the axes of the covariance ellipse (in two dimensions). The σ_{xy} term results from the axes of the covariance ellipse not coinciding with the (arbitrary) coordinate axes for the site location. The relative magnitude of the covariance values for a given characteristic [$C(x,y,z,t = \text{constant})$] could be used as an indicator of spatial variability. The change in center of mass or covariance values for a characteristic over time could be used as indicators of mass transport and temporal variability. The spatial moment approach is not widely used by practitioners and has been mostly limited to research studies.

One of the most widely recognized and extensively studied field sites is the Canadian Forces Base, Borden, Ontario. This site was designed to combine controlled initial conditions and long solute transport distances with detailed three-dimensional descriptions of the hydraulic conductivity fields and a comprehensive monitoring program in order to carefully evaluate dispersive transport models. The site used over 5000 sampling points from which over 19,900 samples were collected over a 3 year period. The inorganic and organic chemical tracers used in the study generally followed the predicted migration patterns based on water level contours and estimated hydraulic characteristics. Dispersive and adsorptive phenomena also followed an expected pattern.

Detailed descriptions of the study and extensive analyses of the results are reported in the literature (Mackay et al., 1986; Roberts et al., 1986; and Curtis et al., 1986).

A more recent study involved a natural-gradient tracer test at the Otis Air Force Base on Cape Cod, Massachusetts. The study was designed to test the relationship between field-scale dispersion and aquifer heterogeneity using conservative (nonretarded) and nonconservative solutes. Water samples were collected about once a month from an array of multilevel sampling devices. Each device includes 15 sampling ports spaced 25 to 76 cm apart to provide vertical profiles of tracer concentrations. The 640 sampling devices are arranged in 71 rows extending about 300 m downgradient from the injection wells. As of December, 1986, 16 sampling rounds had been completed. During each round, as many as 10,000 water samples were collected from 40 to 300 sampling devices to obtain a three-dimensional "snapshot" of the tracer distributions (Le Blanc et al., 1989).

Major conclusions from moments analysis of the Cape Cod tracer test results are (1) longitudinal mixing was the dominant dispersion process, which reached a Fickian limit after 40 m; (2) transverse horizontal and vertical dispersion were relatively small; and (3) the horizontal displacement of the solute cloud was accurately predicted using estimates of hydraulic conductivity, porosity, and measured hydraulic gradient. Results of tracer tests at the Borden and Cape Cods site show uniform velocities of the conservative solute, similar values of longitudinal dispersivity (0.43 m for the Borden site, 0.96 m for Cape Cod), and small values of transverse dispersivity. Results of the Cape Cod tracer test show that solute concentrations are highly variable and difficult to predict on a small scale, but that average characteristics (i.e., moments) can be predicted. It is likely that this general result will be valid for other types of aquifers where the materials heterogeneity is regular (Garabedian et al., 1989).

8.3 GROUND WATER MONITORING AND REMEDIATION

It should be obvious that the processes controlling the subsurface transport and fate of fluids will strongly affect and influence the success of ground water monitoring and/or remediation operations. The design of both ground water monitoring networks and aquifer restoration measures should be based on consideration of the subsurface transport and fate processes that could potentially affect the effectiveness of the design. The implications of the processes described previously in terms of ground water monitoring and remediation are outlined below.

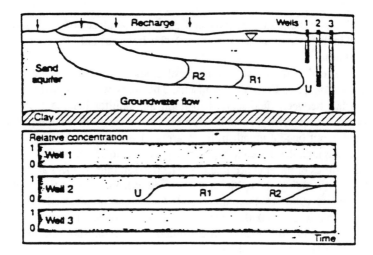

Figure 8.4a.   Contamination from various sources in an unconfined
                aquifer. Continuous source of three dissolved, unretarded
                contaminant (U) and retarded to varying degress (R1,R2).
                (Reprinted with permission from Mackay, D. M., P. V.
                Roberts, and J. A. Cherry, "Transport of Organic Con-
                taminants in Groundwater," *Environ. Sci. Technol.*, Vol.
                19, No. 5, pp. 384–392, Copyright 1985, American Chemi-
                cal Society.)

8.3.1 Monitoring Well Placement

The myriad of processes affecting subsurface transport and fate of miscible and immiscible fluids and their associated constituents would seem to
make ground water monitoring an impossible feat. To consider the almost
infinite possible configurations of monitoring well networks in relation to
aquifer characteristics and subsurface transport and fate processes is truly an
unachievable goal. In fact, the development of even the most basic network
design protocol or flowchart would become bogged down with site-specific
considerations.

The best way to illustrate how to apply the principles described earlier is
to develop some hypothetical scenarios and discuss the implications of some
of the transport and fate processes. Mackay et al. (1985) give an excellent
discussion of the relationships between contaminant source and type, transport
and fate processes, and monitoring well locations. The hypothetical scenarios
used in that discussion are shown in Figure 8.4 and described below.

Figure 8.4(a) shows the arrival of unretarded contaminants (U1) and two
varying (linearly) retarded contaminants (R1 and R2) leached continuously

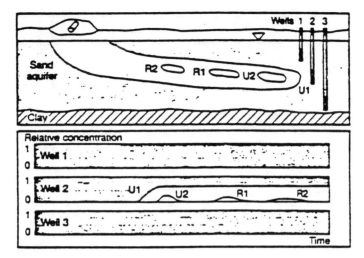

Figure 8.4b. Contamination from various sources in an unconfined aquifer. Continuous source of dissolved, unretarded contaminant (U1) and a pulse source of three dissolved contaminants; unretarded (U2) and retarded (R1, R2). (Reprinted with permission from Mackay, D. M., P. V. Roberts, and J. A. Cherry, "Transport of Organic Contaminants in Groundwater," *Environ. Sci. Technol.*, Vol. 19, No. 5, pp. 384–392, Copyright 1985, American Chemical Society.)

from a surface source. The breakthrough curves from the three monitoring wells show that the contaminants have a definite vertical migration pattern that could result in nondetection (e.g., if only well 1 had been installed). The relative concentration in Well 2 never reaches unity owing to dispersion in the transverse directions.

Figure 8.4(b) shows the arrival of unretarded contaminants (U1) continuously leached; and the arrival of unretarded (U2) and increasingly retarded (R1,R2) contaminants introduced as a pulse or slug. Dispersion reduces the peak relative concentration of all the contaminants; sorption further reduces the peak relative concentrations of the retarded contaminants (proportional to the level of retardation).

Figures 8.4(c) and 8.4(d) show the arrivals of low and high density liquids respectively. It is important to note that none of the three wells is able to detect all of the contaminants. This emphasizes the importance of vertical variation of monitoring well locations.

Figure 8.4(e) shows the overall effects of the abiotic processes and includes transformation of the organic liquids. It is important to note that each of the monitoring wells detects only certain specific contaminants and/or by-products.

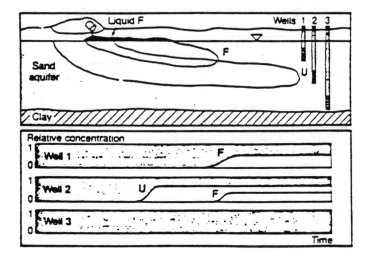

**Figure 8.4c. Contamination from various sources in an unconfined
aquifer. Continuous source of dissolved, unretarded contaminant (U) and a pulse source of organice liquid (F)
that floats on and slowly dissolves into the groundwater.
(Reprinted with permission from Mackay, D. M., P. V.
Roberts, and J. A. Cherry, "Transport of Organic Contaminants in Groundwater," *Environ. Sci. Technol.*, Vol.
19, No. 5, pp. 384–392, Copyright 1985, American Chemical Society.)**

8.3.2 Advances in Remediation Technologies

Improved understanding of the behavior of contaminants in the subsurface
has led to advances in ground water remediation technologies. The most
popular methods for remediating contaminated ground water formations have
traditionally involved removing the contaminated ground water to the surface
for treatment. These "pump-and-treat" schemes involve establishing hydraulic
control over the subsurface through ground water extraction. The ground
water extraction process can influence contaminant migration and adversely
affect the remediation scheme. Two examples in which extraction affects
contaminant transport are discussed below.

8.3.2.1 Pulsed (Cyclic) Pumping

The most common example of how extraction can affect contaminant
transport involves LNAPLs. Consider the hydrocarbon pool floating on the

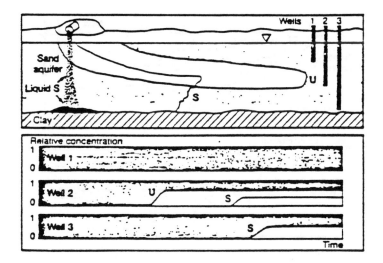

Figure 8.4d. Contamination from various sources in an unconfined aquifer. Continuous source of dissolved, unretarded solutes (U) and two pulse sources: floater (F) and sinker (S). The aquifer is aerobic except within the anaerobic U plume. Dissolved F degrades aerobically to CO_2 and water. Dissolved S degrades anaerobically to by-product SS, which is not readily degraded and is less retarded than S or F. (Reprinted with permission from Mackay, D. M., P. V. Roberts, and J. A. Cherry, "Transport of Organic Contaminants in Groundwater," *Environ. Sci. Technol.,* Vol. 19, No. 5, pp. 384–392, Copyright 1985, American Chemical Society.)

water table shown in Figure 8.5. Much effort has been focused on developing technologies for recovering hydrocarbons. Most of the technologies involve pumping the ground water to establish a cone of depression in which the hydrocarbon pool will accumulate. However, the cone of depression created by pumping will expose clean aquifer material to the hydrocarbons which results in contaminant adsorption and entrapment of a significant volume of immiscible liquid within the pore spaces. These schemes also involve pumping (and exposing) increased volumes of uncontaminated ground water through the contaminated zone for subsequent treatment at the surface, resulting in large volumes of ground water requiring treatment.

Related to the example above is the use of continuous extraction/injection of ground water to remove dissolved contaminants that also adsorb to aquifer materials. The combination of adsorption onto the soil matrix and increased ground water velocities serves to limit the time for desorption of the adsorbed contaminant or diffusion of contaminants from aggregate regions into the aqueous phase. Once the extraction system reaches an acceptable level it is

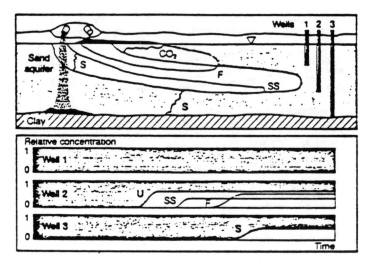

Figure 8.4e. Contamination from various sources in an unconfined aquifer. Continuous source of dissolved, unretarded contaminant (U) and a pulse source of an organic liquid (S) that sinks through and slowly dissolves into the groundwater. (Reprinted with permission from Mackay, D. M., P. V. Roberts, and J. A. Cherry, "Transport of Organic Contaminants in Groundwater," *Environ. Sci. Technol.*, Vol. 19, No. 5, pp. 384–392, Copyright 1985, American Chemical Society.)

turned off. However, if the system were switched back on the aqueous concentrations would be higher than the acceptable level. This rebound phenomenon is depicted in Figure 8.6.

One solution to the rebound phenomenon shown in Figure 8.6 is cyclic pumping. The resting phase of the cyclic operation allows sufficient time for contaminants to diffuse out of low permeability zones or for equilibrium desorption of sorbed contaminants to be reached. Recovery of the water table also allows previously dewatered zones to desorb. Cyclic pumping can result in more efficient contaminant recovery by reducing the large volumes of mildly contaminated water that are produced by continuously pumped systems.

Another advancement in ground water remediation technology is vacuum extraction. The use of vacuum extraction of soil gases increases volatilization of certain contaminants. In fact, vacuum extraction techniques are capable of removing pounds of volatile organic contaminants (VOCs) per day, whereas air stripping of VOCs from ground water typically results in the removal of only a few grams of VOCs per day. Vacuum extraction is more efficient in removing VOCs because it directly addresses the gaseous phase; whereas air stripping is a phase transfer process involving contaminants with a low aqueous solubility, i.e., a much lower mass per unit volume (Keely, 1989).

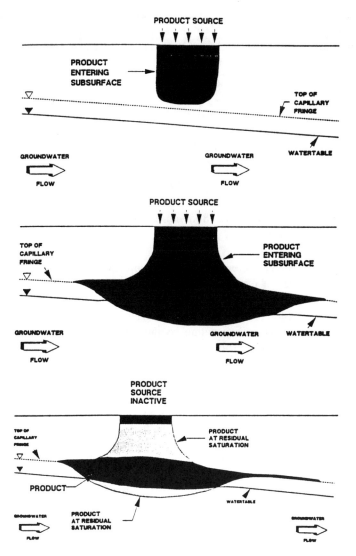

Figure 8.5. Movement of LNAPLs into the subsurface (Palmer and Johnson, 1989).

8.3.2.2 Soil Washing

Conventional pump and treat remediation has proven ineffective in remediating contamination by immiscible and/or strongly hydrophobic contaminants. The basic premise of soil washing or flushing is to introduce chemicals/solvents into the aqueous phase that will remove contaminants by reversing the forces causing the contaminants to be held in the solid matrix.

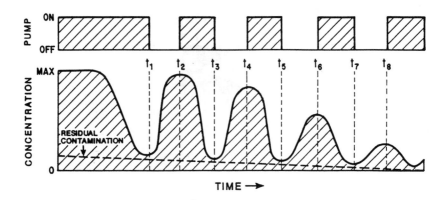

Figure 8.6. **Reduction of residual contaminant mass by pulsed pumping (Keely, 1989).**

The chemicals/solvents proposed for these applications are broadly grouped into the categories of organic solvents, surfactants, and chelating agents.

The proposed remediation processes are often referred to as surfactant flushing, soil washing or solvent extraction. The washing and extraction processes generally involve excavating contaminated soil, mixing with a contaminant extractant, and solids/liquid separation. Surfactant flushing is a modified pump-and-treat scheme that involves injecting an aqueous solution of surfactants and withdrawing the aqueous surfactant contaminant solution. The withdrawn fluid can then be treated at the surface.

Interest in using surfactants for flushing contaminated soils is an outgrowth of the enhanced oil recovery (EOR) research of the petroleum industry. It has been extensively documented that dilute solutions of aqueous surfactants can lower the interfacial tension between hydrocarbon phases and water, thereby enhancing the extraction of oil from porous media. The logical extension of this work is to use surfactants to remove other hydrophobic and slightly hydrophilic organic compounds from contaminated soils (Roy and Griffin, 1988).

Several overview or state-of-the-art assessments regarding surfactant flushing have been conducted (U.S. Environmental Protection Agency, 1984, 1985; Amdurer et al., 1986). All of these studies conclude that the technique is viable, but considerably more research is needed. Roy and Griffin (1988) completed a literature review on surfactant and chelate-induced soil decontamination. Their overall assessment was that the empirical studies had shown that these procedures were viable and warranted further research and development. The major drawback to the procedures was the costs of the reagents and they recommended research to develop schemes for recovering and re-using the reagents, e.g., surfactants.

A recent assessment of surfactant washing studies was completed by Raghavan et al. (1990). The study examined a variety of soil cleaning technologies. The processes reviewed usually involved excavation of the soil, mixing with the cleaning solution, and solid/liquid separation. The study suggests that the technologies were feasible for removal of organics in sandy soils and that separation of the extractant was not widely proven. Recommendations from the study were to expand bench-scale testing to establish operational conditions and to implement more pilot scale studies to assess environmental and economic practicability of the processes. For a current discussion of research related to surfactant-based remediation processes the reader is referred to Sabatini and Knox (1992).

8.4 CASE STUDY — INORGANICS

To use the information presented in the previous chapters to analyze the transport of inorganic constituents in the subsurface, we will draw on information from a contamination episode involving oilfield brine. Brine, like other natural waters, is comprised mainly of the predominant ionic species as shown in Table 8.1. However, brine is much more concentrated; its ionic strength is even higher than that of seawater (from which it originated). Table 8.1 includes the results of analyses of two samples of the same brine taken in response to an alleged contamination episode. Although the two samples were taken from different facilities within the same operation (heater treater and holding tank), the composition of the brine remains essentially constant.

The suspect brine in Table 8.1 was being disposed of through an injection well (Figure 8.7) located on the banks of small creek. Residents in the area relied on shallow wells tapping the alluvial aquifer for their freshwater supplies. The drilling records for these wells show the formation to be predominantly sand or fine sand with the water table encountered around 15 ft below land surface. Natural ground water quality (Table 8.1) would be characterized as hard, but certainly drinkable.

The saltwater (brine) injection well in Figure 8.7 was intended to inject the brine into a deep formation. Although the injection well had operating troubles, as evidenced by its failure of several mechanical integrity tests (MITs), its most probable impact on the shallow alluvial aquifer came in the form of spills at the surface. One spill was of such magnitude that brine ponded over an acre of land. The results of the spill were dead vegetation in the immediate area and dead livestock that drank the ponded water. Subsequent impacts were increased erosion of the denuded soils and, ultimately, ground water contamination.

Approximately two years after the major brine spill, the water in the domestic well located southeast of the spill site became salty. Sampling and analysis of the well water showed chloride concentrations exceeded drinking

Table 8.1 Brine Spill Mixing Data

Chemical Constituents in Equivalents per Million

Parameter	Background-1 Circle	Background-2 Circle[a]	Background-3 Circle[a]	MW1 Square[a]	MW2 Diamond[a]	Brine-1 X^b	Brine-2 X^b
Ca	5.59	5.99	6.19	10.98	7.19	291.96	277.29
Mg	5.51	6.00	5.43	14.40	5.10	190.10	277.29
Na	2.96	1.96	2.09	8.26	5.57	2113.23	1977.94
K	0.01	0.03	0.04	0.06	0.04	12.12	11.30
HCO3	10.41	10.33	10.06	8.11	6.23	3.39	1.93
SO4	0.25	0.52	0.94	1.04	0.62	1.21	0.10
Cl	1.03	0.73	0.42	22.65	9.51	2621.70	2403.49

Percent Reacting Values

Sample	%Ca	%Mg	%(Na+K)	%Cl	%SO4	%HCO3	%Error
Background-1	39.74	39.19	21.07	8.81	2.14	89.05	9.23
Background-2	42.82	42.94	14.24	6.33	4.49	89.17	9.40
Background-3	45.02	39.50	15.47	3.70	8.20	88.09	9.22
MW1	32.57	42.71	24.71	71.22	3.27	25.51	2.89
MW2	40.15	28.50	31.35	58.11	3.82	38.07	4.49
Brine-1	11.20	7.29	81.51	99.82	0.05	0.13	0.37
Brine-2	11.43	6.6	81.98	99.92	0.00	0.08	0.43

[a] Dated June 4, 1989.
[b] Dated January 9, 1989.

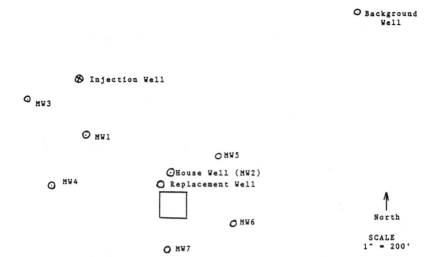

Figure 8.7. Contamination site for inorganics case study.

water standards and were increasing with time. The residents had to transport water for domestic purposes (cooking, bathing, and drinking). Eventually, the salt content caused excessive scaling, corrosion, and discoloration of the pipes and plumbing fixtures within the house; hence, the well was shut off completely.

Table 8.2 is a complete listing of all samples and analyses originally made available. Figure 8.8 is a computer generated trilinear diagram (Piper plot) of the data listed in Table 8.2. It should be noted that much of the data in Table 8.2 and Figure 8.8 are misleading and incomplete in terms of utilizing the trilinear diagram. The values recorded as zero for certain major ionic constituents are indicative of no analysis, not of zero concentration. These samples will plot on the axes of the trilinear diagram and are not of much use. Some of the samples were analyzed just for chloride content simply to show trends. The original brine analyses are also incomplete and of little use for analyzing the effects of mixing; however, they do show the tremendous ionic content of oilfield brines. The lack of complete data sets for individual wells or samples is an unfortunate, yet common, complication when dealing with ground water contamination episodes. Reporting zero values for constituents not analyzed for is a common transcribing error for ground water quality data.

After weeding out the nonusable data, a modest sampling and analysis program was undertaken, the results of which were previously shown in Table 8.1. The "background" values listed in Table 8.1 were taken from three different wells hydraulically upgradient from the spill site. The values for MW1 represent samples taken at the impacted house well and the values for MW2 were taken from a sample at a monitoring well installed at the spill site.

Table 8.2 Original Brine Spill Analyses

Chemical Constituents in Equivalents per Million

Parameter	Background-1 Circle[a]	Background-2 Circle[a]	Background-3 Circle[a]	MW1 Square[a]	MW2 Diamond[a]	Brine-1 X[g]	Brine-2 X[g]
Ca	5.59	5.99	6.19	10.98	7.19	291.96	277.29
Mg	5.51	6.00	5.43	14.40	5.10	190.10	160.08
Na	2.96	1.96	2.09	8.26	5.57	2113.23	1977.94
K	0.01	0.03	0.04	0.06	0.04	12.12	11.30
HCO_3	10.41	10.33	10.06	8.11	6.23	3.39	1.93
SO_4	0.25	0.52	0.94	1.04	0.62	1.21	0.10
Cl	1.03	0.73	0.42	22.65	9.51	2621.70	2403.49

Parameter	MW1 Square[b]	MW2 Diamond[c]	MW1 Square[d]	MW1 Diamond[d]	MW2 Diamond[e]	MW2 Diamond[f]
Ca	15.17	8.13	50.30	24.75	11.68	25.55
Mg	42.36	0.99	0.00	0.00	25.01	0.00
Na	0.00	0.00	0.00	0.00	0.00	0.00
K	0.00	0.00	0.00	0.00	0.00	0.00
HCO3	8.16	6.80	7.60	6.74	0.00	7.54
SO4	0.69	1.46	0.75	0.81	0.85	0.62
Cl	12.69	3.39	17.80	2.76	3.41	3.13

Percent Reacting Values

Sample	%Ca	%Mg	%(Na+K)	%Cl	%SO$_4$	%HCO$_3$	%Error
Background-1	39.74	39.19	21.07	8.81	2.14	89.05	9.23
Background-2	42.82	42.94	14.24	6.33	4.49	89.17	9.40
Background-3	45.02	39.50	15.47	3.70	8.20	88.09	9.22
MW1 (6/89)	32.57	42.71	24.71	71.22	3.27	25.51	2.89
MW2 (6/89)	40.15	28.50	31.35	58.11	3.82	38.07	4.49
MW1 (1/86)	26.37	73.63	0.00	58.92	3.19	37.89	45.51
MW2 (12/86)	89.18	10.82	0.00	29.07	12.52	58.41	12.15
MW1 (1/87)	100.00	0.00	0.00	68.06	2.87	29.08	31.58
MW2 (4/87)	100.00	0.00	0.00	26.81	7.87	65.32	41.18
MW1 (8/87)	31.83	68.17	0.00	79.99	20.01	0.00	79.16
MW2 (12/87)	100.00	0.00	0.00	27.72	5.53	66.75	38.69
Brine-1	11.20	7.29	81.51	99.82	0.05	0.13	0.37
Brine-2	11.43	6.6	81.98	99.92	0.00	0.08	0.43

[a] Dated June 4, 1989.
[b] Dated January 6, 1986.
[c] Dated December 15, 1986.
[d] Dated January 28, 1987.
[e] Dated April 8, 1987.
[f] Dated August 12, 1987.
[g] Dated January 9, 1989.

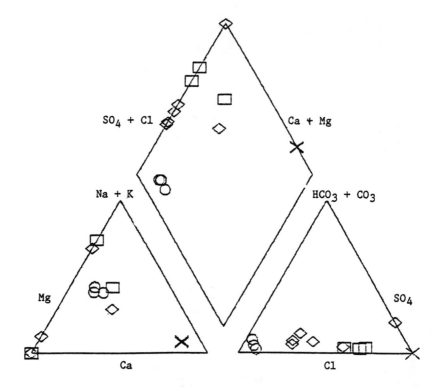

Figure 8.8. **Original brine spill mixing diagram: ○, background sample;**
□, MW1 sample; and ◇, MW2 sample.

The two brine samples were obtained from the saltwater disposal operation,
as noted previously.

The trilinear diagram for the data in Table 8.1 is shown in Figure 8.9. To
analyze the transport and fate of brine through the subsurface the following
assumptions are made:

1.   The natural quality of the impacted residential well was similar
      to the three background wells. This is probably justified given the
      proximity of the four wells and the fact that the three background
      wells show little variation in their composition, i.e., they all plot
      together on the trilinear diagram.
2.   The composition of the brine spilled is similar to the two brine
      analyses shown in Table 8.1. This is probably justified given that
      the analyzed brine is being produced from the same formation as
      when the spill occurred.
3.   Chlorides are a conservative solute that move essentially at the
      same rate as the ground water. This is a widely practiced and

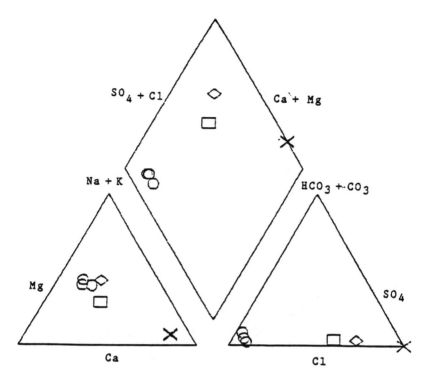

Figure 8.9. **Brine spill mixing diagram: O, background sample; ☐, MW1**
sample; and ◇, MW2 sample.

accepted assumption; hence, we can use chloride values for track-
ing the brine plume. This is especially convenient given the tremendous chloride load in brines.

Figure 8.10 is a reproduction of Figure 8.9 with the addition of three
mixing lines. The mixing lines are drawn from the approximate centroid of
the brine data points to the approximate centroid of the background data
points on all three figures. The mixing lines indicate several obvious trends:

1.   The anionic mixing line shows that MW1 and MW2 are defi-
      nitely a mixture of background water and brine. In fact, the an-
      ionic composition of these wells more closely resembles brine
      than freshwater, i.e., they are physically closer to the brine data
      points.
2.   The cationic mixing line does not show absolute mixing of the
      brine and background waters. The points for MW1 and MW2 do
      fall in between the background and brine centroids, but they are

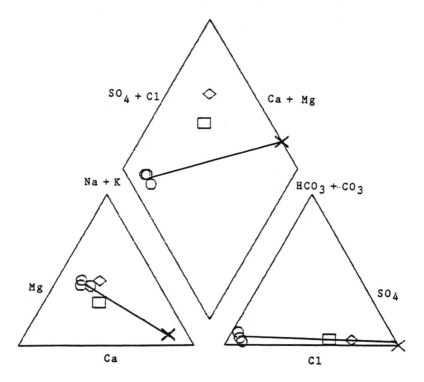

Figure 8.10.   Brine spill mixing diagram with mixing lines: ○,
background sample; ▢, MW1 sample; and ◇, MW2
sample.

off the mixing line. The cationic composition of MW1 and MW2
more closely resembles background quality than brine.

3.   The mixing line on the upper diamond shows that the overall
quality of the water in MW1 and MW2 is not a pure mixture of
background water and brine. Once again, the data points for MW1
and MW2 fall between the two centroids, but off the mixing line.
The two points appear to fall about halfway between the two
extremes.

The above results would seem to be contradictory and inconclusive as to
whether the spilled brine had impacted the house well. However, the mixing
lines shown in Figure 8.10 depict exactly what we would expect given our
knowledge of subsurface transport and fate processes. The mobile, anionic
constituents of the spilled brine have moved freely through the subsurface
and have impacted both MW1 and MW2. The relative position of MW1
(diamond) vs MW2 (square) on the anionic mixing line shows it to be more

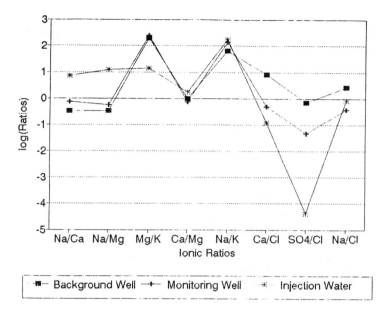

Figure 8.11.   Schoeller diagram of water quality analyses.

highly impacted. This can be attributed to one of two possibilities. First, MW1 may be within the middle of the brine plume with MW2 in the back or the receding end of the plume, i.e., the heavier contamination has moved past MW2. Second, MW1 may be screened deeper in the alluvial aquifer than MW2; hence, the higher density of brine will result in higher concentrations with depth.

The fact that the data points for MW1 and MW2 are not located on the other two mixing lines is exactly what would be expected. The cationic constituents of the brine are much less mobile than the anions. The movement of the cations will be retarded due to adsorption and/or cation exchange reactions. It is important to note that the mixing lines show absolute mixtures of the two end points only in the absence of other reactions.

In order to extend the analysis and more fully document the claims of contamination of the residential well, an ionic ratio (Schoeller) diagram was prepared (Figure 8.11). Plotted on the diagram are the various ionic ratios of the first background well, the house well (MW1), and the first brine sample. Claims of contamination are documented by two major aspects of Figure 8.11. First, all of the data points for MW1 fall in between the background well and the brine sample. This indicates that MW1 could be a mixture of the two fluids. Second, the overall pattern for MW1 more closely follows that of the brine sample than the background well, i.e., the pattern for MW1 rises and falls similar to the brine pattern. The peaks and valleys of MW1 are

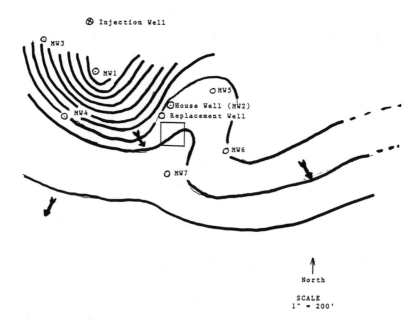

Figure 8.12.   Potentiometric map of site.

attenuated somewhat (especially for cationic ratios). This attenuation is due
to both dilution of the brine (dispersion and mixing) and retardation of the
cationic constituents.

Figure 8.12 depicts the water table of the area as determined from water
level measurements in the house wells and several monitoring wells. The
natural gradient appears to be predominantly to the south. It is hard to conceive
that lateral dispersion from the spill site would be able to solely account for
the increased salt content of the residential well. However, it is important to
note that this potentiometric surface was developed after both residential
wells south of the spill had been abandoned. The flow patterns of the ground
water formation when these two wells were being used would have been
decidedly different. In fact, the residential well labeled (MW1) was pumped
for two days and the chloride levels increased. This indicates that the well
was capturing part of the brine plume.

An interesting exercise in studying the alleged contamination is to use the
recommended statistical procedures for comparing the concentrations of
constituents in different wells. For this case, the concentrations of certain
inorganic constituents in the monitoring wells (MW1 and MW2) will be

Table 8.3 Summary of Statistical Analyses

Well	Method	Parameter[a]			
		Ca	HCO$_3$	SO$_4$	Cl
MW1	Parametric	N	Y	N	Y
	Nonparametric	Y	N	N	Y
	Tolerance limit	Y	Y	N	Y
MW2	Parametric	N	Y	N	N
	Nonparametric	Y	Y	N	N
	Tolerance limit	Y	Y	N	Y

[a] Y — yes, contamination is indicated; N — no, contamination is not indicated.

statistically compared with the background wells. This is similar to the comparison of background and compliance wells at RCRA facilities. The recommended procedures for these situations include analysis of variance (ANOVA) and tolerance limit techniques (U.S. Environmental Protection Agency, 1989).

The recommended procedures for the above techniques include requirements for minimum sample sizes in both the background and compliance wells. Referring to Table 8.2, only four ionic constituents (Ca^{+2}, HCO$_3^-$, SO$_4^{-2}$, Cl$^-$) meet the minimum requirements. Analyzing the concentrations of these constituents in the two compliance wells (MW1 and MW2) vs the concentrations of the three background wells yields some interesting results relative to the claims of contamination. The results of two ANOVA (parametric and nonparametric) procedures and one tolerance limit procedure are outlined in Table 8.3. Detailed calculations for these results are included in Appendix B.

The overall results in Table 8.3 do not present a clear mandate for declaring contamination. However, several subtle points relative to these results need to be noted. First, the evidence of "no contamination" relative to sulfates is misleading. It is common practice in oil and gas industry to chemically reduce the sulfate concentration of brines to prevent bacterial growth. Contamination from brines in this instance would mean a decrease in the monitoring well concentrations below the background well concentrations.

The remaining results depicted in Table 8.3 must be characterized as indicating contamination, but not conclusive proof of contamination. Several factors could be influencing these results. First, the early data for MW2 may be attenuating or masking the effects of contamination. This early time influence can be seen by noting the increasing concentration values for chlorides, the most mobile constituent. Second, the oscillatory pattern of the relatively immobile cation (Ca^{+2}) is indicative of some inconsistency in the sampling and/or analysis program or the effects of some geochemical reaction. The "calcium plume" would be expected to lag behind the "chloride plume" in both monitoring wells. However, detailed scrutiny of the values in Tables 8.1 and 8.2 point to several deficiencies in the analytical data. First, the sample taken on January 28, 1987 was only analyzed for total hardness; not all of which is attributable to calcium. Second, the total hardness (calcium + magnesium) percentages of MW1 and MW2 are lower than the background well percentages; the percentage attributable to the monovalent cations (Na + K) is higher. These combined attributes are paralleling the percentage composition of brine. The statistical analyses using concentration values do not identify these definite trends indicating contamination.

The third factor to consider about the data in Tables 8.1 and 8.2 is the magnitude of certain constituents, especially chloride. The concentrations of chloride in the brine samples are orders of magnitude higher than the monitoring wells, which are an order of magnitude higher than the background wells. However, the ANOVA statistical methods utilized in the previous analyses tend to attenuate the absolute differences in concentrations. Once again, percent concentration is more indicative of the true composition and is not subject to the attenuations noted above. The percentage values listed in Table 8.1b show the composition of the monitoring well waters to be similar to brine. An interesting analysis would be to use the percentage values in the statistical procedures; however, this could only be done with samples that were analyzed for the complete suite of ions.

The above point seems to invalidate the statistical exercise. In fact, very few of the samples listed in Table 8.2 have a complete set of analyses. This void in the data set minimizes the validity of the statistical analyses, but serves the purpose of showing the need to scrutinize all analytical data. In this instance, the graphical analyses and interpretation are more meaningful than the rigorous statistical analyses using incomplete data.

8.5 CASE STUDY — ORGANICS

In previous chapters fundamental processes affecting the transport of reactive chemicals were discussed. The purpose of this section is to present data from a field scale study that demonstrate the cumulative impacts of the individual processes discussed previously. The large-scale study to be discussed was conducted at the Canadian Forces Bases, Borden, Ontario. The study was conducted over a two year period with 5000 sampling points (including multilevel wells) and 19,900 samples analyzed. Two inorganic tracers (chloride and bromide) and five halogenated organic chemicals (bromoform, carbon tetrachloride, tetrachloroethylene, 1,2-dichlorobenzene, and hexachloroethane)

Table 8.4 Injected Solutes and Their Properties

Solute	Injected Concentration (mg/L)	Injected Mass (%)	Octanol-Water Partition Coefficient (K_{ow})	Potential for Biotansformation Aerobic	Anoxic[b]
Tracers					
Chloride ion	892	10,700	—	–	–
Bromide ion	324	3870	—	–	–
Organic solutes					
Bromoform	0.032	0.38	200	–	++
Carbon tetrachloride	0.031	0.37	500	–	+
Tetrachloroethylene	0.030	0.36	400	–	+
1,2-Dichlorobenzene	0.332	4.0	2500	+	–
Hexachloroethane	0.020	.023	4000	?	?

Source: Mackay, D. M., D. L. Freyberg, and P. V. Roberts, *Water Resour. Res.,* Vol. 22, 2017–2029, 1986, Copyright by the American Geophysical Union.

[a]   –, little potential for biotranformation; +, moderate potential for biotranformation; ++, good potential for biotranformation; ?, potential for biotranformation unknown at onset of this work.

[b]   Methanogenic conditions.

were injected as a pulse and allowed to migrate under natural gradient conditions. The solutes injected, the mass and concentration of the injected chemicals as well as several properties of the chemicals are shown in Table 8.4. Detailed monitoring was conducted horizontally, vertically and temporally over 2 year period to monitor migration of the chemicals. Information presented herein is taken from Mackay et al. (1986), Freyberg (1986), Roberts et al. (1986), and Curtis et al. (1986).

The experimental site was located below an inactive sand quarry and downgradient from a landfill, as shown in Figure 8.13. Figure 8.14 shows a cross-section of the experimental zone, indicating that it was located above the zone of contamination from the landfill. Figure 8.15 shows a plan view of the monitoring well locations and a profile view of their multilevel sampling points. The line AA' on Figure 8.15 is parallel to the general ground water flow direction at the site during the study. The sampling plan was designed to obtain spatial and temporal chemical distributions, i.e., spatial distributions of the chemicals in the aquifer at a point in time (spatial, synoptic, plume delineation, snapshot in time) and temporal distributions of the chemicals at a given location (temporal, time series, breakthrough, watching the plume migrate past a point in space).

Migration and growth of the chloride plume over time is demonstrated in

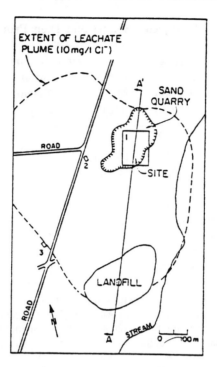

Figure 8.13.   Experimental site (Mackay, D. M., D. L. Freyberg, and P. V. Roberts, *Water Resour. Res.*, Vol. 22, 2017–2029, 1986, Copyright by the American Geophysical Union).

Figure 8.16. The initial condition for the plume indicates a rectangular shape (day 1 on Figure 8.16), as dictated by the injection design. It is observed that the plume migrated in the direction of ground water flow and that the plume followed a linear path. This suggests that the ground water flow direction was fairly uniform over the 2 year period of the study and that significant flow heterogeneities in the horizontal did not occur. After 647 d of transport, the plume apparently encountered a heterogeneity in hydraulic conductivity (velocity), causing a distinct vertical layering and decreased rate of migration of the center of mass of the chloride plume. From a hydrodynamic standpoint, it is observed from Figure 8.16 that the aerial extent of the chloride plume increased (due to dispersion) and concomitantly the concentrations decreased. It is also observed that the plume dispersed more longitudinally (in the direction of flow) than it did transversely (perpendicular to the ground water flow), indicating that the longitudinal dispersivity of the aquifer is greater than the transverse dispersivity. The rate of migration of the centroid of the chloride plume (ground water velocity) was 0.091 m/d. This is similar to values calculated from field slug tests and laboratory tests (0.078 to 0.081 m/ d). All of these observations are in keeping with a conservative chemical and a relatively homogeneous sandy aquifer.

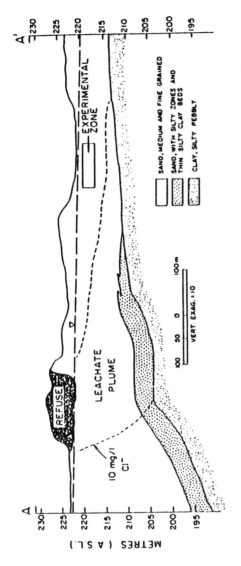

Figure 8.14. Approximate vertical geometry of aquifer along cross section AA' (Mackay, D. M., D. L. Freyberg, and P. V. Roberts, *Water Resour. Res.*, Vol. 22, 2017–2029, 1986, Copyright by the American Geophysical Union).

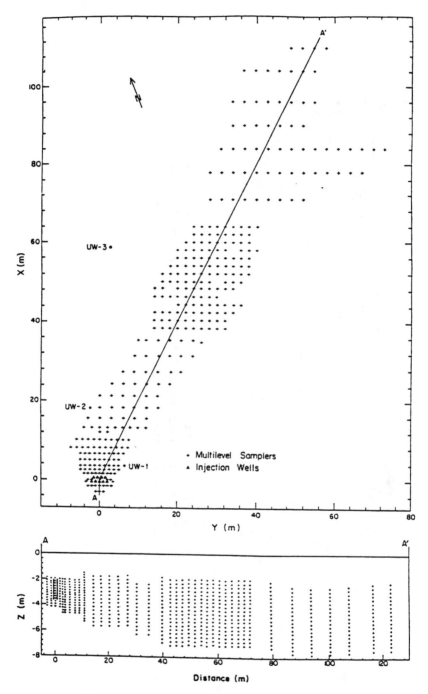

Figure 8.15.   Locations of multilevel samplers and injection wells
              (Mackay, D. M., D. L. Freyberg, and P. V. Roberts, *Water
              Resour. Res.,* Vol. 22, 2017–2029, 1986, Copyright by the
              American Geophysical Union).

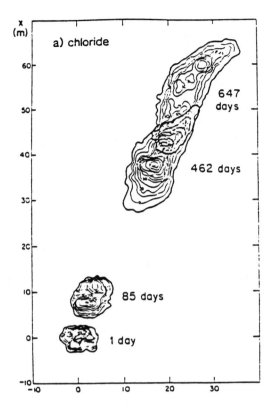

Figure 8.16. Vertically averaged concentration distribution of chloride (Mackay, D. M., D. L. Freyberg, and P. V. Roberts, *Water Resour. Res.*, Vol. 22, 2017–2029, 1986, Copyright by the American Geophysical Union).

Migration of the organic chemicals was also analyzed with respect to temporal (time series, breakthrough) and spatial (synoptic, plume delineation) distributions. The sampling points monitored for breakthrough of the chemicals are show in Figure 8.17 with the breakthrough curves for chloride, carbon tetrachloride and tetrachloroethylene shown in Figure 8.18. It is observed that the breakthrough of each chemical is distinct, with the chloride appearing first, followed by carbon tetrachloride and tetrachloroethylene. The breakthrough of the bromoform was very similar to that of carbon tetrachloride. Thus, classical chromatographic separation of the chemicals was observed for these chemicals (this is demonstrated spatially in Figure 8.19). Breakthrough for dichlorobenzene and hexachloroethane was not complete (mass was not conserved), with only small concentrations observed in an erratic fashion. The fate of these compounds will be discussed subsequently.

Figures 8.18A and 8.18C correspond to breakthrough curves for points a and c shown in Figure 8.17. It is observed from Figure 8.17 that point c is

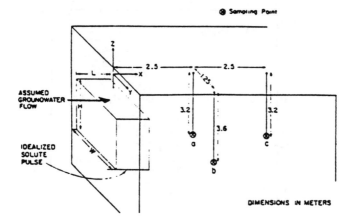

Figure 8.17.   Configuration of the injected pulse and time series sam-
               pling points (Roberts, P. V., M. N. Goltz, and D. M.
               Mackay, *Water Resour. Res.*, Vol. 22, 2047–2058, 1986,
               Copyright by the American Geophysical Union).

downgradient from point a. As expected, the breakthrough of each component occurs later at point c than at point a. It is also observed that the
maximum relative concentration is less in the downgradient well for each
component. This reflects the increased dispersion that has occurred in the
additional travel time from point a to c. The areas below the breakthrough
curves at a given point (a or c) are observed to be approximately equal for
each component. However, the maximum concentration realized decreases
and the spreading (standard deviation) of the breakthrough curve increases as
sorption (retardation) increases. This reflects the increased spreading (dispersion) resulting from sorptive transport (it takes the plume a longer time to
pass the observation point). Comparing the breakthrough curves at two points,
it is observed that the area below the breakthrough curve at point c is less
than at point a for a given component. This observation demonstrates that the
increased dispersion during migration between points a and c has resulted in
dilution of the plume.

   Analysis of the spatial data (plume delineation as a function of time)
results in a plot of distance traveled by the plume center of mass as a function
of time, as shown in Figure 8.20. The slope of Figure 8.20 (change in distance with respect to time) corresponds to the velocity of migration of the
chemical of interest. It is observed that the chloride plot is linear (constant
velocity) and the slope is greatest (greatest mobility). Decreasing mobility
(increasing retardation) is observed in the order: carbon tetrachloride,
bromoform, tetrachloroethylene, 1,2-dichlorobenzene, and hexachloroethane.
The plots for the sorbing chemicals are seen to be nonlinear, suggesting that
the mobility decreased with time.

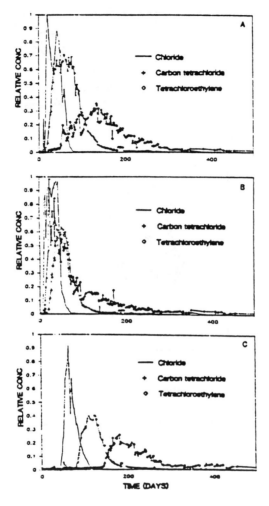

Figure 8.18. Breakthrough curves for chloride, carbon tetrachloride and tetrachloroethylene (Roberts, P. V., M. N. Goltz, and D. M. Mackay, *Water Resour. Res.*, Vol. 22, 2047–2058, 1986, Copyright by the American Geophysical Union).

Figure 8.21 shows values of retardation factors vs time for the organic chemicals, indicating that the retardation factors increased with time (or distance of migration). For purpose of this discussion, four possible explanations for this variation in retardation factors will be delineated: (1) a directional trend of increasing sorption affinity of the aquifer solids, which happens to coincide with the direction of plume migration, (2) nonlinear equilibrium sorption behavior that would result in higher values of the distribution coefficient as dispersion of the plume reduces the solution phase concentrations,

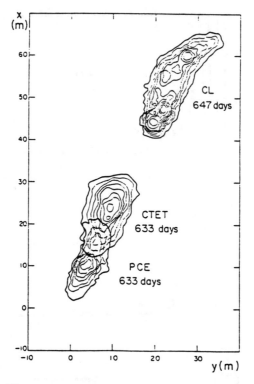

**Figure 8.19.   Chromatographic separation of plumes (Roberts, P. V., M.
N. Goltz, and D. M. Mackay, *Water Resour. Res.*, Vol. 22,
2047–2058, 1986, Copyright by the American Geophysical
Union).**

(3) nonsingular sorption-desorption behavior (hysteresis of desorption), and
(4) gradual increases in the distribution coefficients as a result of slow approach to sorption equilibrium due to intraparticle or intraaggregate diffusion.
The first explanation would account for increased sorption due to increases
in the organic fraction (sorptive capacity) of the porous medium. No such
trend in f_{oc} at the field site was observed by the researchers. The second
explanation is based upon the fact that as the plume migrates and dilutes, the
aqueous phase concentrations will decrease. If sorption is described by a
nonlinear isotherm, the slope of the isotherm effectively increases as the
aqueous phase concentrations decrease (the plume is operating at a point on
the isotherm that is more efficient in sorbing the chemical and thus has a
higher slope or distribution coefficient). This could account for increases in
the retardation factor. However, the researchers report that laboratory isotherm studies indicated that the linear isotherm was appropriate for the concentration ranges evaluated in the field study. The third explanation suggests
that hysteresis of desorption is evidenced by the chemical. This would cause

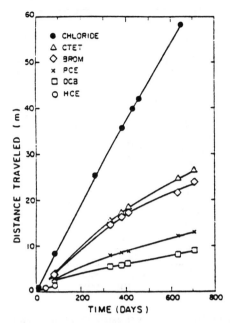

Figure 8.20.   Distances traveled by plume centers of mass in the horizontal plane (Roberts, P. V., M. N. Goltz, and D. M. Mackay, *Water Resour. Res.*, Vol. 22, 2047–2058, 1986, Copyright by the American Geophysical Union).

the chemical to desorb more slowly than it sorbed and effectively reduce the mobility of the plume. However, the researchers again report that the isotherm studies in the laboratory showed no evidence of hysteresis of desorption. The researchers thus suggest that the fourth explanation is responsible for the variations in the retardation factors, although admitting that the evidence is not conclusive. The researchers report that laboratory studies do indicate the slow approach to complete sorption, suggesting rate limiting sorption. The increased plume size (and thus time for the plume to pass) and the decreased aqueous phase concentrations (and thus less mass required to diffuse into inter- or intraparticle sites) could both act to enhance the approach to equilibrium sorption.

Several practical implications of the retardation factors being a function of the extent of plume migration can be considered. If one is concerned with the time of appearance of a contaminant plume downgradient, the actual time of appearance will be greater than that estimated if the retardation factor is taken to be a constant when in actuality it increases. From a natural migration perspective, the estimate will be conservative (the plume will appear after the estimate). However, monitoring should be conducted before and after the estimates to assure the plume does not go undetected. From a remediation perspective, additional ground water will be pumped and additional time will

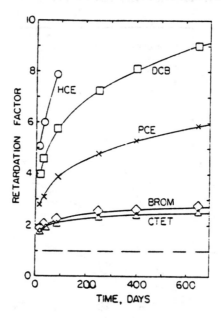

Figure 8.21.   Retardation factors estimated from synoptic sampling data (Roberts, P. V., M. N. Goltz, and D. M. Mackay, *Water Resour. Res.*, Vol. 22, 2047–2058, 1986, Copyright by the American Geophysical Union).

be required to flush the contaminants from the subsurface as the plume is diluted.

The researchers compared retardation factors developed from estimation methods, derived from laboratory batch tests and determined from the field test (from both spatial and temporal data). Table 8.5 summarizes retardation factors determined from these methods. The predicted values of retardation factors were developed from the regression equation of Schwarzenbach and Westall (1981) (see Table 3.1). It is observed that these estimated values are less than those measured in the laboratory or determined from field data. The researchers attribute these lower estimates of retardation factors to the low f_{oc} of the porous medium (0.02%); they suggest that sorption to the mineral surfaces was significant and accounted for the increased values of retardation factors. It is also observed that the estimates predicted bromoform to be the least sorbed with carbon tetrachloride and tetrachloroethylene to be similarly sorbed; while laboratory and field studies showed carbon tetrachloride and bromoform to experience similar levels of sorption and to be less sorbed than tetrachloroethylene. This demonstrates the need to recognize the level of accuracy possible with estimation procedures. Agreement was good between retardation factors from the laboratory and the temporal field data for carbon tetrachloride, bromoform and tetrachloroethylene, with equilibrium labora-

Table 8.5 Comparison of Retardation Factors

	Predicted[a]	Batch Experiments	Field Data Temporal	Field Data Spatial
CTET	1.3	1.9 ± 0.1	1.6 ± 1.8	1.8 ± 2.5
BROM	1.2	2.0 ± 0.2	1.5 ± 1.8	1.9 ± 2.8
PCE	1.3	3.6 ± 0.3	2.7 ± 3.9	2.7 ± 5.9
DCB	2.3	6.9 ± 0.7	1.8 ± 3.7	3.9 ± 9.0
HCE	2.3	5.4 ± 0.5	4.0	5.1 ± 7.9

Source: Curtis, G. P., P. V. Roberts, and M. Reinhard, *Water Resour. Res.,* Vol. 22, pp. 2059–2067, 1986, Copyright by the American Geophysical Union.
[a] Calculated from the regression by Schwarzenbach and Westall (1981) with f_{oc} = 0.02%.

tory results being on the upper end of the temporal field results. The spatial field data resulted in a wider range and increasing magnitude of retardation factors, with the greatest values corresponding to increased times of migration (travel distance, plume dispersion and dilution, etc.). This indicates the need to interpret laboratory and field data appropriately and recognize the degree of accuracy possible when generating estimates of future plume migration.

Mass balance relationships between the mass of chemical detected in the subsurface at various times vs the mass of chemical injected are summarized in Figure 8.22 (with the mass detected including estimates of mass of chemical sorbed to the subsurface media). As can be seen from Figure 8.22, the data suggests that transformations (losses) were realized for the following chemicals, and that the losses increased in the following order: bromoform, 1,2-dichlorobenzene and hexachloroethane. The disappearance of the chemicals was observed to be virtually immediate for hexachloroethane, with lag periods of approximately 100 d for bromoform and 1,2-dichlorobenzene. The disappearance of chemicals, once begun, can be described by a first order decay process, with the following first order rate constants: hexachloroethane (0.02 d⁻¹), 1,2-dichlorobenzene (0.004 d⁻¹), and bromoform (0.003 d⁻¹). Biotransformation is a likely process to account for the loss of mass in this study. However, the loss of 1,2-dichlorobenzene along with bromoform and hexachloroethane was puzzling to the researchers. The chemicals were selected based on evidence that 1,2-dichlorobenzene would transform under aerobic conditions and bromoform, carbon tetrachloride and tetrachloroethylene would transform under anoxic conditions (hexachloroethane was subsequently shown to transform under aerobic conditions). Dissolved oxygen concentrations in water samples at the site were generally in the range of 1 to 5 mg/L, with

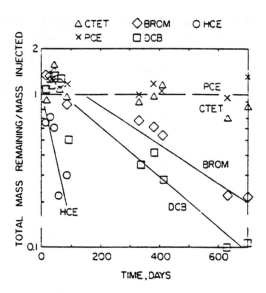

Figure 8.22. Relative total mass inferred from synoptic sampling data (Roberts, P. V., M. N. Goltz, and D. M. Mackay, *Water Resour. Res.*, Vol. 22, 2047–2058, 1986, Copyright by the American Geophysical Union).

values occasionally observed at or below detection limits. The fact that chemicals from each group transformed, but not all chemicals within the aerobic group transformed, is puzzling. The researchers report that conclusive evidence is not available to explain these observations.

In summary, the results of this field scale study serve to reinforce the fundamental concepts discussed in previous chapters. Also, interpretation of the data provides greater insight into the characterization of the subsurface. It should be noted that this field study was conducted under near ideal conditions — relatively homogeneous conditions, simple subsurface hydrogeology, discrete pulse of contaminants, etc. Generally, field sites will be more complicated than at the Borden site. However, the same fundamental principles will apply.

REFERENCES

Amdurer, M., R. T. Fellman, J. Roetzer, and C. Russ "Systems to Accelerate In-Situ Stabilization of Waste Deposits," EPA/540/2-86/002, U.S. Environmental Protection Agency, Washington, D.C. (1986).

Curtis, G. P., P. V. Roberts, and M. Reinhard "A Natural Gradient Experiment on Solute Transport in a Sand Aquifer: 4. Sorption of Organic Solutes and its Influence on Mobility," *Water Resour. Res.* 22(13, December):2059–2067 (1986).

Davis, S. N., D. J. Campbell, H. W. Bentley, and T. J. Flynn "Ground Water Tracers," Cooperative Agreement CR-810036, U.S. Environmental Protection Agency, Ada, OK (1985).

Freyberg, D. L. "A Natural Gradient Experiment on Solute Transport in a Sand Aquifer: 2. Spatial Moments and the Advection and Dispersion of Nonreactive Tracers," *Water Resour. Res.* 22(13):2031–2046 (1986).

Garabedian, S. P. et al. "Natural Gradient Tracer Test in Sand and Gravel; Results of Spatial Moments Analysis," in *Toxic Waste-Ground Water Contamination Program Third Technical Meeting* (U.S. Geological Survey, 1989).

Guven, O. R., W. Falta, J. Molz, and J. G. Melville "Analysis and Interpretation of Single-Well Tracer Tests in Stratified Aquifers," *Water Resour. Res.* 21:676–684 (1985).

Guven, O. R., W. Falta, J. Molz, and J. G. Melville "A Simplified Analysis of Two-Well Tracer Tests in Stratified Aquifers," *Ground Water* 24:68–82 (1986).

Keely, J. F. "Performance Evaluations of Pump-and-Treat Remediations," EPA/540/4-89/005, U.S. Environmental Protection Agency, Ada, OK (1989).

LeBlanc, D. R., S. P. Garbedian, W. W. Wood, K. M. Hess, and R. D. Quadri "Natural Gradient Tracer Test in Sand and Gravel: Objective, Approach, and Overview of Tracer Movement," in *Toxic Waste-GroundWater Contamination Program Third Technical Meeting,* (Washington, D.C.: U.S. Geological Survey, 1989).

Mackay, D. M., P. V. Roberts, and J. A. Cherry "Transport of Organic Contaminants in Groundwater," *Environ. Sci. Technol.* 19(5):384–392 (1985).

Mackay, D. M., D. L. Freyberg, and P. V. Roberts "A Natural Gradient Experiment on Solute Transport in a Sand Aquifer 1: Approach and Overview of Plume Movement," *Water Resour. Res.* 22(13, December):2017–2029 (1986).

Molz, F. J., J. G. Melville, O. Guven, R. D. Crocker, and K. T. Matteson "Design and Performance of Single-Well Tracer Test at the Mobile Site," *Water Resour. Res.* 22:1497–1502 (1985).

Molz, F. J., O. Guven, J. G. Melville, and J. F. Keely "Performance and Analysis of Aquifer Tracer Tests With Implications for Contaminant Transport Modeling," EPA/600/2-86/062, U.S. Environmental Protection Agency, Ada, OK (1986).

Palmer, C. D. and R. L. Johnson "Physical Processes Controlling the Transport of Contaminants in the Aqueous Phase," in *Transport and Fate of*

Contaminants in the Subsurface, EPA/625/4-89/019, U.S. Environmental Protection Agency, Ada, OK (1989).

Raghavan, R., E. Coles, and D. Dietz "Cleaning Excavated Soil Using Extraction Agents: A State-of-the-Art Review," EPA/600/52-89/034, U.S. Environmental Protection Agency, Cincinnati, OH (1990).

Roberts, P. V., M. N. Goltz, and D. M. Mackay "A Natural Gradient Experiment on Solute Transport in Sand Aquifer 3: Retardation Estimates and Mass Balances for Organic Solutes," *Water Resour. Res.* 22(13, December):2047–2058 (1986).

Roy, W. R. and R. A. Griffin "Surfactant and Chelate Induced Decontamination of Soil Materials: Current Status," Open File Report No. 21, Environmental Institute for Waste Management Studies, Tuscaloosa, AL (1988).

Sabatini, D. A. and R. C. Knox, Eds., *Transport and Remediation of Subsurface Contaminants: Colloidal Interfacial, and Surfactant Phenomena,* ACS Symposium Series 491, American Chemical Society, Washington, D.C., 1992.

Schwarzenbach, R. P. and J. Westall "Transport of Nonpolar Organic Compounds from Surface Water to Groundwater: Laboratory Sorption Studies," *Environ. Sci. Technol.* 15:1360–1367 (1981).

U.S. Environmental Protection Agency "Statistical Analysis of Ground-Water Monitoring Data at RCRA Facilities: Interim Final Guidance," EPA/530/SW-89/026, Washington, D.C. (1989).

U.S. Environmental Protection Agency "Remedial Actions at Waste Disposal Sites," EPA/625/6-85/006, Washington, D.C. (1985).

U.S. Environmental Protection Agency "Review of In-Place Treatment Techniques for Contaminated Surface Soils. Volume 1: Technical Evaluation," EPA/540/2-84/0039, Washington, D.C. (1984).

Appendix A
PHYSICAL, CHEMICAL, AND FATE
DATA FOR PRIORITY POLLUTANTS

Pollutant	CAS #	Formula	Formula Weight	Henry's Constant (atm-m^3/mol)	log K_{ow}	log K_{oc}
Acenaphthene	83-32-9	C12H10	154.21	1.70E-04	4.13	1.25
Acenaphthylene	208-96-8	C12H8	152.20	2.00E-04	4.07	3.68 E
Acetone	67-64-1	C3H6O	58.08	3.97E-05	-0.24	-0.43 E
Acrolein	107-02-8	C3H4O	56.06	4.40E-06	-0.09	-0.28 E
Acrylonitrile	107-13-1	C3H3N	53.06	1.10E-04	0.25	-1.13 E
Aldrin	309-00-2	C12H8Cl6	364.92	2.67E-05	5.52	2.61
Anthracene	120-12-7	C14H10	178.24	6.50E-05	4.45	4.27
Benzene	71-43-2	C6H6	78.11	5.40E-03	2.12	1.94
Benzidine	92-87-5	C12H12N2	184.24	3.88E-11	1.63	1.60 E
Benzo[a]anthracene	56-55-3	C8H12	228.30	2.30E-06	5.90	6.14
Benzo[b]fluoranthene	205-99-2	C20H12	252.32	1.20E-05	6.57	5.74
Benzo[k]fluoranthene	207-08-0	C20H12	252.32	1.04E-03	6.85	6.64 E
Benzoic acid	65-85-0	C7H6O2	122.12	7.02E-08	1.88	2.26
Benzo[ghi]perylene	191-24-2	C22H12	276.34	1.40E-07	7.10	6.89 E
Benzo[a]pyrene	50-32-8	C20H12	252.32	2.40E-06	6.00	6.00 E
Benzyl alcohol	100-51-6	C7H8O	108.14	NDA	1.10	1.98 E
Benzyl butyl phthalate	85-68-7	C19H20O4	312.37	1.30E-06	4.78	2.18
alpha–BHC	319-84-6	C6H6Cl6	290.83	5.30E-06	3.77	3.28
beta–BHC	319-85-7	C6H6Cl6	290.83	2.30E-07	3.96	3.46
delta–BHC	319-86-8	C6H6Cl6	290.83	2.50E-07	4.14	3.28
Bis(2-chloroethoxy)methane	111-91-1	C5H10Cl2O2	173.04	3.78E-07	1.26	2.06 E
Bis(2-chloroethyl)ether	111-44-4	C4H8Cl2O	143.01	1.30E-05	1.35	1.15
Bis(2-chloroisopropyl)ether	108-60-1	C6H12Cl2O	171.07	1.10E-04	2.58	1.79
Bis(2-ethylhexyl)phthalate	117-81-7	C24H38O4	390.57	1.10E-05	4.65	5.00
Bromodichloromethane	75-27-4	CHBrCl2	163.83	2.12E-04	1.88	1.79
Bromoform	75-25-2	CHBr3	252.73	5.46E-04	2.34	2.34
4-Bromophenyl phenyl ether	101-55-3	C12H9BrO	249.20	1.00E-04	5.15	4.94 E
2-Butanone	78-93-3	C4H8O	72.11	4.66E-05	0.28	0.09 E
Carbon disulfide	75-15-0	CS2	76.13	1.33E-02	2.00	2.47 E
Carbon tetrachloride	56-23-5	CCl4	153.82	2.40E-02	2.78	2.62
Chlordane	57-74-9	C10H6Cl8	409.78	4.80E-05	6.00	5.36
4-Chloroaniline	106-47-8	C6H6ClN	127.57	1.07E-05	2.30	2.75

Pollutant	Solubility (mg/l)	Specific Density	Vapor Pressure (mm Hg)	Oxidation singlet 1/M*hr	Oxidation peroxy 1/M*hr	Hydrolysis neutral: kn 1/hr	Biotrans kb ml/cell*hr
Acenaphthene	3.47E+00	1.0242	1.55E-03	<3600	8.00E+03	0	3.00E-09
Acenaphthylene	3.93E+00	0.8988	2.90E-02	4.00E+07	5.00E+03	0	3.00E-09
Acetone	1.00E+06	0.7906	2.70E+02	NDA	NDA	NDA	NDA
Acrolein	4.00E+05	0.8419	2.20E+02	1.00E+07	3.40E+03	0	3.00E-09
Acrylonitrile	7.90E+04	0.8032	1.00E+02	1.00E+08	3.60E+01	0	3.00E-09
Aldrin	2.70E-02	1.7000	6.00E-06	<3600	5.00E+03	3.00E-06	3.00E-09
Anthracene	4.50E-02	1.2600	1.95E-04	5.00E+08	2.20E-05	0	3.00E-09
Benzene	1.75E+03	0.8680	9.52E+01	<<360	<<1	0	1.00E-07
Benzidine	4.00E+02	1.2500	5.00E-04	4.00E+07	1.10E+08	0	1.00E-10
Benzo[a]anthracene	1.20E-02	1.2740	2.20E-08	5.00E+08	2.00E+04	0	1.00E-10
Benzo[b]fluoranthene	1.40E-02	NDA	5.00E-07	4.00E+07	5.00E+03	0	3.00E-12
Benzo[k]fluoranthene	5.50E-04	NDA	9.59E-11	4.00E+07	5.00E+03	0	3.00E-12
Benzoic acid	3.40E+03	1.3160	1.00E+00	NDA	NDA	NDA	NDA
Benzo[ghi]perylene	2.60E-04	NDA	1.01E-10	<360	<360	0	3.00E-12
Benzo[a]pyrene	3.90E-03	1.3510	5.60E-09	5.00E+08	2.00E+04	0	3.00E-12
Benzyl alcohol	3.50E+04	1.0424	1.00E+00	NDA	NDA	NDA	NDA
Benzyl butyl phthalate	2.76E+00	1.1200	8.60E-06	<<360	2.80E+02	0	3.00E-09
alpha-BHC	1.63E+00	1.8700	2.50E-05	<3600	6.00E+00	0	1.00E-10
beta-BHC	2.40E-01	1.8900	2.80E-07	<3600	6.00E+00	0	1.00E-10
delta-BHC	3.14E+01	1.8700	1.70E-05	<3600	6.00E+00	0	1.00E-10
Bis(2-chloroethoxy)methane	8.10E+04	1.2339	1.00E+00	<<360	5.20E+01	4.00E-06	3.00E-12
Bis(2-chloroethyl)ether	1.02E+04	1.2199	1.20E+00	<<360	2.40E+01	4.00E-06	3.00E-09
Bis(2-chloroisopropyl)ether	1.70E+03	1.1080	8.50E-01	<<360	2.00E+00	4.00E-06	1.00E-10
Bis(2-ethylhexyl)phthalate	3.00E-01	0.9873	2.00E-07	<<360	7.20E+00	0	4.00E-12
Bromodichloromethane	4.50E+03	1.9860	5.00E+01	<<360	2.00E-01	5.76E-08	1.00E-10
Bromoform	3.10E+03	2.8899	5.00E+00	<<360	5.00E-01	2.50E-09	1.00E-10
4-Bromophenyl phenyl ether	NDA	1.4208	1.50E-03	<<360	<<1	0	3.00E-09
2-Butanone	2.70E+05	0.8051	7.75E+01	NDA	NDA	NDA	NDA
Carbon disulfide	2.10E+03	1.2600	3.00E+02	NDA	NDA	NDA	NDA
Carbon tetrachloride	8.00E+02	1.5880	9.00E+01	<360	<<1	NDA	1.00E-10
Chlordane	5.60E-02	1.6000	1.00E-05	<3600	3.00E+01	0	3.00E-12
4-Chloroaniline	3.90E+03	1.4290	1.50E-02	NDA	NDA	NDA	NDA

Pollutant	CAS #	Formula	Formula Weight	Henry's Constant (atm-m³/mol)	log K_{ow}	log K_{oc}
Chlorobenzene	108-90-7	C6H5Cl	112.56	3.93E-03	2.84	2.10
p-Chloro-m-cresol	59-50-7	C7H7ClO	142.59	1.78E-06	3.03	2.89 E
Chloroethane	75-00-3	C2H5Cl	64.52	1.00E-02	1.43	0.54 E
2-Chloroethyl vinyl ether	110-75-8	C4H7ClO	106.55	2.50E-04	1.28	0.82
Chloroform	67-66-3	CHCl3	119.39	3.23E-03	1.90	1.64
2-Chloronaphthalene	91-58-7	C10H7Cl	162.62	6.12E-04	4.07	3.93 E
2-Chlorophenol	95-57-8	C6H5ClO	128.56	8.28E-06	2.15	2.56 E
4-Chlorophenyl phenyl ether	7005-72-3	C12H9ClO	204.66	2.20E-04	4.08	3.60 E
Chrysene	218-01-9	C18H12	228.30	7.26E-20	5.61	5.39 E
Cresol	1319-77-3	C7H8O	108.14	1.10E-06	1.97	2.70
P,p'-DDD	72-54-8	C14H10Cl4	320.05	2.16E-05	5.99	4.64 E
p,p'-DDE	72-55-9	C14H8Cl4	319.03	2.34E-05	5.77	5.39
p,p'-DDT	50-29-3	C14H9Cl5	354.49	4.89E-05	6.19	5.38
Dibenz[a,h]anthracene	53-70-3	C22H14	278.36	7.33E-09	6.36	6.22
Dibenzofuran	132-64-9	C12H8O	168.20	NDA	4.17	4.00 E
Dibromochloromethane	124-48-1	CHBr2Cl	208.28	9.90E-04	2.08	1.92
Di-n-butyl phthalate	84-74-2	C16H22O4	278.35	6.30E-05	4.57	3.14
1,2-Dichlorobenzene	95-50-1	C6H4Cl2	147.00	1.90E-03	3.40	2.26
1,3-Dichlorobenzene	541-73-1	C6H4Cl2	147.00	3.60E-03	3.38	2.23
1,4-Dichlorobenzene	106-46-7	C6H4Cl2	147.00	3.10E-03	3.39	2.20
3,3-Dichlorobenzidine	91-94-1	C12H10Cl2N2	253.13	4.50E-08	3.51	3.30 E
Dichlorodifluoromethane	75-71-8	CCl2F2	120.91	3.00E+00	2.16	2.56 E
1,1-Dichloroethane	75-34-3	C2H4Cl2	98.96	5.45E-03	1.79	1.48
1,2-Dichloroethane (EDC)	107-06-2	C2H4Cl2	98.96	9.80E-04	1.48	1.15
1,1-Dichloroethylene	75-35-4	C2H2Cl2	96.94	1.80E-02	1.84	1.81
trans-1,2-Dichloroethylene	156-60-5	C2H2Cl2	96.94	7.20E-03	2.09	1.77
2,4-Dichlorophenol	120-83-2	C6H4Cl2O	163.00	6.66E-06	3.13	2.94 E
1,2-Dichloropropane	78-87-5	C3H6Cl2	112.99	2.94E-03	2.28	1.71
cis-1,3-Dichloropropylene	10061-01-5	C3H4Cl2	110.97	1.30E-03	1.41	1.52
trans-1,3-Dichloropropylene	10061-02-6	C3H4Cl2	110.97	1.30E-03	1.41	1.55
Dieldrin	60-57-1	C12H8Cl6O	380.91	2.00E-05	4.53	4.08
Diethyl phthalate	84-66-2	C12H14O4	222.24	8.46E-07	2.35	1.84

Pollutant	Solubility (mg/l)	Specific Density	Vapor Pressure (mm Hg)	Oxidation singlet 1/M*hr	Oxidation peroxy 1/M*hr	Hydrolysis neutral: kn 1/hr	Biotrans kb ml/cell*hr
Chlorobenzene	4.88E+02	1.1095	1.00E-01	<<360	<1	0	3.00E-09
p-Chloro-m-cresol	3.85E+03	NDA	5.00E-02	7.00E+05	1.00E+07	0	3.00E-09
Chloroethane	5.70E+03	0.8978	1.01E+03	<<360	<1	7.20E-04	NDA
2-Chloroethyl vinyl ether	1.50E+04	1.0484	2.68E+01	1.00E+10	3.40E+01	4.00E-06	1.00E-10
Chloroform	8.00E+03	1.4890	1.60E+02	<<360	7.00E-01	2.50E-09	NDA
2-Chloronaphthalene	6.74E+00	1.2656	1.70E-02	<3600	1.60E+03	4.00E-07	1.00E-10
2-Chlorophenol	2.80E+04	1.2600	1.42E+00	7.00E+05	1.00E+07	0	1.00E-07
4-Chlorophenyl phenyl ether	3.30E+00	1.2026	2.70E-03	<<360	<1	0	1.00E-07
Chrysene	1.80E-03	1.2740	6.30E-09	1.00E+06	1.00E+03	0	1.00E-10
Cresol	3.10E+04	NDA	2.40E-01	NDA	NDA	NDA	NDA
p,p'-DDD	2.00E-02	1.4760	1.02E-06	<3600	1.60E+03	4.00E-07	1.00E-10
p,p'-DDE	4.00E-02	NDA	6.49E-06	<3600	1.20E+05	6.60E-07	3.00E-12
p,p'-DDT	5.00E-03	1.5600	1.90E-07	<3600	3.60E+03	6.80E-06	3.00E-12
Dibenz[a,h]anthracene	5.00E-03	1.2820	1.00E-10	5.00E+08	1.50E+04	NDA	3.00E-12
Dibenzofuran	1.00E+01	1.0886	NDA	NDA	NDA	NDA	NDA
Dibromochloromethane	4.00E+03	2.4510	7.60E+01	<<360	5.00E-01	2.88E-08	1.00E-10
Di-n-butyl phthalate	1.30E+01	1.0460	1.00E-05	<<360	1.40E+00	0.00E+00	3.20E-08
1,2-Dichlorobenzene	1.00E+02	1.3064	1.00E+00	<<360	<1	0	1.00E-10
1,3-Dichlorobenzene	1.23E+02	1.2881	2.28E+00	<<360	<1	0	1.00E-10
1,4-Dichlorobenzene	7.90E+01	1.2475	6.00E-01	<<360	<1	0	1.00E-10
3,3-Dichlorobenzidine	3.11E+00	NDA	1.00E-05	4.00E+07	4.00E+07	0	3.00E-12
Dichlorodifluoromethane	2.80E+02	1.3500	4.87E+03	0	0	0	NDA
1,1-Dichloroethane	5.50E+03	1.1757	1.82E+02	<<360	1.00E+00	1.15E-07	NDA
1,2-Dichloroethane (EDC)	8.69E+03	1.2530	6.40E+01	<360	<1	1.80E-09	1.00E-10
1,1-Dichloroethylene	2.25E+03	1.2180	5.91E+02	<1E+08	3.00E+00	0	NDA
trans-1,2-Dichloroethylene	6.30E+03	1.2565	2.65E+02	<1E+05	6.00E+00	0	NDA
2,4-Dichlorophenol	4.50E+03	1.4000	8.90E-02	7.00E+05	1.00E+07	0	1.00E-07
1,2-Dichloropropane	2.70E+03	1.1590	4.20E+01	<<360	1.00E+00	7.20E-04	1.00E-10
cis-1,3-Dichloropropylene	2.70E+03	1.2240	2.50E+01	<1E+08	4.40E+01	4.20E-04	1.00E-10
trans-1,3-Dichloropropylene	2.80E+03	1.1994	2.50E+01	<1E+08	4.40E+01	4.20E-04	1.00E-10
Dieldrin	2.00E-01	1.7500	1.80E-07	<3600	3.00E+01	5.40E-06	3.00E-12
Diethyl phthalate	8.96E+02	1.1175	2.00E-03	<<360	1.40E+00	0	1.00E-07

Pollutant	CAS #	Formula	Formula Weight	Henry's Constant (atm-m³/mol)	log K_{ow}	log K_{oc}
2,4-Dimethylphenol	105-67-9	C8H10O	122.17	6.55E-06	2.42	2.07 E
Dimethyl phthalate	131-11-3	C10H10O4	194.19	4.20E-07	1.56	1.63
4,6-Dinitro-o-cresol	534-52-1	C7H6N2O5	198.14	1.40E-06	2.12	2.64E
2,4-Dinitrophenol	51-28-5	C6H4N2O5	184.11	1.57E-08	1.54	1.25 E
2,4-Dinitrotoluene	121-14-2	C7H6N2O4	182.14	8.67E-07	1.98	1.79 E
2,6-Dinitrotoluene	606-20-2	C7H6N2O4	182.14	2.17E-07	2.00	1.79 E
Di-n-octyl phthalate	117-84-0	C24H38O4	390.57	1.41E-12	9.20	8.99 E
1,2-Diphenylhydrazine	122-66-7	C12H12N2	184.24	4.11E-11	2.94	2.82 E
alpha-Endosulfan	959-98-8	C9H6Cl6O3S	406.92	1.01E-04	3.55	3.31 E
beta-Endosulfan	33213-65-9	C9H6Cl6O3S	406.92	1.91E-05	3.62	3.37 E
Endosulfan sulfate	1031-07-8	C9H6Cl6O4S	422.92	NDA	3.66	3.37 E
Endrin	72-20-8	C12H8Cl6O	380.92	5.00E-07	4.56	3.92
Endrin aldehyde	7421-93-4	C12H8Cl6O	380.92	3.86E-07	5.60	4.43 E
Ethylbenzene	100-41-4	C8H10	106.17	6.60E-03	3.13	2.20
Ethylene Dibromide (EDB)	106-93-4	C2H4Br2	187.87	6.73E-04	1.76	1.64
Ethylene Oxide	75-21-8	C2H5O	44.05	7.56E-05	-0.22	0.34
Fluoranthene	206-44-0	C16H10	202.26	1.69E-02	5.22	4.62
Fluorene	86-73-7	C13H10	166.22	2.10E-04	4.18	3.70
Formaldehyde	50-0-0	HCHO	30.05	9.87E-07	0.00	0.56
Heptachlor	76-44-8	C10H5Cl7	373.32	2.30E-03	4.40	4.34 E
Heptachlor epoxide	1024-57-3	C10H5Cl7O	389.32	3.20E-05	3.65	4.32 E
Hexachlorobenzene	118-74-1	C6Cl6	284.78	1.70E-03	5.23	3.59
Hexachlorobutadiene	87-68-3	C4Cl6	260.76	1.03E-02	4.78	3.67 E
Hexachlorocyclopentadiene	77-47-4	C5Cl6	272.77	1.60E-02	5.04	3.63
Hexachloroethane	67-72-1	C2Cl6	236.74	2.50E-03	3.58	3.34
2-Hexanone	591-78-6	C6H12O	100.16	1.75E-03	1.38	2.13 E
Indeno[1,2,3-cd]pyrene	193-39-5	C22H12	276.34	2.96E-20	7.70	7.49 E
Isophorone	78-59-1	C9H14O	138.21	5.80E-06	1.67	1.49 E
Kepone	143-50-0	NDA	490.60	NDA	2.00	4.74
Lindane	58-89-9	C6H6Cl6	290.83	3.25E-06	3.70	3.03
Malathion	121-75-7	C10H19PS2O6	330.36	NDA	2.89	2.68 E
Methoxychlor	72-43-5	C16H15Cl3O2	345.66	NDA	4.40	4.95

Pollutant	Solubility (mg/l)	Specific Density	Vapor Pressure (mm Hg)	Oxidation singlet 1/M*hr	Oxidation peroxy 1/M*hr	Hydrolysis neutral: kn 1/hr	Biotrans kb ml/cell*hr
2,4-Dimethylphenol	6.20E-03	0.9650	9.80E-02	4.00E+06	1.10E+08	0	1.00E-07
Dimethylphthalate	4.00E+03	1.1905	2.00E-03	<<360	5.00E-02	0	5.20E-06
4,6-Dinitro-o-cresol	1.28E+02	NDA	5.00E-05	3.00E+04	5.00E+05	0	3.00E-09
2,4-Dinitrophenol	5.60E+03	1.6830	1.49E-05	3.00E+04	5.00E+05	0	3.00E-09
2,4-Dinitrotoluene	2.70E+02	1.3790	5.10E-03	<<360	1.44E+02	0	1.00E-07
2,6-Dinitrotoluene	3.00E+02	1.2833	1.80E-02	<<360	1.44E+02	0	3.10E-10
Di-n-octyl phthalate	3.00E+00	0.9780	1.40E-04	<<360	1.40E+00	0	1.00E-10
1,2-Diphenylhydrazine	2.21E+02	1.1580	2.60E-05	4.00E+07	1.00E+09		1.00E-10
alpha-Endosulfan	5.30E-01	1.7450	1.00E-05	<3600	3.60E+04	2.60E-03	3.00E-09
beta-Endosulfan	2.80E-01	1.7450	1.00E-05	<3600	3.60E+04	2.90E-03	3.00E-09
Endosulfan sulfate	NDA	NDA	NDA	<3600	2.00E+01	1.30E-04	1.00E-10
Endrin	2.30E-01	1.6500	7.00E-07	<3600	2.00E+01	0	1.00E-10
Endrin aldehyde	2.60E-01	NDA	2.00E-07	<3600	3.10E+03	0	3.00E-09
Ethylbenzene	1.52E+02	0.8669	7.08E+00	<<360	7.20E+02		3.00E-09
Ethylene Dibromide (EDB)	4.30E+03	2.1792	1.17E+01	NDA	NDA	NDA	NDA
Ethylene Oxide	1.00E+06	0.8824	1.31E+03	NDA	NDA	NDA	NDA
Fluoranthene	2.40E-01	1.2520	5.00E-06	<3600	<360	0	1.00E-10
Fluorene	1.69E+00	1.2030	7.00E-04	<360	3.00E-03	0	3.00E-09
Formaldehyde	4.00E+05	0.8150	1.00E+01	NDA	NDA	NDA	NDA
Heptachlor	1.80E-01	1.6600	3.00E-04	3.00E+10	2.50E+03	3.00E-02	NDA
Heptachlor epoxide	3.50E-01	NDA	2.60E-06	<3600	2.00E+01	0	3.00E-12
Hexachlorobenzene	6.00E-03	1.5691	1.09E-05	<<360	<<1	0	3.00E-12
Hexachlorobutadiene	3.23E+00	1.5542	1.50E-01	<1000	6.00E+00		1.00E-10
Hexachlorocyclopentadiene	1.10E+00	1.7119	8.10E-02	<1000	1.20E+01	2.00E-03	1.00E-10
Hexachloroethane	5.00E+01	2.0910	8.00E-01	0	0	0	1.00E-10
2-Hexanone	3.50E+04	0.8113	2.00E+00	NDA	NDA	NDA	NDA
Indeno[1,2,3-cd]pyrene	6.20E-02	0.0620	1.00E-10	5.00E+08	2.00E+04	0	3.00E-12
Isophorone	1.20E+04	0.9225	3.80E-01	<1E+07	2.25E+02	0	3.00E-09
Kepone	9.90E-03	NDA	NDA	NDA	NDA	NDA	NDA
Lindane	7.50E+00	1.5691	6.70E-05	<3600	6.00E+00	0	1.00E-10
Malathion	1.45E+02	1.2076	4.00E-05	NDA	NDA	NDA	NDA
Methoxychlor	4.00E-02	1.4100	NDA	NDA	NDA	NDA	NDA

Pollutant	CAS #	Formula	Formula Weight	Henry's Constant $(atm\text{-}m^3/mol)$	log K_{ow}	log K_{oc}
Methyl bromide	74-83-9	CH3Br	94.94	2.00E–01	1.10	1.92 E
Methyl chloride	74-87-3	CH3Cl	50.48	8.82E–03	0.90	1.40 E
Methylene chloride	75-09-2	CH2Cl2	84.93	2.00E–03	1.30	0.94
2-Methylnaphthalene	91-57-6	C11H10	142.20	NDA	4.11	3.93
4-Methyl-2-pentanone	108-10-1	C6H12O	100.16	1.49E–05	1.09	0.79 E
2-Methylphenol	95-48-7	C7H8O	108.14	1.23E–06	1.93	1.34
4-Methlyphenol	106-44-5	C7H8O	108.14	7.92E–07	1.92	1.69
Naphthalene	91-20-3	C10H8	128.18	4.60E–04	3.36	3.11
2-Nitroaniline	88-74-4	C6H6N2O2	138.13	9.72E–05	1.79	1.43 E
3-Nitroaniline	99-09-2	C6H6N2O2	138.13	NDA	1.37	1.26 E
4-Nitroaniline	100-01-6	C6H6N2O2	138.13	1.14E–08	1.39	1.08 E
Nitrobenzene	98-95-3	C6H5NO2	123.11	2.45E–05	1.84	1.95
2-Nitrophenol	88-75-5	C6H5NO3	139.11	3.50E–06	1.78	1.57 E
4-Nitrophenol	100-02-7	C6H5NO3	139.11	3.00E–05	1.91	2.33
N-Nitrosodimethylamine	62-75-9	C2H6N2O	74.09	1.43E–01	0.06	1.41 E
N-Nitrosodiphenylamine	86-30-6	C12H10N2O	198.22	2.33E–08	3.13	2.76 E
N-Nitrosodi-n-propylamine	621-64-7	C6H14N2O	130.19	6.92E–06	1.31	1.01 E
PCB-1016	12674-11-2	various	avg = 258	7.50E–04	4.38	4.70 E
PCB-1221	11104-28-2	various	avg = 192	3.24E–04	2.80	2.44 E
PCB-1232	11141-16-5	various	avg = 221	8.64E–04	3.20	2.83 E
PCB-1242	53469-21-9	various	avg = 261	5.60E–04	5.58	3.71 E
PCB-1248	12672-29-6	various	avg = 288	3.50E–03	6.11	5.64 E
PCB-1254	11097-69-1	various	avg = 327	2.70E–03	6.47	5.61
PCB-1260	11096-82-5	various	avg = 370	7.10E–03	6.91	6.42 E
Pentachlorophenol	87-86-5	C6HCl5O	266.34	2.10E–06	5.01	2.95
Phenanthrene	85-01-8	C14H10	178.24	3.90E–05	4.52	4.36
Phenol	108-95-2	C6H6O	94.11	2.70E–07	1.46	1.43
Pyrene	129-00-0	C16H10	202.26	1.09E–05	5.09	4.81
Styrene	100-42-5	C8H8	104.15	2.61E–03	3.16	2.87 E
TCDD (Dioxin)	1746-01-6	C12H4Cl4O2	321.98	5.40E–23	6.20	6.66
1.1.1.2-Tetrachloroethane	630-20-6	C2H2Cl4	167.85	3.81E–04	NDA	1.73
1.1.2.2-Tetrachloroethane	79-34-5	C2H2Cl4	167.85	3.80E–04	2.56	2.07

Pollutant	Solubility (mg/l)	Specific Density	Vapor Pressure (mm Hg)	Oxidation singlet 1/M*hr	peroxy 1/M*hr	Hydrolysis neutral: kn 1/hr	Biotrans kb ml/cell*hr
Methyl bromide	1.32E-04	1.6755	1.63E+03	<<360	1.00E-01	1.44E-03	NDA
Methyl chloride	7.25E+03	0.9159	3.79E+03	NDA	NDA	NDA	NDA
Methylene chloride	2.00E-04	1.3266	3.49E+02	<<360	2.00E-01	1.15E-07	NDA
2-Methylnaphthalene	2.46E+01	1.0058	NDA	NDA	NDA	NDA	NDA
4-Methyl-2-pentanone	1.70E+04	0.8000	1.50E+01	NDA	NDA	NDA	NDA
2-Methlyphenol	2.50E+04	1.0273	2.40E-01	NDA	NDA	NDA	NDA
4-Methlyphenol	2.30E+04	1.0178	4.00E-02	NDA	NDA	NDA	NDA
Naphthalene	3.00E+01	1.1450	5.40E-02	<360	<1	0	1.00E-07
2-Nitroaniline	1.26E+03	1.4400	8.10E+00	NDA	NDA	NDA	NDA
3-Nitroaniline	1.10E+03	0.9011	NDA	NDA	NDA	NDA	NDA
4-Nitroaniline	8.00E+02	1.4240	1.50E-03	NDA	NDA	NDA	NDA
Nitrobenzene	1.90E+03	1.2050	1.50E-01	<<360	<<1	0	3.00E-09
2-Nitrophenol	2.10E+03	1.4950	2.00E-01	2.00E+05	2.00E+06	0	3.00E-09
4-Nitrophenol	1.32E+04	1.4790	1.00E-04	2.00E+05	2.00E+06	0	1.00E-07
N-Nitrosodimethylamine	1.00E+06	1.0059	8.10E+00	<3600	<3600	0	3.00E-12
N-Nitrosodiphenylamine	3.51E+01	NDA	1.00E-01	<3600	<3600	0	1.00E-10
N-Nitrosodi-n-propylamine	9.90E+03	0.9160	4.00E-01	<3600	<3600	0	3.00E-12
PCB-1016	4.20E-01	1.3300	4.00E-04	<<360	<1	0	3.00E-09
PCB-1221	5.90E-01	1.1500	6.70E-03	<<360	<1	0	3.00E-09
PCB-1232	1.45E+00	1.2400	4.60E-03	<<360	<1	0	3.00E-09
PCB-1242	2.40E-01	1.3920	1.00E-03	<<360	<1	0	3.00E-09
PCB-1248	5.40E-02	1.4100	4.94E-04	<<360	<1	0	3.00E-09
PCB-1254	5.00E-02	1.5050	7.71E-05	<<360	<1	0	3.00E-09
PCB-1260	8.00E-02	1.5660	4.05E-05	<<360	<1	0	3.00E-09
Pentachlorophenol	2.00E+01	1.9780	1.10E-04	<7000	1.00E+05	0	3.00E-09
Phenanthrene	1.00E+00	1.1790	6.80E-04	<360	<36	0	1.60E-07
Phenol	8.20E-04	1.0576	2.00E-01	7.00E+05	1.00E+07	0	3.00E-06
Pyrene	1.35E-01	1.2710	2.50E-06	5.00E+08	2.20E+04	0	1.00E-10
Styrene	3.00E+02	0.9060	5.00E+00	NDA	NDA	NDA	NDA
TCDD (Dioxin)	2.00E-04	1.8270	6.40E-10	<360	<1	0	1.00E-10
1,1,1,2-Tetrachloroethane	2.90E+03	NDA	5.00E+00	NDA	NDA	NDA	NDA
1,1,2,2-Tetrachloroethane	2.90E+03	1.5953	5.00E+00	<<360	2.00E+00	1.20E-07	3.00E-12

Pollutant	CAS #	Formula	Formula Weight	Henry's Constant (atm-m³/mol)	log K_{ow}	log K_{oc}
Tetrachloroethylene (PCE)	127-18-4	C2Cl4	165.83	1.53E-02	2.60	2.42
Toluene	108-88-3	C7H8	92.14	6.70E-03	2.65	2.18
Toxaphene	8001-35-2	C10H10Cl8	413.82	6.30E-02	3.30	3.18 E
1,2,4-Trichlorobenzene	120-82-1	C6H3Cl3	181.45	2.32E-03	4.11	3.16
1,1,1-Trichloroethane	71-55-6	C2H3Cl3	133.40	1.62E-02	2.47	2.18
1,1,2-Trichloroethane	79-00-5	C2H3Cl3	133.40	7.40E-04	2.18	1.75
Trichloroethylene (TCE)	79-01-6	C2HCl3	131.39	9.10E-03	2.53	2.03
Trichlorofluoromethane	75-69-4	CCl3F	137.37	1.10E-01	2.53	2.20
2,4,5-Trichlorophenol	95-95-4	C6H3Cl3O	197.45	1.76E-04	3.85	3.51 E
2,4,6-Trichlorophenol	88-06-2	C6H3Cl3O	197.45	9.07E-08	3.06	3.03
Vinyl acetate	108-05-4	C4H6O2	86.09	4.81E-04	0.73	0.45 E
Vinyl chloride	75-01-4	C2H3Cl	62.50	1.22E+00	0.60	0.39 E
o-Xylene	95-47-6	C8H10	106.17	5.27E-03	2.95	2.11
m-Xylene	108-38-3	C8H10	106.17	7.00E-03	3.20	3.20
p-Xylene	106-42-3	C8H10	106.17	7.10E-03	3.18	2.31

Pollutant	Solubility (mg/l)	Specific Density	Vapor Pressure (mm Hg)	Oxidation singlet 1/M*hr	Oxidation peroxy 1/M*hr	Hydrolysis neutral: kn 1/hr	Biotrans kb ml/cell*hr
Tetrachloroethylene (PCE)	1.50E+02	1.6227	1.40E+01	<<100	6.00E+00	0	1.00E-10
Toluene	5.15E+02	0.8669	2.20E+01	<<360	1.44E+02	0	1.00E-07
Toxaphene	7.40E-01	1.6000	3.30E-05	<3600	3.00E+00	0	3.00E-12
1,2,4-Trichlorobenzene	3.00E+01	1.4542	4.00E-01	<<360	<<1	5.60E-05	1.00E-10
1,1,1-Trichloroethane	1.55E+03	1.3390	1.00E+02	<<360	1.00E+00	1.70E-04	NDA
1,1,2-Trichloroethane	4.50E+03	1.4397	1.90E+01	<<360	3.00E+00	1.20E-07	3.00E-12
Trichloroethylene (TCE)	1.10E+03	1.4642	5.78E+01	<1000	6.00E+00	0	1.00E-10
Trichlorofluoromethane	1.10E+03	1.4870	6.87E+02	0	0	0	NDA
2,4,5-Trichlorophenol	1.19E-03	1.6780	2.20E-02	NDA	NDA	NDA	NDA
2,4,6-Trichlorophenol	8.00E+02	1.4901	1.70E-02	7.00E+04	1.00E+06	0	3.00E-09
Vinyl acetate	2.00E+04	0.9317	8.30E+01	NDA	NDA	NDA	NDA
Vinyl chloride	1.10E+03	0.9106	2.58E+03	<1E+08	3.00E+00	0	NDA
o-Xylene	1.52E+02	0.8802	1.00E+01	NDA	NDA	NDA	NDA
m-Xylene	1.58E+02	0.8642	8.29E+00	NDA	NDA	NDA	NDA
p-Xylene	1.98E+02	0.8610	8.76E+00	NDA	NDA	NDA	NDA

Sources:

1. Montgomery, J. H. and L. M. Welkom, *Groundwater Chemicals Desk Reference*, Lewis Publishers, Chelsea, Michigan, 1990.

2. Mabey, et al., "Aquatic Fate Process Data for Organic Priority Pollutants", EPA-440/4-81-014, Prepared by SRI International, EPA Contract Nos. 68-01-3867 and 68-03-2981, Dec. 1982, prepared for Monitoring and Data Support Division, Office of Water Regulations and Standards, Washington, DC.

3. *Superfund Public Health Evaluation Manual*, EPA/540/1-86/060 (OSWER Directive 9285.4-1), Oct. 1986, Office of Emergency and Remedial Response, Office of Solid Waste and Emergency Response, U.S. Environmental Protection Agency, Washington, DC.

4. Lyman, W. J., W. F. Reehl, and D. H. Rosenblatt, *Handbook of Chemical Property Estimation Methods*, McGraw-Hill, New York, 1982.

5. Suflita, J. M., "Microbial Ecology and Pollutant Biodegradation in Subsurface Ecosystems," in *Transport and Fate of Contaminants in the Subsurface*, EPA/625/4-89/019, Sept. 1989, U.S. Environmental Protection Agency, Center for Environmental Research Information, Cincinnati, OH.

6. Ellington, J. J., F. E. Stancil, Jr., and W. D. Payne, "Measurement of Hydrolysis Rate Constants for Evaluation of Hazardous Waste Land Disposal: Volume 1. Data on 32 Chemicals", EPA/600/S3-86/043, Apr. 1987, U.S. Environmental Protection Agency, Environmental Research Laboratory, Athens, GA.

Notes:

Oxidation and biotransformation data were taken from reference 2. These data are for aquatic systems and should be used for comparative purposes; their direct application to subsurface systems is cautioned. For biotransformation information for specific subsurface sytems, the reader is referenced to Chapter 4 and Tables 4.11 and 4.12. Hydrolysis data were taken from references 2 and 6. All other values were taken from reference 1. NDA means no data available; E means value was estimated in original reference, and a value of 0 means this was reported in the original reference. Where more than one reliable value or a range of values was reported, the median value is given in this appendix. All of the data should be used with caution due to accuracy and general applicability concerns. Site specific data are always preferable.

The second-order biotransformation constant, k_b, is given in ml/cell*hr. The rate of biotransformations, R_B, is given by the expression

$$R_B = k_b \, [B] \, [C]$$

where [B] is the microbial population in cells/ml and [C] is the concentration of the chemical. When the population count, [B], can be assumed to be constant, $k_b*[B]$ gives a first-order rate constant (reference 2). Typical microbial population count values, which range from 0.036×10^6 to 49×10^6 cells/ml, can be found in reference 5.

Many of the listed chemicals can be named in several ways. In this appendix, the chemicals are named as listed in reference 1.

Appendix B

EXAMPLE STATISTICAL CALCULATIONS

Table 1 One-Way Parametric ANOVA Table for Chloride Data

Source of Variation	Sums of Squares (SS)	Degrees of Freedom	Mean Squares (MS)	F
Between Wells	624,677	2	12,339	27
Error (within wells)	104,096	9	11,566	
Total	728,773	11		

Tabulated F statistic, $F_{2,9,.95}$ = 4.26.

Calculated F > Tabulated F, therefore reject hypothesis of equal well means at the 5% level, i.e., contamination is present.

Bonferroni t-statistics for chloride data:

n_{up} = 3 (number of background observations)

$\overline{x}_{up}$ = 25.8 (mean of background chloride concentration)

MW1: $\overline{x}_i - \overline{x}_{up}$ = 628–25.8 = 602.2 mg/l

MW2: $\overline{x}_i - \overline{x}_{up}$ = 163–25.8 = 137.2 mg/l

MW1: SE = $[MS_{error} (1/n_{up} + 1/n_i)]^{1/2} = [11,566(1/3 + 1/3)]^{1/2}$ = 87.81

MW2: SE = $[11,566(1/3 + 1/6)]^{1/2}$ = 76.05

Tabulated t-statistic, $t_{9,.975}$ = 2.26

MW1: D = SE × t = 87.81 × 2.26 = 198.45

MW2: D = SE × t = 76.05 × 2.26 = 171.87

MW1: 602.2 > 198.45, therefore contamination

MW1: 137.2 < 171.87, therefore no contamination

Table 2 One-Way Nonparametric ANOVA for Chloride Data

Background	MW1	MW2
36.5 (3)	803 (12)	337 (9)
26 (2)	450 (10)	120 (6)
15 (1)	631 (11)	98 (4)
		121 (7)
		111 (5)
		190 (8)
$n_1 = 3$	$n_2 = 3$	$n_3 = 6$
$R_1 = 6$	$R_2 = 33$	$R_3 = 39$
$R_1 = 2$	$R_2 = 11$	$R_3 = 6.5$

N = 12 (Total Observations)

K = 3 (Number of Wells)

Calculated H statistic = 9.33

Tabulated Chi-Squared Statistic, $\chi^2_{2,.95}$ = 5.991

Calculated H > Tabulated Chi-Squared, therefore reject the hypothesis of no contamination at the 5% level.

The critical difference is calculated for each monitoring well:

MW1:

$z = 1.96$

$C_2 = 1.96[12(12 + 1)/12]^{1/2}[1/3 + 1/3]^{1/2} = 5.77$

$R_2 - R_1 = 11 - 2 = 9$

9 > 5.77, therefore, contamination indicated

MW2:

$C_3 = 1.96[12(12 + 1)/12]^{1/2}[1/3 + 1/6]^{1/2} = 4.997$

$R_3 - R_1 = 6.5 - 2 = 4.5$

4.5 < 4.997, therefore, no contamination is indicated

Tolerance Interval:

Background Calcium:

$\bar{x}$ = 118.7 mg/l

s = 6.11 mg/l

n = 3

k = 7.655

TL = $\bar{x}$ + ks = 118.7 + 7.655(6.11) = 165.5 mg/l

If any of the monitoring wells have a calcium value greater than this tolerance limit of 165.5 mg/l, then it is considered to be statistically significantly greater and therefore contamination is indicated. Based upon this test, both MW1 and MW2 exhibit contamination.

Table 3 Summary Statistics for Data

		Background	MW1	MW2
Ca	$\bar{x}$	118.7 mg/l	510.7	350
	s	6.11	432.8	190
HCO_3	$\bar{x}$	626	485	423
	s	11	19	32
SO_4	$\bar{x}$	27.3	40	43
	s	16.6	9.1	14.8
Cl	$\bar{x}$	25.8	628	163
	s	10.8	177	91

It is readily apparent that the CV ($\bar{x}$/s) for all of the data is <1.

INDEX